ORGANIC SPECTROSCOPY THROUGH SOLVED PROBLEMS

ORGANIC SPECTROSCOPY THROUGH SOLVED PROBLEMS

Dr K S Mukherjee MSc, PhD
Formerly Principal, Siksha Bhavan [Dean, Faculty of Science]
Visva-Bharati, Santiniketan, West Bengal, India

and

Bodhisattwa Mukhopadhyay MSc
Senior Executive, Production
Fresenius Kabi Oncology Ltd
Kalyani, Nadia, West Bengal, India

New Central Book Agency (P) Ltd
LONDON
HYDERABAD ERNAKULAM BHUBANESWAR
NEW DELHI KOLKATA PUNE GUWAHATI

NCBA

REGD OFFICE
8/1 Chintamoni Das Lane, Kolkata 700 009, India
email: ncbapvtltd@eth.net

OVERSEAS
NCBA (UK) Ltd, 149 Park Avenue North, Northampton, NN3 2HY, UK
email: ncbauk@yahoo.com

EXPORT
NCBA Exports Pvt. Ltd, 212 Shahpur Jat, New Delhi 110 049
email: ncbaexports@ncbapvtltd.com

BRANCHES

2/27 Ansari Road, 1st Floor
Daryaganj, **New Delhi** 110 002
email: ncbadel@ncbapvtltd.com

Shop Nos. 3 & 4, Vinayak Towers
681/B Budhwar Peth, Appa Balwant Chowk
Pune 411 002
email: ncbapune@ncbapvtltd.com

Shop No. 15, Bharati Towers, Ground Floor
Forest Park
Bhubaneswar 751 001
email: ncbabub@ncbapvtltd.com

House No. 3-1-315, 1st Floor
Nimboliadda, Kachiguda, **Hyderabad** 500 027
email: ncbahydb@ncbapvtltd.com

GSS Shopping Complex, 1st Floor
Opposite College Post Office, Convent Road
Ernakulam 682 035
0484-4021054/53

Radhamayee Complex (Second Floor)
Opp. Axis Bank, Panbazar, Jaswanta Road
Guwahati 781 001
email: ncbaguwahati@ncbapvtltd.com

ORGANIC SPECTROSCOPY THROUGH SOLVED PROBLEMS • Dr K S Mukherjee, Bodhisatta Mukhopadhyay

First Published: August 2013
Reprinted: January 2015
Reprinted: 2018

PUBLISHER AND TYPESETTER
New Central Book Agency (P) Ltd
8/1 Chintamoni Das Lane, Kolkata 700 009

PRINTER
New Central Book Agency (P) Ltd
Web-Offset Division, Dhulagarh, Sankrail, Howrah

TECHNICAL EDITOR
Dipan Roy

PROJECT TEAM
Swapan Sarkar, Prodip Baidya and Pradip Biswas

ISBN: 978 81 7381 877 6

Price ₹ 350.00

CONTENTS

Preface ix

CHAPTER 1. INTRODUCTION 1-9

(i) Spectroscopy, spectrometry and spectrum 1
(ii) Nature of electromagnetic radiation 1
(iii) Different regions of electromagnetic radiation 4
(iv) Classification of spectroscopic technique 4
(v) Differentiation between atomic and molecular spectra 4
(vi) Objective of molecular spectroscopy 5
(vii) Fate of molecules on excitation 5
(viii) Different types of molecular spectroscopy 6
(ix) Expression of absorption data in spectroscopy 6
(x) Fourier transform spectroscopy 7
(xi) Signal to Noise Ratio, Width of spectral lines and Heisenberg uncertainty principle 8

CHAPTER 2. INFRARED SPECTROSCOPY 10-44

(i) Cause of Infrared spectrum and nature of transition 10
(ii) Vibrational motion 10
(iii) Infrared, vibrational and rotational spectroscopy 11
(iv) Information obtainable from infrared spectroscopy 11
(v) Nature of molecular vibration 11
(vi) Fundamental molecular vibration 12
(vii) Essential requirement for IR absorption 15
(viii) Selection rules for vibrational transition 16
(ix) Coupled vibrations, Overtones and Fermi resonance 16
(xi) Basis of classical theory of vibrational interactions 18
(xi) Sampling technique for IR analysis 18
(xii) Instrumentation for IR spectrum 19

(xiii) FTIR spectroscopy 20
(xiv) Scanning of infrared spectrum 20
(xv) Sketch of IR spectrum, presentation of data, functional group and fingerprint region 21
(xvi) Chart for characteristics group frequencies 23
(xvii) Factors influencing the position of frequencies 25
(xviii) Important methods of approach for interpretation of IR spectrum 28
(xix) Applications of IR spectroscopic technique 29

CHAPTER 3. ULTRAVIOLET SPECTROSCOPY 45-91
(i) Ultraviolet spectroscopy, its definition and range 45
(ii) Nature of transition in UV-spectroscopy 45
(iii) UV-spectrum 45
(iv) Measurement of UV-spectrum 46
(v) Different regions of UV-spectrum 48
(vi) Electronic transition 48
(vii) Information obtainable from UV-measurement, Chromophore and Auxochrome 49
(viii) Selection rules governing electronic transition 52
(ix) Absorption bonds 53
(x) Absorption laws 56
(xi) Bathochromic, Hypsochromic, Hyperchromic and Hypochromic shift 59
(xii) Factors affecting the position of UV-bands 60
(xiii) Woodward-Fieser rules 64
(xiv) Applications of UV-spectroscopy 66
(xv) Behaviour of benzene and its derivatives in UV-radiation 73

CHAPTER 4. NUCLEAR MAGNETIC RESONANCE SPECTROSCOPY 92-153
(i) Nuclear magnetic resonance and basic theory 92
(ii) Angular momentum and Zeeman effect 94
(iii) Quantisation, Saturation and Relaxation 95
(iv) Spin active property 96
(v) Relative number of nuclei in the higher and lower energy state 96
(vi) Proton magnetic resonance 97
(vii) Measurement of ^{1}HNMR spectrum 97
(viii) Instrumentation 98
(ix) Internal standard 101
(x) Solvents 102

(xi) Chemically equivalent protons 102
(xii) Field sweep and frequency sweep 105
(xiii) Chemical shift 106
(xiv) Characteristic Chemical shifts 108
(xv) Factors controlling the chemical shifts 109
(xvi) Shielding and deshielding effect 111
(xvii) Splitting of signals and coupling constant 115
(xviii) Order of NMR spectrum 124
(xix) Analysis of Second order NMR spectra 124
(xx) Experimental techniques for the simplification of non-first order NMR spectra 139
(xxa) Shift reagent 140
(xxb) Double resonance 143
(xxc) Spin tickling 145
(xxd) N.O.E. 145
(xxe) 2D-NMR 147
(xx1) Applications of NMR technique 152

CHAPTER 5. CARBON-13 NMR (CMR) SPECTROSCOPY 154-173
(i) ^{13}CNMR and its measurement 154
(ia) Off resonance technique 157
(ii) Characteristics of ^{13}CNMR Chemical shift, factors controlling Chemical shift and chart of ^{13}CNMR Chemical shift 157
(iii) DEPT Spectra 166

CHAPTER 6. MASS SPECTROSMETRY 174-225
(i) Mass spectrometry and its difference from other spectroscopic techniques... 174
(ii) Basic principle of mass spectrometry 174
(iii) Significance of unipositive ion in mass spectrometry 177
(iv) Importance of mass spectral technique in Organic Chemistry 177
(iva) Molecular ion 179
(ivb) Base peak 179
(ivc) Squasi molecular ion peak 179
(v) Factors controlling the formation molecular and fragment ions 179
(vi) Instrumentation for mass spectral analysis 180
(via) E.I. Method 181
(vib) Chemical ionisation method 182

(vic) FDI method 183
(vid) FAB technique 183
(vie) MALDI technique 184
(vif) T.O.F. Method 185
(vii) Determination of exact molecular formula from mass value 187
(viii) Nitrogen rule 188
(ix) Metastable ions 189
(x) General mode of fragmentation 192
(xa) Allylic cleavage 194
(xb) retro-Diels-Alder cleavage 195
(xc) α-Cleavage 196
(xd) Hydrogen transfer reaction 198
(xe) Ortho effect 199
(xf) Mclafferty rearrangment 200
(xi) Important features of mass spectra of hydrocarbons 202
(xii) Characteristic features of mass spectra of hydroxy compounds 205
(xiii) Characteristic features of mass spectra of carbonylic compounds 206
(xiv) Interpretation of mass spectral data 212

CHAPTER 7. **OPTICAL ROTATORY DISPERSION (ORD) AND CIRCULAR DICHROISM (CD)** 226-246
(i) ORD and CD 226
(ii) Measurement of ORD and CD 229
(iii) Cotton effect 231
(iv) Types of ORD Curves 234
(iva) Plain Curves 234
(ivb) Anomalous Curves 235
(v) Octant rule 235
(vi) Haloketone rule 237
(vii) Applications of ORD and CD 238
(viii) Benzoate sector rule 245

CHAPTER 8. **SOLUTIONS OF SOME COMPOSITE SPECTRAL PROBLEMS** 247-266

BIBLIOGRAPHY 267-268

PREFACE

Spectroscopic technique is one of the most powerful and versatile tools at the hands of scientist—particularly to Organic Chemists—for solving a wide range of complicated problems of Organic Chemistry involving molecular structure and sample analysis whose classical approach has no answer. Rapid development of this technique—both in instrumentation and interpretation in recent years has revolutionised our approach towards the problems of structure determination of complex organic molecules. As such, a study on this technique has been recognised as an essential part in the curriculum of Chemistry course in all universities.

Undoubtedly, a meaningful and indispensable method of understanding any subject—particularly like this technique—is to deal with a wide range of well-designed problems. Students can acquire thorough command over the subject only through the solution of problems by the application of the knowledge they have gained in the classroom study. With that end-objective this manual has been prepared to meet the needs of the students. Each class of spectroscopy has been presented in the form of a sufficient number of solved problems; and their answers have been provided in simple and lucid language which will help the students for deeper understanding of the subject. In addition, a separate chapter on composite problems has been provided at the end where structure determination of unknown compounds has been ascertained by the use of different spectral data to help the students in proper application of this technique. We are sure the book will be highly useful to both students and the teachers alike.

In preparing the manuscript of this manual a large number of books on this subject by national and international authors have been thoroughly consulted. We are grateful to them.

We record our profound regards to late Sri Shibsankar Mukhopadhyay, Former Assistant Teacher, Ramkrishna Sikshapith, Tumbuni, Birbhum, and Dr Durgasankar Mukhopadhyay, Former Head, Department of Bengali, Calcutta University, for their constant encouragement. Appreciation and thanks are due to (Mrs) Gouri Mukherjee and (Mrs) Sarada Mukhopadhyay for their lively and untiring inspiration during the preparation of the manuscript.

We are extremely grateful to the authorities and the production staff of New Central Book Agency (P) Ltd for publishing this book.

17 June, 2013
Phuler Madhu
Santiniketan
Birbhum

K S Mukherjee
Bodhisattwa Mukhopadhyay

CHAPTER 1

INTRODUCTION

1. What do you mean by the terms spectroscopy, spectrometry and spectrum?

➜ Spectroscopy is a subject which deals with the interaction of electromagnetic radiation with the matter. It is obvious that during such interaction different kinds of changes in the matter occur. As a result there may be variation in the intensity of electromagnetic radiation. The instruments by which this variation in intensity of electromagnetic radiation is measured are known as spectrometers.

The change in intensity of electromagnetic radiation is expressed as a graph. This graph is a plot of change of intensity against one of the parameters (wavelength or frequency) of the electromagnetic radiation. This graph is termed as the spectrum of the matter undergoing interaction with the electromagnetic radiation.

Spectrometry, in principle, is different from the usual spectroscopy. Spectroscopic techniques depend upon the changes in the intensity of electromagnetic radiation while in spectrometry, as in mass spectrometry, the photographic plate is calibrated to measure the various value of mass and charge ratios (m/z).

2. Discuss the nature of electromagnetic radiation.

➜ It is a kind of energy which has two oscillating fields (vectors) electrical and magnetic. This radiation has following characteristic features:

(i) It has tremendous velocity.

(ii) It can move both in the presence and absence of any medium. Its velocity in vacuum is approximately 3×10^{10} cm/sec. Light is an electromagnetic radiation.

(iii) Another important characteristic of this radiant energy is that it exhibits dual character—exhibiting both wave and particle character. Phenomena like interference, refraction, diffraction, etc. can be best interpreted in terms of wave nature while photo-electric effect can be properly explained on the basis of particle nature. However, majority of phenomena may be explained on the basis of wave nature.

(iv) Both the fields electrical and magnetic of this radiation oscillate in a plane perpendicular to each other and their resultant (direction of propagation) is also perpendicular to both the fields, i.e. the three components electric, magnectic and resultant are perpendicular to each other. On the basis of these characteristics this radiation may be pictured as below:

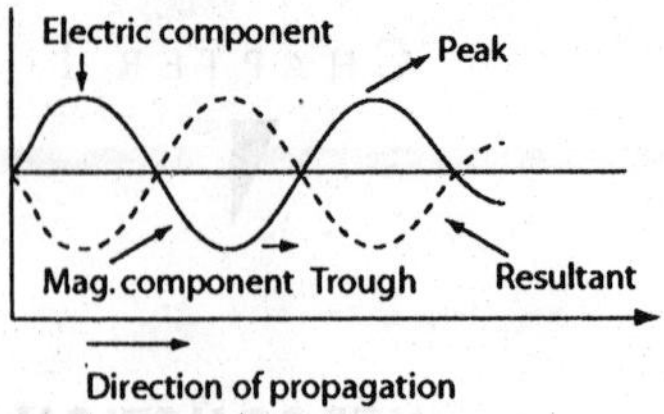

The above picture clearly reveals that this radiation is wave in nature. Thus electromagnetic radiation may be regarded as an wave, i.e., as electromagnetic wave.

(v) This electromagnetic radiation is expressed by certain parameters as mentioned below:

(a) *Wavelength*: Wavelength is the distance between two consecutive peaks or troughs. It is indicated by λ (lambda) which is expressed by the following units.

(i) Å (angstrom); 1 Å = 10^{-8} cm.

(ii) μ (micrometer), 1 $\mu = 10^{-4}$ cm.

(iii) mμ (millimicrometer), 1 m$\mu = 10^{-7}$ cm.

It is also known as nanometer, i.e., mμ = nm.

(b) *Frequency*: It is the number of waves passing through a point in a second (unit time) and is indicated by ν (nu).

(c) *Wave number*: It is just the inverse of wavelength and is represented by $\bar{\nu}$ which is equal to $\frac{1}{\lambda}$.

These three parameters are related to each other in the following manner:

$$\nu \propto \frac{1}{\lambda}$$

$$\text{or,} \quad \nu = \frac{c}{\lambda},$$

where c is a constant and is the velocity of electromagnetic radiation, i.e., light.

$$\text{or,} \quad \nu = c.\frac{1}{\lambda} = c\bar{\nu} \quad (\because \bar{\nu} = \frac{1}{\lambda})$$

The above relationships may be represented in the tabular form as under:

Quantity	Relationship	Unit
Wavelength (λ)	$\lambda = \frac{1}{\bar{\nu}} = \frac{c}{\nu}$	Å, μ, μm, nm
Wave number	$\bar{\nu} = \frac{1}{\lambda} = \frac{\nu}{c}$	cm^{-1}
Frequency	$\nu = \frac{c}{\lambda} = c\bar{\nu}$	S^{-1} (Hertz, Hz)

(vi) According to Quantum theory, electromagnetic radiation consists of a stream of energy particles known as *photons*, which travel in the direction of propagation of electromagnetic radiation with the velocity of light. Each photon has certain amount of energy. For this reason photon is also termed as quanta of energy. The energy, E, of the photon is related to the frequency of the radiation (Plank's equation) by the following equation,

$$E \propto \nu$$

$$\text{or,} \quad E = h\nu$$

where h is Plank's constant (6.625×10^{-27} erg/second, or 6.625×10^{-34} joule/second).

$$\text{or,} \quad E = h.\frac{c}{\lambda} \ (\because \nu = \frac{c}{\lambda})$$

The energy E of the photon is called quantum of energy. Thus, energy of electromagnetic radiation is inversely proportional to the wavelength but does not depend upon the intensity of radiation.

A few problems may be cited in this context to show how the relationships between different parameters and energy content may suitably be applied.

Problem 1: Convert 1500Å to cm, cm^{-1}, erg and Hz.

➡*Solution*: $\lambda = 1500\ \text{Å} = 1.5 \times 10^3\ \text{Å}$

or, $1.5 \times 10^3 \times 10^{-8}\ \text{cm} = 1.5 \times 10^{-5}\ \text{cm}$

Now $1.5 \times 10^{-5}\ \text{cm} = \dfrac{1}{1.5\times10^{-5}}\ \text{cm}^{-1} = 0.66 \times 10^5\ \text{cm}^{-1}$

$$0.66 \times 10^5\ \text{cm}^{-1} = 0.66 \times 10^{+5} \times 1.985 \times 10^{-10}\ \text{erg/mole}$$
$$= 0.66 \times 1.985 \times 10^{-5}\ \text{erg/mole.}$$

Again, $\nu = \dfrac{c}{\lambda} = \dfrac{3\times10^{10}\ \text{cm}}{\lambda} = \dfrac{3\times10^{10}\ \text{cm}}{1.5\times10^{-5}\ \text{cm}} = 2 \times 10^{15}\ \text{Hz}$

Problem 2: A local radio station transmists (i) at 90 MHz on high frequency and (ii) at 800 kHz on medium frequency. Calculate the wavelength of these transmission.

➡*Solution*: The broadcast frequencies are given.

We know, $\nu = \dfrac{c}{\lambda} = \dfrac{3\times10^{10}\ \text{cm}}{\lambda}$

Again, 1 Megahertz = 1000,000 Hz

(i) $\nu = 90\ \text{MHz} = 90 \times 1000{,}000\ \text{Hz} = \dfrac{3\times10^{10}}{\lambda}\ \text{cm}$

$(\therefore \nu = \dfrac{c}{\lambda}, c = 3 \times 10^{10}\ \text{cm or, } \lambda = \dfrac{c}{\nu})$

or, $\lambda = \dfrac{3\times10^{10}}{90\times1000000}\ \text{cm} = \dfrac{10^3}{3}\ \text{cm}$

$= 3.33 \times 100\ \text{cm} = 333\ \text{cm}$

(ii) $\nu = 800\ \text{KHz} = 800 \times 10^3\ \text{Hz}$

$$\therefore \lambda\left(\frac{c}{\nu}\right)$$

$$\lambda = \frac{3\times10^{10}}{800\times10^{3}}\ \text{cm} = 0.37 \times 10^5\ \text{cm} = 3.7 \times 10^4\ \text{cm}$$

3. Indicate the different regions of electromagnetic radiation (Electromagnetic Spectrum) or range.

➜ The whole range of electromagnetic radiation although has certain properties in common but the different portions are sufficiently different. So the different portions of electromagnetic radiation require different techniques for handling them. On this basis, the whole range of electromagnetic radiations has been divided into several regions depending on the wavelength as shown in the table below:

Regions	Cosmic rays	Gamma rays	X-rays	UV rays	Visible	Near IR	Far IR	Micro-wave	Radio-frequency
Wave length	10^{-12} cm	10^{-10} – 10^{-9} cm	10 nm	200 – 400 nm	400 – 800 nm	1 – 2.5 m	2.5 – 25 m	1 – 100 nm	1 cm – 15 m
Type of molecular spectrum	Mossbanur (Nuclear transitions)		Core orbitals (Electronics)	Electronic (valency orbitals)		Rotational and Vibrational		Rotational	NMR (spin-orientation)

4. How are spectroscopic techniques classified?

➜ When electromagnetic radiation passes from vacuum to the surface of the matter the electrical field (vector) of the radiation interacts with the matter. As a result radiation may either be absorbed or emitted. Hence spectroscopy may be classified into two major groups — (i) absorption spectroscopy and (ii) emission spectroscopy, and the corresponding spectra are respectively termed as absorption and emission spectra.

Again, the matter which interacts with the electromagnetic radiation may be an atom or a moelcule. Consequently, spectroscopy may further be divided into two classes— (i) atomic spectroscopy and (ii) molecular spectroscopy. The corresponding spectra are as usual called atomic and molecular spectra.

Thus, spectroscopy, in general, are classified into four categories as mentioned below:

(i) Atomic absorption spectroscopy
(ii) Atomic emission spectroscopy
(iii) Molecular absorption spectroscopy
(iv) Molecular emission spectroscopy.

5. How would you differentiate between atomic and molecular spectra?

➜ The atoms in a molecule can rotate and vibrate with respect to each other. These rotations and vibrations have discrete energy states. The transition from ground state rotational and vibrational levels to the corresponding levels in the excited electronic state can occur. As a result broad spectra known as band spectra are obtained.

But in the case of atom, the species neither vibrate nor rotate as a molecule does. So the atom does not have rotational or vibrational energy levels. Hence the transition between energy levels of an atomic species results in very sharp lines.

In fact, whether the radiation absorbed or emitted by a species is from a molecule or an atom is ascertained by the fact that a molecule furnishes a *band spectrum* and the atom a *line spectrum*.

6. What is the objective of molecular absorption spectroscopy?

➜ In molecular absorption spectroscopic technique electromagnetic radiation is allowed to pass through a molecule when energy associated with the radiation is absorbed by the molecule. After absorption of energy (electromagnetic radiation) by the molecule two things may occur.

(i) The absorbed energy (radiation) may be stored by molecule which is utilised for the cleavage of bonds, i.e., *photochemical reaction occurs*.

(ii) By absorbing energy molecule is excited and is promoted to the energetically higher level. After excitation the molecule returns to its normal (ground) state by emitting the absorbed energy. The energy is emitted as bands. It is obvious that since the bands are emitted from the molecule they must have the characteristics of the molecule which absorbs the energy. For this reason these bands are known as *molecular spectra*. These spectra are collected in a suitable device and are analysed to get the information regarding the characteristics of the molecule which has absorbed the electromagnetic radiation (energy). This is, in fact, the subject matter of molecular absorption spectroscopy.

7. What actually happens when molecule is excited?

➜ It is quite obvious that during excitation i.e., during absorption of electromagnetic radiation, the energy of the molecule is increaased and hence its energy levels are changed. There are several types of energy levels in a molecule. So during absorption of electromagnetic radiation these different types of energy levels are changed. Broadly, there are two types of energy levels in a molecule, viz., (i) kinetic and (ii) potential energy. It may be pointed out that kinetic energy levels are degenerate and hence practically kinetic energy levels are not changed. So only the potential energy levels are changed. Now the question arises how many potential enrgy levels are there? In order to find out the answer to this question let us consider a diatomic molecule AB. Such a molecule consists of two nuclei corresponding to the atoms A and B and their accompanying electrons. The electrons of these atoms can exist in a number of energy levels. When excited i.e.when energy is absorbed, transition of an electron from low energy orbit to high energy orbit occurs and on returning to the low energy level spectral lines are emitted. If this electronic transition had been the only mode of absorption or emission of energy by the molecule, the electronic spectrum of a molecule would be a line spectrum similar to that of atoms. But in fact spectral bands are obtained. This suggests that molecule can utilise the absorbed energy in other ways as well. The absorbed energy may set the molecule as a whole rotating or to cause the vibration of the atoms A and B relative to one another within the molecule.

It is thus evident from the above discussion that there may be three types of potential energy levels such as—(i) rotational, (ii) vibrational and (iii) electronic levels. Under normal condition these energy states are quantised. All these three energy levels mainly originate due to the arrangement of electrons. Hence spectra obtained from the transitions of these levels are termed as electronic spectra.

It may be mentioned in this connection that when electrons or the nuclei of certain elements are subjected to a strong magnetic field, additional quantised energy levels are produced due to their magnetic properties and hence spectra may be obtained due to transition between lower and higher levels of these additional quantised levels. Such spectroscopy is known as *induced spectroscopy* which are of two types:

(i) Electron Spin Resonance (ESR), which involves a change in the direction of the resultant electron spin in the magnetic field

(ii) Nuclear Magnetic Resonance (NMR) which involves the change in the direction of the nuclear spin in the presence of magnetic field.

8. Discuss in short the different types of molecular absorption spectroscopic techniques applied in organic chemistry and the information obtainable from each of these techniques.

➜ Although almost all parts of electromagnetic spectra are used for studying matter in organic chemistry, however, primarily the following types of molecular spectroscopy are commonly used.

(i) *Infrared spectroscopy*: In this spectroscopy electromagnetic radiation from mid infrared region is allowed to pass through organic molecule. In this technique vibrational and rotational transitions of the molecule occurs. The wavelength range is 4000 to 650 cm^{-1} and the corresponding energy range is 1-35 kcal/mole. Generally information reagrding functional groups are obtained.

(ii) *Ultraviolet and visible spectroscopy*: This branch of spectroscopy deals with the transitions between electronic energy levels of a molecule brought by the absorption of ultraviolet or visible radiation. The wavelenght range is 200-400 nm and energy range is 35-150 kcal/mole. Information regarding the nature of π-electron and non-bonding electron systems may be obtained.

(iii) *Nuclear magnetic resonance* (NMR) *spectroscopy*: In this technique interaction of electromagnetic radiation with spin-active nuclei like ^1H, ^{13}C is involved. In this study electromagnetic radiation from radio wave region (ν, 60-600 MHz) is applied and hydrogen and carbon atoms in different environments can be detected and counted.

(iv) *Mass spectrometry*: This technique is based on the principle that molecules in the gaseous phase are converted by bombarding with electron beams into parent and fragment ions which are sorted into their mass numbers. This techniques is frequently used to determine the relative molecular mass (molecular weight) and to deduce the molecular structures from the fragments produced.

9. How the absorption data are expressed in spectroscopic techniques?

➜ In molecular absorption spectroscopic techniques, electromagnetic radiation of a particular frequency for a particular spectroscopy is allowed to pass through a sample and the amount of transmitted radiation is detected. The instrument (spectrometers) compares the intensity of the transmitted light (radiation) with that of incident light. Automatic instruments gradually and continuously change the frequency and an automatic recorder plots a graph of absorption against frequency or wavelength. This graph is the desired spectrum. The particular position in the spectrum is specified by some convenient units.

However, because of the great difference in the wavelength of different region it is not convenient to use the same unit for the all the spectroscopic methods. In the infrared regions (4000–650 cm^{-1}), the wavelengths are expessed in micrometres or as the reciprocal of wavelength ($\frac{1}{\lambda}$) in centimetres termed as wave number ($\overline{\nu}$) while in uv (200-400 nm) and visible (400-800 nm) wavelengths are expressed as nanometres (nm). In NMR absolute frequencies are used rather than the wave numbers. Thus a wavelength of 6 meters corresponds to a frequency of 6×10^7 Hz or 60 Mhz where Hz represents hertz, defined as cycles per second and Mhz is megahertz (10^6 Hz).

10. Write a brief account of Fourier transform spectroscopy?

➜ Fourier transform spectroscopy is a process of producing electromagnetic spectra. The conventional method of producing spectra as a plot of frequency versus intensity is a very slow process. In such conventional method, each point of the spectrum has to be recorded separately – the spectrometer is set to start reading at one end, the frequency is swept smoothly across the whole span of the spectrum and the detector signal is monitored and recorded. On the other hand, Fourier transform technique is a rapid process. It is based on the fact that the intensity of electromagnetic radiation is related to two parameters, time and frequency. Hence spectral data may be plotted either as intensity versus time or intensity versus frequency. Spectra obtained in both the cases furnish the same kind of information and in fact, one form is the mathematical reciprocal of the other. The former (intensity vs. time) is referred as *time domain spectra* while latter is known as *frequency domain* spectra.

In principle, Fourier transform technique is applicable to both emission and absorption spectroscopies, but it is relatively easier to visualise Fourier transform spectroscopy in terms of emission and that is why in most of the cases it is expressed in terms of emission.

In this process the sample is subjected to a strong broad electromagnetic radiation band of frequencies called *pulse* of the required region for a short time (10-100 msec) when all the absorvable functionalities and nuclei present in the molecule absorb their respective frequencies. Then the radiation is stopped for a short time (during which measurement is carried out). The molecules or nuclei relax to the original equilibrium state and emit electromagnetic radiation at the same frequency and intensities as the absorbed radiation. These intensities are detected by a detector. The whole process is repeated several times. The data are presented as total intensity versus time graph as shown below. This graph is known as *time domain spectrum*. In such graph intensity of peaks or signals is maximum after the pulse is stopped and diminishes rapidly with time.

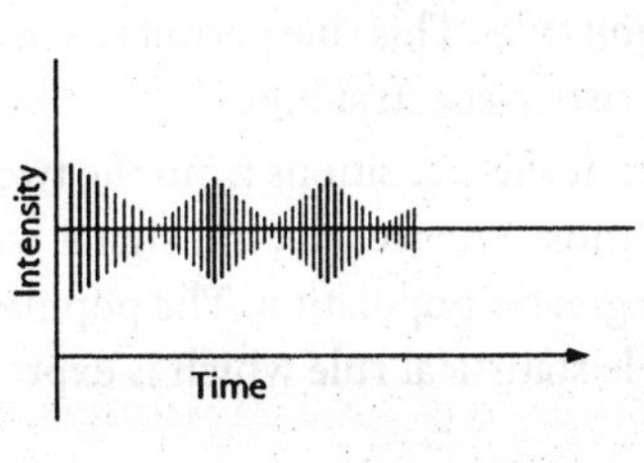

Time domain spectrum

This time domain spectrum consisting of the sum of the intensities of all radiations of different frequencies at different time is practically of no help in interpretation. Fortunately there is a simple and quite general method to resolve the different frequencies along with their intensities for different functionalities. This is a mathematical process developed by French mathematician Jean Baptiste Fourier in the early 1800 and is usually known as *Fourier Transform* after his name. It is a process of integration of the complex waveform which is very conveniently carried out by computers.

It may be mentioned that although Fourier transform spectrum is confined to time domain signals and monitoring the detector as a function of time but it can equally be utilised in terms of interference phenomenon as is observed in IR measurement. The FT IR-spectroscopy is based on the principle of interference. That is why sometimes it is also termed as interferometric infrared spectroscopy.

11. Anotate the following terms.

(a) *Signal to Noise Ratio*: Since in modern spectrometers electronic amplification procedure is used to magnify the signals, every recorded spectrum has a background of randam fluctuation due to spurious electronic signals generated in the amplifying equipments. These fluctuations are known as noise. In order to identify the spectral peak, its intensity must be three to four times more than that of the noise fluctuations. By using computer properly such intensity may be maintained.

(b) *Width of Spectral Lines*: In most of the cases of spectral measurements, broad peaks instead of sharp lines are obtained. Several factors are responsible for this broadening of peaks as noted in the sequel.

(i) *Collisions*: In liquids and gasses, atoms or molecules are in continious motion and collide each other. As a result some deformations of the particles occur and so energies particularly of the outer electrons in each are perturbed. This causes widening in the UV and visible spectral lines. Similarly, vibrational and rotational spectra are broadened due to this collision.

(ii) *Heisenberg uncertainty principle*: The uncertainity principle of Heisenberg is also responsible for broadening of the spectral lines. According to this principle, the energy of a particular state of a system is uncertain.

(iii) In liquids and gases since the motion of the particle is random shifting of frequencies to both high and low occurs i.e., *Doppler Effect* occurs. So the spectral line is broadened.

(c) *Intensity of spectral lines*: Intensity of specral lines is primarily controlled by three factors as discussed below:

(i) *Transition probability*: Generally the process of transition probability is the deduction of *selection rules*. This rule permits us to decide between which levels transition will give rise to spectral line.

(ii) *Population of states*: If the transitions from the two levels to a third are equally probable, then the most intense spectral signal will definitely arise from the level which has the greater population. The population of a set of energy levels is guided by a simple statistical rule which is expressed as

$$\frac{N_{upper}}{N_{lower}} = \exp(-\Delta E/KT)$$

where N is the total number of molecules distributed between two levels, T is the temperature in kelvin, *K* is a constant and $\Delta E = E_{upper} - E_{lower}$. The constant *K* is known as Boltzmann distribution and it has a value of 1.38×10^{-23} J/kelvin.

(iii) *Path-length of the sample*: The intensity of a signal depends upon the path-length through which electromagnetic radiation traverses. The relationship between path-length (l), concentration (C) and the intensities of incident (I_0) and emergent (I) radiations is usually expressed by *Lambert-Beer* law. This law is expressed as

$$\log\left(\frac{I_0}{I}\right) = \varepsilon.\,c.\,l$$

or, $A = \varepsilon.\,c.\,l$ where $\log\left(\frac{I_0}{I}\right) = A$, is the *absorbance* or *optical density* and ε (epsilon) is a constant known as *molar extinction coefficient*. It is purely a number and its value is very high and is generally expressed as log, i.e., as log ε.

12. (a) Convert 10 joules into Hz.

➻ We know $E = h\nu$ or, $\nu = \frac{E}{h} = \frac{10\text{J}}{6.6262\times10^{-34}\text{ Js}} = 1.5092 \times 10^{34}$ Hz

(b) Convert 10 Mhz into wavelength and wave number.

$$\lambda = \frac{c}{\nu} = \frac{3\times10^{10}\text{ cm/s}}{10\times10^{6}\text{ Hz}} = 3 \times 10^{3}\text{ cm}$$

Again, $\bar{\nu} = \frac{1}{\lambda} = \frac{1}{3\times10^{3}}\text{ cm}^{-1}$

(c) The wavelength of radiation absorbed has been found to be 121.567 nm. Express this in frequency (Hz).

We know, $\nu = \frac{c}{\lambda} = \frac{3\times10^{10}\text{ cm/s}}{121.567\text{ nm}} = \frac{3\times10^{8}\text{ m/s}}{121.567\times10^{-9}\text{ m}}$

$= 2.466 \times 10^{15}$ Hz (approx).

CHAPTER 2

INFRARED SPECTROSCOPY

1. What happens when electromagnetic radiation of infrared region is passed through a molecule?

or, What causes an infrared spectrum? What transition is involved and how is it induced?

➻ In infrared spectroscopy, electromagnetic radiation of infrared region is passed through the molecule. The wavelength range of infrared region is 0.25–50 μ or 4000–650 cm^{-1} and its energy range is 1–35 kcal/mole. By this amount of energy transition of the molecule from one vibrational energy level to another occurs. The transition is induced by the absorption of energy of infrared light of appropriate frequency (ν) which matches with the energy gap between the two levels.

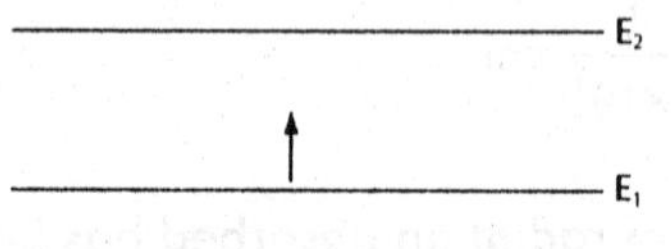

$$\Delta E = E_2 - E_1 = h\nu$$

2. Explain vibrational motion or what is meant by vibrational motion?

➻ The vibration of the molecule means vibration of the atoms of the molecule and it is a to and fro motion, just like a simple harmonic motion. It has been observed that under ordinary condition molecules are in a state of constant vibration with the individual bond having its characteristic stretching and bending frequencies and can absorb energy. The vibration of two atoms joined together by a chemical bond may be illustrated with the help of a spring and ball model. It is known that every spring has a vibrational frequency.

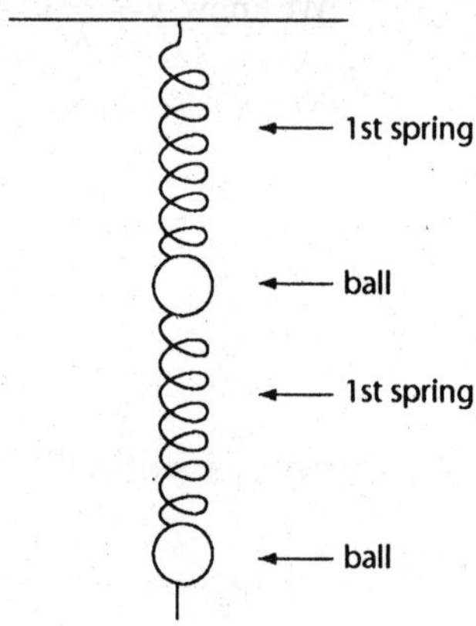

Spring and ball model

Let us consider a ball connected to a spring which is held hanging by a hook and is stretched to a fixed distance and then released as shown in the figure. The ball starts moving to and fro with a certain frequency and amplitude, i.e., a simple harmonic motion is generated (Hook's Law). Now if another ball with a spring connected to it is attached to the first ball and the first ball is stretched to the same fixed distance and then released as before it is observed that this time amplitude and frequency become different from those of previous one.

From the above observations it is evident that the vibrational motion in between two balls is dependent upon the following factors.

(i) strength of the spring
(ii) masses of balls attached to it
(iii) the system as a whole.

Now if spring is substituted by chemical bond and the balls are replaced by atoms, we can conclude that the molecular vibration is associated with the following factors:

(i) strength of the bond
(ii) masses of the atoms connected to the bond
(iii) the nature of the molecule as a whole.

3. Explain why infrared spectroscopy is called rotational and vibrational spectroscopy and why the infrared (IR) peaks are not very sharp?

➜ It is known that in IR spectroscopy the energy range is 1–35 kcal/mole. This energy range is much higher than that of rotational transitions for which energy required is 0.0001–0.1 kcal/mole. So in IR spectroscopy along with vibrational transitions, rotational transitions also occur. For this reason infrared spectroscopy is also termed as rotational-vibrational spectroscopy. It may be mentioned that since along with vibrational transitions rotational transitions also occur in this spectroscopy, the peaks in IR spectrum are not so sharp like peaks in rotational spectra. In fact, the IR peaks are due to the large number of both rotational and vibrational transitions.

4. What information can be obtained from IR spectral measurement?

➜ Since the molecular vibration depends upon the strength of the bond it is logical to assume that information regarding the nature of chemical bonding, particularly nature of functional group may be obtained.

5. Describe the nature of molecular vibrations in IR spectroscopic technique?

➜ It is known that vibrational motion of a diatomic molecule is quantised. This means that the atoms of the molecule can vibrate at certain frequencies, and transition from one level

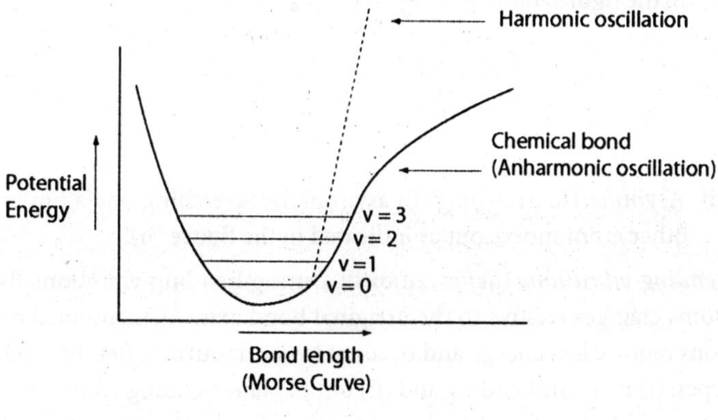

to another can occur only when energy equal to the difference between two vibrational energy levels is absorbed by the molecule. At ordinary temperature most of the molecules of a sample exist in the lowest energy level and the potential energy diagram approximates that of a harmonic oscillator, i.e., the potential energy diagram is a symmetrical parabola. But at higher temperature deviation occurs as shown in the figure in page 11. (Morse curve). The chemical bonds are, therefore, anharmonic oscillators. If sufficiently large amount of energy is added, bond may break. However, for small oscillations (say, $v = 0$ to $v = 1$) the behaviour may remain harmonic. In absorption of radiation from IR region of particular frequency vibrational transition occurs. The transition from ground state ($v = 0$) to the first excited state ($v = 1$) absorbs light most strongly and give rise to intense bands known as *fundamental vibration.* But the transition from the grand state to the second ($v = 2$) or third ($v = 3$) excited state give rise to weak bands known as *overtones.* It may be mentioned that since IR absorption bands due to fundamental vibration are most intense. they are the most important for chemical study. However, in IR spectroscopy most of the overtones are found in near infrared region beyond 4000 cm^{-1}. Such absorption are very weak.

6. Discuss in detail different types of fundamental molecular vibrations.

➡ It is known that a molecule may be regarded as a system of balls (atoms) and springs (bonds). When such a system (molecule) absorbs radiation, it is set into vibrations as a result of which deformations of bonds occur. The deformations of bond may be due to two factors—(i) stretching and (ii) bending of the bond. Consequently there are two types of fundamental molecular vibrations which are termed as—(a) Stretching vibrations (b) Bending vibrations.

We know that to stretch a spring more energy is required than to bend it, so bending vibrations require less energy and occur at shorter wave numbers (1600–650 cm^{-1}) than the stretching vibrations (4000–1600 cm^{-1}).

(a) *Stretching Vibrations*: These vibrations are rhythmic back and forth movements of atoms along the covalent bond axis and in this process the distance between the atoms increases or decreases but the atoms always remain in the same bond axis. As mentioned before these vibrations require higher energy and occur at higher frequency. There are two types of stretching vibrations:

(i) *Symmetric stretching*: In this type of stretching the movement of the atoms with respect to a particular atom in the molecule occur in the same direction as shown in the figure (i).

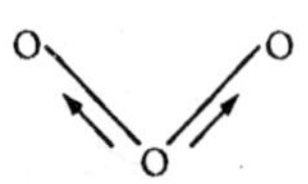

(i)

(ii)

(ii) *Asymmetric stretching*: In asymmetric stretching one atom moves in while the other atom moves out as indicated in the figure (ii).

(b) *Bending vibrations* (*deformations*): During bending vibrations the position of the atoms changes relative to the orriginal bond axis. As mentioned earlier, these vibrations require less energy and occur at lower frequrncy. Bending vibrations are of two types: (i) in-plane bending and (ii) out of plane bending vibration.

(i) *In-plane bending vibration*: A bending vibration in which the atoms remain in the same plane (known as nodal plane of the system), is called in-plane bending vibration. There are two kinds of in-plane bending vibrations:

(a) In-plane *scissoring* bending: In this type, two atoms approach each other, i.e., swing in concert towards opposite directions (Figure a).

(a) (b)

(b) *In-plane rocking deformation*: In this type the movement of atoms takes place in the same direction (Figure b).

(ii) *Out-of-plane bending vibration*: When the atoms bend out of nodal plane, the bending mode is called out-of-plane bending. They are of two types.

(a) *Out-of, lane wagging deformation*: In this kind of bending two atoms move up and below the plane with respect to central atom (Figure c)

(c) (d)

(b) *Out-of-plane twisting deformation*: In this type, one atom moves up the plane while the other moves down the plane with respect to the central atom (Figure d).

It may be mentioned that for all these different types of fundamental vibrations energy may be absorbed and hence transitions may occur. Again, these different types of vibrations may absorb energy separately or jointly. So there are possibilities of appearance of large number of IR bands in the the spectrum.

7. What do you mean by number of fundamental vibration in polyatomic molecules? How are they related to number of degrees of freedom in the molecule?

➜ The term fundamental vibrations according to quantum mechanical treatment, means the first vibrantional transitions from the zeroth vibrational level (ground state) to the first vibrationally excited state, i.e., $v_0 \rightarrow v_1$. A polyatomic molecule may exhibit a large number of fundamental vibrations. The number of fundamental vibrations is directly related to the number of degrees of freedom in the molecule. In a polyatomic molecule no atom loses its degrees of freedom. Thus, the number of degrees of freedom in a polyatomic molecule is the total number of degrees of freedom of all the atoms present in it. The number of degrees of freedom of any atom means the total number of co-ordinates required to locate the atom in the space. We know that to specify any atom in the space three co-ordinates (x, y, z) are necessary. So any atom in a molecule has three degrees of freedom. Thus, for a polyatomic molecule having 'n' number of atoms, the total number of degrees of freedom is $3n$. These are also called translational degrees of freedom.

When atoms combine to form a molecule although there is no loss in the total number of degrees of freedom but some of the translational motions are converted to rotational and vibrational motions. Thus, in n-atomic molecule the total number of degrees of freedom ($3n$) consists of translational, rotational and vibrational degrees of freedom.

$3n$ = Translational + Rotational + Vibrational degrees of freedom

or, Vibrational degrees of freedom =

$\equiv 3n$ – (Rotational + Translational degrees of freedom)

Now when the molecule as a whole moves its motion which is translational, can be described in terms of the motion of its centre of gravity and this requires only three co-ordinates. Thus, out of $3n$ degrees of freedom only three are translational for the molecule.

However, number of rotational degrees of freedom depends upon the nature of the molecule. The nature of the molecule may be linear or non-linear. When the molecuel is linear in nature, its number of degrees of freedom is only two. This is because the rotation along the axis of linearity (internuclear axis) produces non-effective change in rotational level. Hence the total number of fundamental vibrations for linear molecule is

$$3n - 2 - 3 = 3n - 5 \quad \text{(i)}$$

But when the molecule is non-linear in nature, the molecule is free to undergo rotaional motion along the three co-ordinates and so the number of rotational degrees of freedom is three. Therefore, the total number of fundamental vibrations for the non-linear molecule is

$$3n - 3 - 3 = 3n - 6 \quad \text{(ii)}$$

These 3n–5 or 3n–6 vibrational modes are also called 'normal modes' of vibration. Out of these, there are (n–1) stretching and rest are bending or deformation modes.

Based on the above expressions (i & ii), number of fundamental vibrations for a few simple molecules are given in the table below:

No. of atoms (n) in a molecule	Degrees of Freedom (3n)				
	Transitional	Rotational	Vibrational (Fundamental)		
				Stretching (n – 1)	Bending
(a) Linear Molecule (3n–5)					
1. O=C=O (n=3)	3	2	$3 \times 3 - 5 = 4$	2	2
2. HCl (n = 2)	3	2	$3 \times 2 - 5 = 1$	1	–
3. H–C≡C–H, (n = 4)	3	2	$3 \times 4 - 5 = 7$	3	4
4. H–C≡N, (n = 3)	3	2	$3 \times 3 - 5 = 4$	2	2
(b) Non-linear Molecule					
1. C_6H_6 (n = 12)	3	3	$3 \times 12 - 6 = 30$	11	19
2. H_2O (n = 3)	3	3	$3 \times 3 - 6 = 3$	2	1
3. NH_3 (n = 4)	3	3	$3 \times 4 - 6 = 6$	3	3

It may be remembered that ($3n$–5) and ($3n$–6) are the theoretical number of fundamental vibrations. In practice, they are seldom obtained.

8. What is the essential requirement for IR absorption?

➜ Although theoretically a large number of fundamental vibrations may occur but all transitions may not have equal intensity, even some of the vibrations may not occur at all. It has been observed that the most strong transitions (intense bands) occur for those which are associated with the change of dipole moment—greater the dipole moment changes more intense becomes the absorption of energy and hence more intense becomes the IR absorption band. It must be mentioned in this connection that during the absorption of IR energy, there must be appreciable change in the dipole moment of the molecule and this is the essential requirement for IR absorption. It has been found that some compounds appear to have dipole moment but they do not produce any observable IR band. This is because in such compounds there is no appreciable change in dipole moment during absorption of IR energy. Thus,

(a) 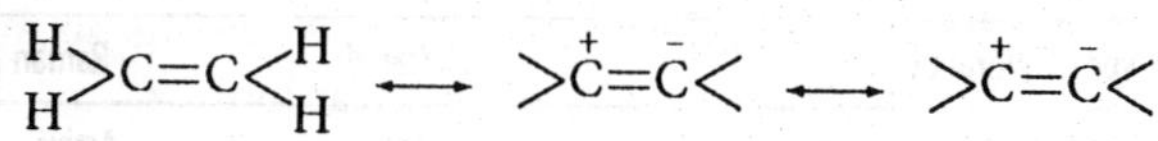

So there is no net change in dipole moment.

(b) 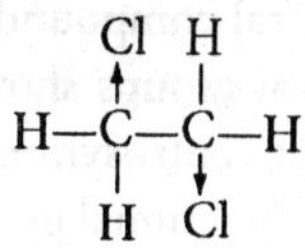 In this compound there is no change in dipole moment.

(c) 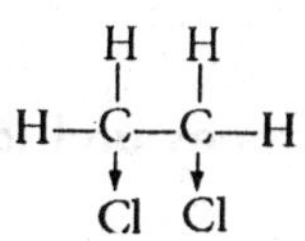 In this compound, appreciable change of dipole moment occurs.

The molecules in which appreciable change in dipole moment occurs during IR absorption are known as *IR-active compounds* such as in (c). But the molecules in which there is no appreciable change in dipole moment during IR absorption are termed as *IR-inactive compounds* as in (a) and (b). Only IR active compounds show absorption. The IR inactive compounds may produce observable bands in Raman Spectra, i.e., they are Raman active.

9. Explain why change in dipole moment is essential for IR absorption.

➜ Electromagnetic radiation consists of two component vectors—electrical and magnetic. Again, any molecule may be looked upon as an assembly of electrons which are electrical in nature. So it is expected that the electrical component of electromagnetic radiation (IR) plays the most important role in the absorption of energy. Due to dynamic character of electrons as well as the continuous vibration of a bonding system any bond may be regarded as a dipole (incipient). The electrical component of electromagnetic radiation attracts the positive pole of such dipole and hence the position of the bonding atoms are changed. As a result bond may be expanded. The expansion of the bond becomes smooth if it has the inherent tendency to expand. The inherent tendency for expansion of the bonds depends upon the difference in the power of attraction of the electron pair of the bonding atoms. So greater the difference in attraction of electrons (i.e., dipole moment) more is the tendency of the bond to expand and hence smoother becomes the absorption of IR energy and so more intense becomes the IR band. For this reason, change in dipole moment is the essential factor for IR absorption.

10. State the selection rules for vibrational transition.

→ It has been established that infrared light is absorbed when a change in dipole moment of the molecule takes place. Total symmetry about a bond will eliminate certain absorption bands so that the number of absorption bands does not coincide exactly with the number of fundamental vibrations and the molecule displays somewhat simplified spectrum. Thus, some of the funadamental vibrations are infrared active while others are not. This is governed by the selection rules which may be stated as below:

(i) In a molecule with a centre of symmetry, the vibrations symmetrical about the centre of symmetry are inactive in the infrared but active in the Raman.

(ii) The vibrations, without a centre of symmetry, are active in infrared but inactive in the Raman.

The above two selection rules may be tabulated as shown below:

Sl. No.	Type of Vibration	Infrared	Raman
1.	Centrosymmetric	Inactive	Active
2.	Non-centrosymmetric	Active	Inactive

It may be mentioned that the scrutiny of various bands in chemical compounds reveals that in general, those bonds which are associated with functional groups show appreciable dipole moment change because functional groups are not centrosymmetric. So they are infrared active. Thus from IR absorptions nature of the functional groups may be ascertained.

11 What do you mean by (a) Coupled Vibrations, (b) Overtones and (c) Fermi Resonance? How do they originate?

→ (a) *Coupled Vibrations*: In some cases it has been observed that two different vibrational levels have nearly the same energy. In such situation vibrational coupling takes place between two bonds which are reasonably close to each other in the molecule. The coupling vibrations may both be fundamental, or fundamental vibration may couple with the overtone of some other vibrations. The first one, i.e., when coupling takes place between two fundamental vibrations, the vibration is termed as *coupled vibration* and the corresponding absorption band is known as *coupled vibrational band*.These bands are very weak compared to fundamental vibrations. Thus for a methyl group, the three hydrogen atoms have the same mass and the C—H bonds are all of similar strength. So the vibrational mode is coupled and appears as symmetric and antisymmetric CH_3 stretching vibrations.

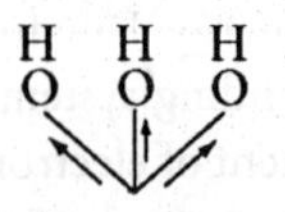

Symmetric Stretching Vibrations

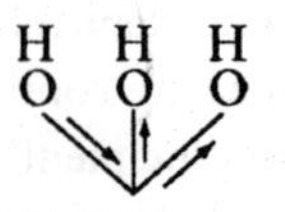

Asymmetric Stretching Vibrations

Coupled vibrations within a group of three atoms as in the case of methylene group, —CH_2; nitrogroup —NO_2 and amino group —NH_2 are very common. Each of these groups shows two bands corresponding to symmetric and antisymmetric (asymmetric) stretching vibrations. In general, asymmetric stretching vibrations are of higher frequency.

(b) *Fermi Resonance*: When vibrational coupling takes place by the interaction between a fundamental vibration of one bond with the overtone of some other vibration, a type of resonance occurs. This type of resonance is called *Fermi Resonance.* It causes a decrease in the intensity of fundamental vibration and a large increase in the intensity of the overtone. This phenomenon was first observed by Enrico Fermi in the case of carbon dioxide.

O=C=O — Asymmetric Stretching (IR-active)

O=C=O — Symmetric Stretching (IR-inactive)

Another example of Fermi Resonance is provided by aldehyde in which C—H stretching absorption generally appears as a doublet (~2820 cm^{-1} and 2720 cm^{-1}) due to interaction between C—H stretching (fundamental) and overtone of C—H deformation (bending).

Similarly, the appearance of twin peaks at 1746 cm^{-1} and 1750 cm^{-1} due to carbonyl group of cyclopentanone can be explained on the basis of Fermi Resonance.

(c) *Overtones*: When a molecule absorbs infrared radiation of appropriate energy besides fundamental vibrations, other bands also occur correspondings to the excitation to the second, third or fourth excited states. These bands are known as *overtones.* They are weak but are important in characterisation of certain classes of compounds, particaularly, benzene derivatives. Thus aliphatic and aromatic aldehydes may be distinguished from the presence of overtone bands in the range 2000–1665 cm^{-1}.

12. Explain that the number of fundamental vibrations of non-linear polyatomic molecule is 3n–6 where n is the number of atoms. What about the linear molecule?

➔ See the answer against question no. 7.

13. What is meant by 'normal modes of vibrations'? Find out the number of normal modes for HCl, CO_2, NH_3, C_2H_2 and H_2O.

➔ See the answer against question no. 7.

14. The molecule of CO_2 is linear and hence the number of normal modes of vibration is 4. But in IR spectrum only two fundamental bands are observed. Explain.

➔ Carbon dioxide (O=C=O) is a linear triatomic molecule, hence it has 3×3–5 = 4 fundamental modes—two stretching and two bending. One of the stretching is symmetric and the other is asymmetric as shown below:

O=C=O — Symmetric Stretching

O=C=O — Asymmetric Stretching

It should be remembered that two C=O bonds are polar but since they are in opposite directions, the resultant dipole moment is zero. If the bonds are stretched to equal extents (symmetric stretching) the dipole moment will remain zero, so this mode of vibration will be IR inactive (no band in the IR spectrum will appear).

On the other hand, in the asymmetric stretching, one bond is stretched and the other is compressed. So though in equilibrium position normal dipole moment is zero, during this mode of vibration it will continuously increase and decrease and the antisymmetric stretching vibration will be IR active.

The two bending vibrations are as shown below:

Arrows indicate the direction of motion of atoms

In one, carbon and oxygen atoms move in XY-plane and in other, they move in the XZ-plane. Both these bending vibrations are identical in frequency. Both these vibrations are IR active but since they have same frequency they produce only one band in IR spectrum. Such vibrations are termed as *degenerate*. Thus in IR spectrum of CO_2, only two—one stretching (antisymmetric) and one bending bands are observed.

15. What is the basis of classical theory of vibrational interactions?

➻ *Hints*: Interactions between the electrical component of electromagnetic radiation and electrical dipolar motions within the molecules are the basis of classical theory of vibrational interactions (vide answer question no. 9).

16. Explain: Some of the fundamental vibrations are infrared active while others are not.

➻ See answer question no. 8.

17. Explain the fact that ethylene and ethyne unlike propene and propyne have no carbon-carbon multiple bond stretching vibrations.

➻ *Hints*: Ethylene and ethyne are symmetrical molecules and so there is no change in dipole moment when excited. But propene and propyne are not symmetrical and hence there is change in dipole moment on excitation.

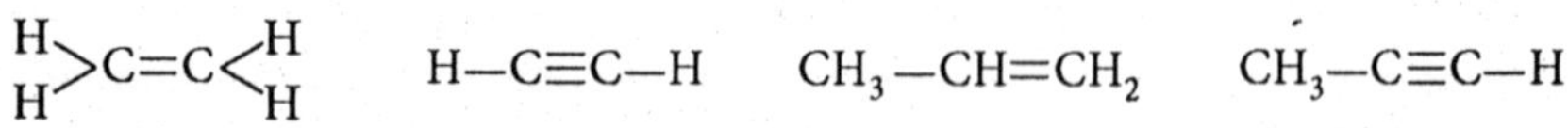

Ethylene Ethyne Propen Propyne

18. Explain: Cis-1,2-Dichloro ethylene is IR active with respect to C=C frequency whereas trans-1,2-dichloroethylene is not.

➻ *Hints*: We know that *cis*-isomer is not symmetrical and hence dipole moment change occurs on absorption of IR energy and so is IR active. On the other hand, trans isomers are highly symmetrical and so remain forbidden.

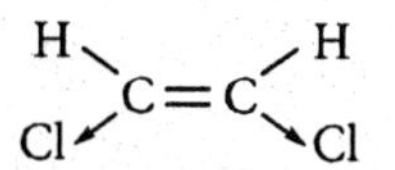

cis-1,2-Dichloroethylene

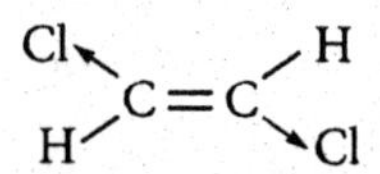

trans-1,2-Dichloroethylene

19. Discuss in short the sampling technique for infrared analysis.

➻ Infrared spectra can be taken for solids, liquids and gases. The sample is taken in a cell which must be IR inactive. The most commonly used cells are made of metallic halides, particularly alkali metal halides. The spectrum of a solid sample is frequently best determined as an alkali halide commonly, potassium bromide, 'pellel'. About 1mg of the sample and 100–200 mg of alkali halide are ground together finely, dried to remove moisture and pressed at high temprature under pressure into a small disc of 10 mm diameter and 1–2 mm thickness. A clear pellet is obtained which contains a solid solution in potassium bromide. The solid samples can also be taken by depositing a thin film of the solid by evaporation of its solution. Solids can also be examined in crystalline form by dilutions in a *mull.* In this case about 5 mg of the sample (solid) is ground to a very fine dispersion with a drop of suitable mulling agent. The spectrum of the mixture is then determined after being placed between two alkali halide plates. The most commonly used mulling agent is nujol—a mixture of high molecular weight liquid paraffin hydrocarbons. However, it exhibits absorption in the regions 3000–2850, ~1450 and ~1375 cm^{-1}. So when nujol is used no information can be derived about the absorption of the sample in these regions.

In the case of liquid sample, spectrum may be obtained without solvent interferences by determination as a liquid film. For this purpose, a small drop of the liquid is placed on alkali metal halide plate, and another alkali metal halide plate is placed on the top of the drop and the plates are placed in a holder and the spectrum is determined.

It is frequently desirable to determine the spectrum of a sample in solution. In infrared, solvents used, generally strongly absorb in at least a few regions of the spectrum. In order to minimise the solvent absorption, concentrated solutions and cells with short path-lengths are used. The most commonly used cells for solution spectra have a path-length of 0.1 mm. For such cells a 10% solution (20 mg in 0.2 ml solvent) generally produces a good spectrum.

The spectra of gases are obtained by taking the sample in a gas cell. The walls of the cell are made of alkali metal halide. Alkali metal halide windows allow the cells to be placed directly in the path of the sample beam. The gas cell is usually 10 cm long.

It is oftem desirable to determine the spectra of a sample under different conditions of physical state or concentration. In very dilute solution or in vapour phase, effects owing to molecular association are minimised whereas in the solid phase or in concentrated solution molecular association becomes important. Effects due to hydrogen bonding are frequently studied in this way.

20. Discuss in brief the instrumentation for infrared spectrum (scanning of infrared spectrum).

➜ The modern double-beam infrared spectrophotometer consists of the following main parts.

1. The source of IR radiation.
2. The monochromators.
3. The sample cells.
4. Detectors.
5. Amplifier.
6. Recorder.

1. *The source of IR radiation*: The most important source of infrared light for scanning the spectrum is *Nernst glower.*It is a rod of sintered mixture of oxides of Zirconium, Ytterium and Erbium. The rod is heated electrically to 1800 °C to produce IR radiation. A rod of silicon carbide known as *Glober* can also be heated electrically to produce IR radiations.
2. *Monochromators*: The source radiations emit radiations of various frequencies. As the sample in IR spectroscopy absorbs only at certain frequencies, it is necessary to select

desired frequencies from the radiation source and reject the radiations of other frequencies. This is achieved by means of monochromators. There are two types of monochromators—prisms or gratings type. Gratings give better resolution than do prisms.

3. As in answers against question 19.
4. *Detectors*: There are various types of detectors. The commonly used detector is Thermopile.

After detection necessary amplification and recording are done.

A simplified working of the double beam instrument may be shown as below:

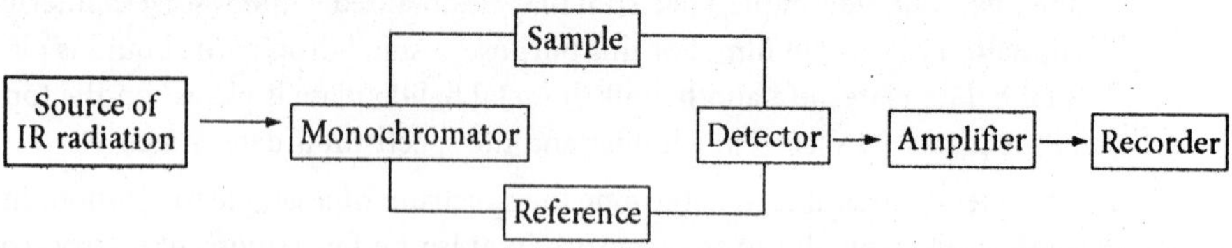

21. Write a short account on Fourier Transform Infrared Spectroscopy (FTIR spectroscopy).

➻ Fourier transform infrared spectroscopy is a technique of obtaining high quality infrared spectra by a mathematical conversion of an interference pattern into a spectrum. The FTIR spectroscopy is based on the principle of interference and that is why FTIR spectroscopy is also termed as interferometric IR spectroscopy.

In FTIR technique the IR radiation consisting of all the wavelengths (4000–650 cm^{-1}) is split into two beams by passing this radiation through a half silvered mirror (beam splitter). One beam is transmitted to plane mirror placed at a fixed length (static mirror) and reflected there back to the beam splitter and other beam is transmitted to another plane mirror of variant path-length (moving mirror) kept perpendicular to the previous mirror and from it is reflected back to the beam splitter. Both these beams are then combined to get a modulated beam. In this modulated beam there will be more light energy (*constructive interference*) or less light energy (destructive interference) for a given wave length. (When the difference of path-lengths between two mirrors is an intergral number of wavelengths, the two emergent beams recombine with *constructive interference*, and high intensity radiation reaches the detector but when the path-lengths differ by exactly one half of a wavelength *destructive interfernce* occurs and low intensity radiation reaches the detector.)

The modulated beam is passed through the sample, digitised and then Fourier transformed by a computer to give the IR spectra.

22. Mention with reasons the special precautions to be taken during the scanning of infrared spectrum.

➻ It has been observed that most of the infrared spectra exhibit a band around 2350 cm^{-1}. This band is due to carbon dioxide (CO_2). In order to avoid the appearance of this band either a carbon dioxide absorbent should be kept within in the instrument or the room should be well ventilated.

The sample must be perfectly dry because water absorbs strongly in the regions 3700 cm^{-1} and 1630 cm^{-1}. These absorptions may obscure absorptions of the substance under investigation and may lead to erroneous assignments. Finger should not come in contact with the optical surfaces. Care should be taken to prevent any contamination, i.e., the sample under scanning must be reasonably pure, otherwise undesirable bands may appear and make the assignment of bands difficult.

Frequently, it is desirable to scan the sample in solution. Most of the common solvents strongly absorb in at least few regions. In order to avoid such undesirable bands, concentrated solution should be used and the solvent should be indicated. CCl_4, CS_2 and $CHCl_3$ should be prefrably used.

The spectrophotometer should be calibrated so that the bands are observed at their proper frequencies or wave length. For this purpose, a calibration line should be recorded on the IR paper. Usually a polystyrene film is commonly used for this purpose. The calibration is done using one of the several absorption bands (3025 cm^{-1}, 2850 cm^{-1}, 1603 cm^{-1}, 1495 cm^{-1} and 906 cm^{-1}) of polystyrene. It is a common practice to record infrared spectra alongwith polystyrene band at 1603 cm^{-1} marked on as a check of frequency accuracy.

23. (a) Give a sketch of a typical IR spectrum of an organic compound. (b) Mention how the data are presented. (c) What do you mean by functional group and finger print regions.

➡ (a) A sketch of a typical IR spectrum may be shown as below:

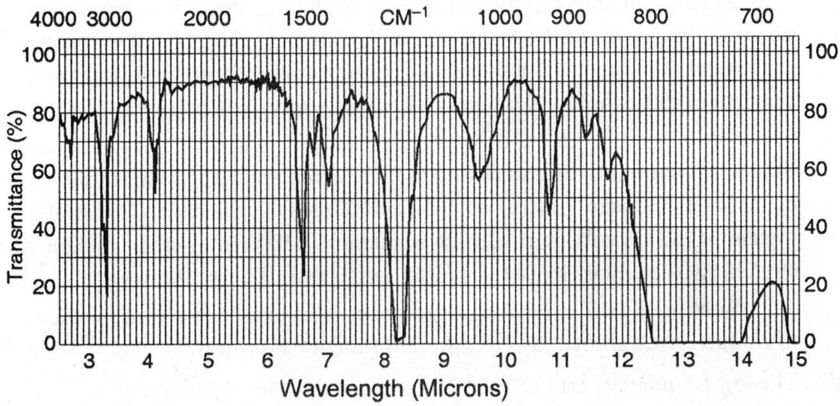

(b) *Presentation of infrared absorption bands*: The different absorption bands appearing in infrared region are presented in a graph which is a plot of intensity versus wavelength (μm) or wave number (cm^{-1}) as shown in the figure above. This is the desired IR spectrum of the compound under the study. It is obvious that the different vibrations do not absorb same amount of energy and hence the intensity of different bands in the IR spectrum may be different. So the intensity of bands in an infrared-spectrum is presented as strong (s), medium (m) and weak (w). The positions of bands are given in wave number (cm^{-1}) and the position of absorption band ν is expressed as ν_{max} cm^{-1}.

(c) *Interpretation of spectral data*: For rational interpretation of IR absorption bands to have a logical conclusion regarding the nature of the molecule under study, the entire range of infrared region (4000–650 cm^{-1}) has been divided into two most important areas,

(i) 4000–1500 cm^{-1} and (ii) 1500–650 cm^{-1}.

(i) The high frequency portion (4000–1500 cm^{-1}) of the spectrum is known as the *functional group region.* The characteristic stretching frequencies for common functional groups found in organic compounds occur in this portion of the spectrum. It may be remembered that absorption frequency of a particular functional group may vary within its specified range due to some changes in structural environment. Thus, under ideal condition >C=O group stretching vibrational frequency appear around 1700 cm^{-1} but if there is an unsaturation in conjugation with with >C=O group absorption band shifts to around 1600–1650 cm^{-1}. Similarly, free O–H group shows a band due to O–H stretching at about 3500–3600 cm^{-1} but this absorption band becomes broader and appears at about 3300 cm^{-1} when the hydroxyl group is H–bonded. Stronger the hydrogen bonding, lower is the stretching frequency. It may be mentioned that the absence of absorption band in the assigned ranges for various functional groups can usually be used as the evidence for the absence of such fundamental groups in the molecule. Further any conclusion reached after examination of a particular band in the functional group region should be confirmed, where possible, by the examination of the other portions of the spectrum. Thus, the assignment of a carbonyl band to analdehyde should be confirmed by the appearance of a band or a pair of bands in the region 2900 cm^{-1} – 2690 cm^{-1} of the spectrum due to C–H stretching vibrations of the aldehyde group. Similarly, the assignment of carbonyl band to an ester should be confirmed by the appearance of a strong band in the C–O stretching region at (1300 – 1100 cm^{-1}).

Again, if phenolic or alcoholic O–H stretching absorption appears in the functional group region of the spectrum, the position of the C–C–O absorption band in the region1250 cm^{-1} – 1000 cm^{-1} very often makes it possible to assign the O–H absorption to alcohols and phenols with highly specific structures.

Once again, it has been observed that more polar bonds are generally stronger than less polar bonds and so absorb at higher frequencies. As far example, C–H, 3000 cm^{-1}; N–H, 3300 cm^{-1} and O–H 3550 cm^{-1} (free OH). Further, C–H stretching of the system –C≡C–H occurs at highest frequency (3300 cm^{-1}) when compared to C–H stretching frequencies in C=C–H (3080 cm^{-1}) and alkanes (3000 cm^{-1}).

By applying this knowledge infrared spectra may be analysed for the assignment of characteristic absorptions to different functional groups.

(ii) The low frequency part (1500 cm^{-1} – 650 cm^{-1}) to the right hand portion of the spectrum contains many absorptions and is usually very complex because both bending vibrations and absorption due to C–C, C–O and C–N stretching vibrations occur in this region. In general in a molecule there are many more bending vibrations than stretching vibrations, but this region is highly rich in absorption bands and shoulders. For this reason, it is frequently called *fingerprint region* because the possibility that two compunds containing the same functional group will have the same absorptions in this region is exceedingly small. Although similar molecules many show similar spectra in the region 4000 cm^{-1} – 1500 cm^{-1}, there will always be discernible differences in the fingerprint region. Thus, if two pure samples give different infrared spectra particularly in this region the compound must be different. On the other hand, if they give the super imposable spectra then they represent the same compound. Perhaps, this is the most powerful function of infrared spectroscopy, i.e., to establish conclusively the identity of two samples.

A few examples may be cited in this context to show the importance of fingerprint region.

Androsterone (A) and epiandrosterone (B), the two naturally occuring steroidal compounds differ only in the stereochemistry of OH group at C_3. The IR spectrum of both exhibit strong absorption bands in the functional group frequency region around 3600 – 1950 and 1735 cm^{-1} for OH, CH and >C=O stretching vibrations respectively. But the fingerprint regions of their spectra are completely different.

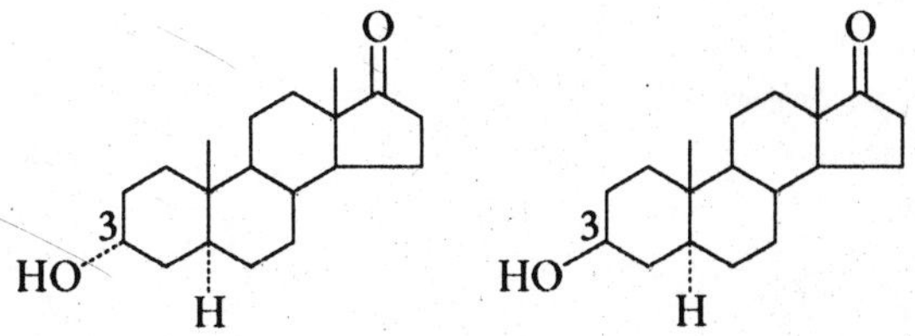

(A) Andosterone (C_3–OH axial)

(B) *epi*-Andosterone (C_3–OH is equatorial)

The class of organic compounds can also be ascertained from the careful scrutiny of fingerprint regions. Thus aromatic compounds frequently show numerous bands in the finger print region than their aliphatic counterpart.

Again, the presence of mono and disubstitution at ortho, para and meta positions in benzene are detected by most characteristic absorptions in this region. A band in the region 750 cm^{-1} – 700 cm^{-1} shows monosubstitution in benzene.

It is, therefore, evident that fingerprint region is very sensitive and changes even with minor chemical or stereochemical changes in a molecule. So this region is of immense importance, in interpreting the infrared spectra.

24. Present a chart containing characteristic group frequencies of some of the most common functional groups occurring in organic compounds.

➧ In a complex molecule there may be many vibrational absorptions in the entire IR region. Among these, absorption occuring in the range of 4000 cm^{-1} – 1500 cm^{-1} are usually termed as *functional group frequencies*. This is because almost all the functional groups found in organic compounds appear in this region of the IR spectrum. It may be mentioned that although the precise frequencies or wavelength at which a particular group absorbs is dependent upon its environment within the molecule, and on its physical state, the ranges are quite well defined. For this reason, these frequencies are regarded as *characteristic group frequencies* and are of great importance to the chemists for identification of the functional groups in organic compounds. Characteristic group frequencies of some common functional groups occuring in organic compounds are presented below. The ranges, given for group absorptions have been assigned from the examination of many compounds containing the groups.

Characteristic infrared absorptions of some common functional groups

Functional Group	V_{max} Range (cm^{-1})
A. Hydrocarbon	
1. C–H stretching	
a. Alkane	2960 – 2850

Contd.

Functional Group	V_{max} Range (cm^{-1})
b. Alkene	3040 – 3010
c. Alkyne	~ 3300
d. Aromatic	~ 3030
2. C–H bending	
a. Alkane	1480 – 1340
b. Alkane, *gem*-dimethyl	1380 – 1375
c. Alkane, *tert*-butyl	1395 – 1360
d. Alkene	995 – 790
e. Alkyne	~ 630
f. Aromatic	880 – 750
3. C—C Multiple bond stretching	
a. Alkene	1680 – 1600
b. Alkyne	2260 – 2100
c. Aromatic	1600 – 1450
B. Carbonyl Group	
1. Ketone stretching	
a. Saturated acyclic	1725 – 1700
b. Saturated cyclic	
(i) 6-membeed	1725 – 1700
(ii) 5-membered	1750 – 1740
c. α, β-unsaturated acyclic	1680 – 1660
d. α, β-unsaturated cyclic	
(i) 6-membered and higher	1680 – 1660
(ii) 5-membered	1725 – 1708
2. Aldehyde	
a. Carbonyl stretching	1740 – 1720
b. C—H stretching	2900 – 2820 2770–2700
c. Aromatic aldehyde	1715–1690
3. Caboxylic Acids	
a. Carbonyl stretching	1710 – 1700
b. Hydroxyl stretching	around 3300
4. Ester stretching	1750 – 1735
5. Amide	
a. Carbonyl stretching	1690 – 1650
b. N—H stretching	3500 – 3350
6. Anhydride stretching	1850 – 1800 and 1790 – 1740
C Alcohols and Phenols	
a. O—H stretching	
(i) Free O—H	3650 – 3590
(ii) Inter molecularly hydrogen bonded	3550 – 3450
(iii) Intra molecularly hydrogen bonded	3400 – 3200
b. O—H Bending and C—O stretching	
(i) alcohol	1050 – 1200
(ii) phenol	1200

Functional Group	V_{max} Range (cm^{-1})
	1410 – 1310
D. Amines	
a. N–H stretching	3500 and 3300
b. N–H bending	1650 – 1590
c. C–N stretching	~ 1300
E. Miscellaneous groups	
a. Nitrile (–C≡N)	2280 – 2200
b. Nitro compounds (aromatic)	1570 – 1500 and 1370 – 1300
c. Nitro compounds (aliphatic)	1570 – 1550 and 1380 – 1370
d. Halogen compounds, C–X stretching	
(i) C–Cl	800 – 600
(ii) C–Br	600 – 500
(iii) C–I	500
e. Sulphur compounds	
(i) S–H stretching	2600 – 2550
(ii) C=S stretching	1200 – 1050
(iii) S–O stretching	1430 – 1030

25. Discuss different factors which influence the positions of absorption frequencies from their normal values.

➜ It is known that the most essential requirement for infrared absorptions is the change in dipole moment during absorption of electromagnetic radiation of infrared region. Again, dipole moment is associated with the strength of the bond. So it is expected that the factors on which the strength of the bond depends will influence the positions of absorption frequencies. The followings are some of the most common factors controlling the frequencies of a particular functional group.

1. *Inductive effect*: Inductive effect offered by the groups attached to functional group shifts significantly their absorption frequencies. The shifts are prominant in carbonyl compounds like ketone, aldehyde, carboxylic acid, anhydride, ester, amide etc. The direction of the shift depends upon the direction in which the inductive effect is offered, when electron releasing groups such as alkyl, amino are attached to the functional group tend to shift stretching frequency to lower wave number (higher wavelength). Thus,

Functional group	Frequency (ν_{max})
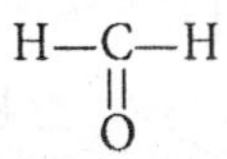	1750 cm^{-1}
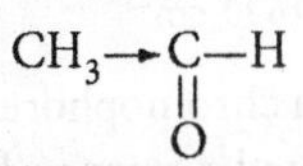	1725 cm^{-1}
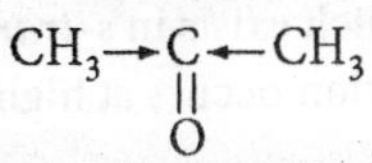	1715 cm^{-1}

Functional group	Frequency (V_{max})
$CH_3\rightarrow C(=O)\leftarrow OH$	1700 cm^{-1}
$CH_3\rightarrow C(=O)\leftarrow NH_2$	1680 cm^{-1}
$CH_3\rightarrow C(=O)\leftarrow OR$	1735 cm^{-1}

On the other hand, when electrons withdrawing groups such as chlorine is attached to the functional group frequency of the functional group is shifted to higher wave number (lower wavelength). Thus,

Compound	Frequency (V_{max})
$CH_3-C(=O)-CH_3$	1715 cm^{-1}
$Cl\leftarrow CH_2-C(=O)-CH_3$	1725 cm^{-1}
$Cl\leftarrow CH_2-C(=O)-CH_2$	1750 cm^{-1}

2. *Conjugation or Resonance*: Conjugation of a functional group with an olefinic bond or aromatic ring results in the delocalisation of the π-electrons and reduces the double character of the functional group as well as the unsaturation attached to it. This causes a lowering of frequency of the functional group. The lowering of absorption frequencies is attributed to the resonance as shown below.

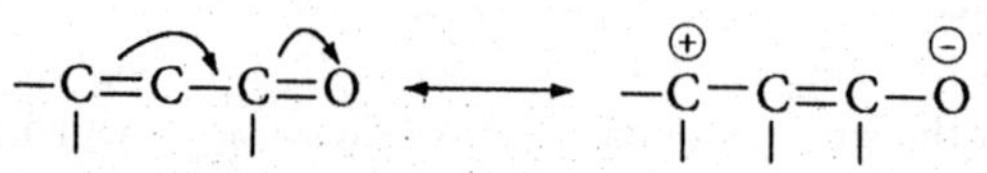

Introduction of a second double bond in conjugation with the functional group results in an additional shift of ~ 15 cm^{-1}. The effect of conjugation is illustrated by the carrbonyl stretching frequency in the following compounds.

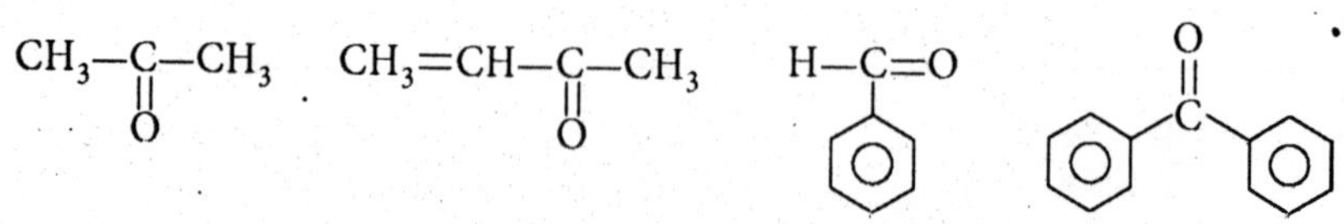

1715 cm^{-1} 1690 cm^{-1} 1690 cm^{-1} 1670 cm^{-1}

The effect of conjugation becomes maximum when chromophores are co-planar. The steric effects disturb the co-planarity of the conjugated system and reduce the effect of conjugation as is observed in benzal acetone which exists in s-*trans* (planar) and s-*cis* (non-planar) forms. In s-*cis* form C=O absorption occurs at higher frequency.

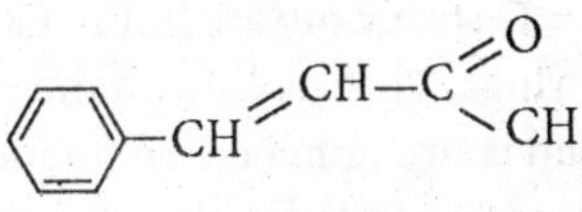

s-trans

ν_{max} 1674 cm^{-1}

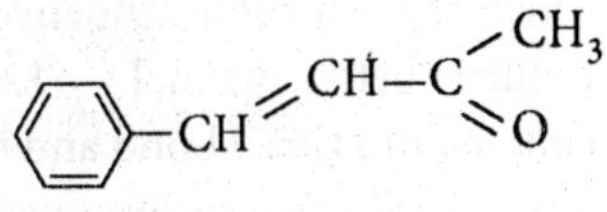

s-cis

ν_{max} 1699 cm^{-1}

3. *Physical state*: Physical state of the compound can shift the frequency of the group. Thus C=O stretching of acetone in vapour phase occurs at 1742 cm^{-1} whereas in the liquid phase, dipole-dipole association of the type $>\overset{\delta+}{C}=\overset{\delta-}{C}\cdots>\overset{\delta+}{C}=\overset{\delta-}{O}$ lowers the frequency to 1715 cm^{-1}. Similar is the case with acetic acid. In the vapour phase it exhibits carbonyl frequency at 1760 cm^{-1} whereas hydrogen bonded dimerisation in liquid state shifts the frequency to 1700 cm^{-1}.
4. *Solvent effect*: Intermolecular hydrogen bonding with hydroxylic solvents such as methyl alcohol shifts the absorption frequency slightly to lower wave number. For example, a pure sample of ethylmethyl ketone absorbs at 1725 cm^{-1} whereas a 10% solution of the ketone in methanol absorbs at 1706 cm^{-1}.
   The O—H frequency of phenol is lowered by 50 cm^{-1} – 100 cm^{-1} when the spectrum is taken in benzene solution compared to carbon-tetrachloride solution. The reason for this lowering is the formation of hydrogen bonds between the aromatic π-bonds (behave as Lewis base) and acidic hydrogen atoms of phenol.
5. *Hydrogen bonding*: Hydrogen bonding brings about significant lowering of frequency. Stronger the hydrogen bonding greater is the shift towards lower wave number than the normal value. Two types of hydrogen bonding can be easily distinguished. Generally, intermolecular hydrogen bond give rise to broad bands whereas intramolecular hydrogen bond exhibits sharp bond. Intermolecular hydrogen bonding is *concentration dependent*.The cause of shifting of frequency to lower wave number is due to the fact that the original O—H bond is lengthened due to hydrogen bonding. As a rersult small amount of energy will be required to stretch such a bond (O—H) and so absorption takes place at lower wave number, i.e.,

$$R-O-H\cdots\underset{\displaystyle H}{\underset{|}{O}}-R$$

   As nitrogen atom is less electronegative than oxygen atom, hydrogen bonding in amines is weaker than that in alcohols and so the frequency shifts in amines are less. Hydrogen bonding is also observed in enols and carboxylic acids.

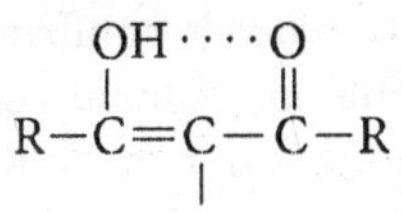

Enol

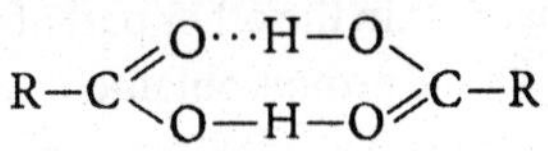

Carboxylic acid

6. *Effect of strain*: Effect of strain is mainly observed in cyclic compounds particularly in cyclic carbonyl compounds. The ring strain causes large shift of C=O stretching to higher frequency. This povides a very good method of differentiating the size of the ring ketones. Thus,

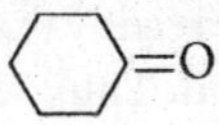

ν_{max} 1725 cm^{-1}

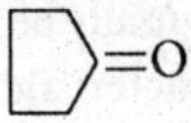

ν_{max} 1770 cm^{-1}

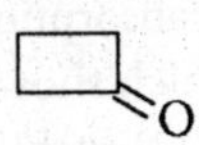

ν_{max} 1780 cm^{-1}

The reasons for this influence of strain on C=O stretching are (i) C—CO—C angle is reduced below normal (~120°) in strained rings. This leads to an increase in the S-character in C=O bond and hence C=O bond is strengthened. So frequency of C=O is increased (ii) in cyclic system a steric hindrance is created by the adjacent C—C bond and so the C=O group is less available and hence more energy is to be required. As a result C=O stretching frequency is increased. Similar effects are observed in lactones. Thus,

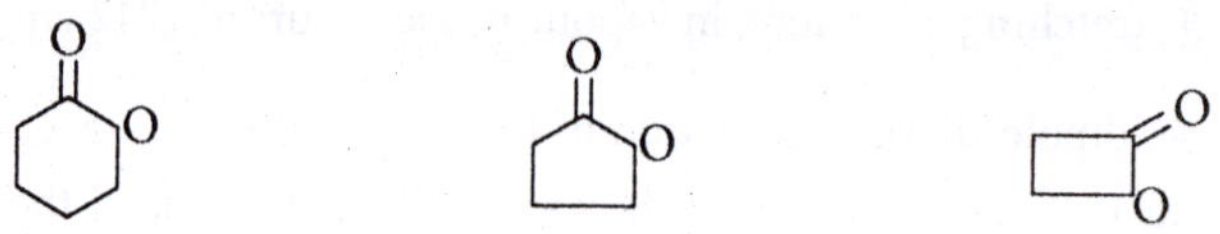

ν_{max} 1735 cm^{-1} ν_{max} 1770 cm^{-1} ν_{max} 1800 cm^{-1}

7. *Coupled vibrations and Fermi Resonance*: Due to coupled vibrations there may be substantial shift in the frequency of a particular group. Fermi Resonance is another cause of shifting of group frequency. Coupled vibrations are often observed in CH_2 (methylene) group whereas in anhydrides and aldehydes Fermi Resonance is found to occur.

26. Mention some of the important methods of approach for the interpretation of infrared spectrum of an unknown compound.

➜ In fact, there is no standard method for the interpretation of infrared spectrum. However, some common practices developed from experience have been found to be valuable. Such common methods may be enumerated as below.

1. Always more weightage is given to negative evidence. The absence of an absorption band for a particular functional group in the specified region is a clear indication of the absence of that functional group in the molecule under study. Thus, if there is no absorption band in the region 1750–1650 cm^{-1}, the carbonylic (>C=O) type of functionalities (aldehyde, ketone, ester, lactone, etc.) must be absent in the compound.
2. During interpretation always higher frequency end of the spectrum should be taken as the starting point. This is because, most of the stretching frequencies occur in this region which are highly informative. In this region the characteristic stretching frequencies of most of the functional groups occuring in organic compounds are observed. The strong bands in this region are picked out and using correlation chart functional groups present in the molecule are identified. Next, presence or absence of band(s) in the other part of the spectrum is searched to confirm the identity of the functional groups obtained. As for example, if the spectrum of a compound shows a strong band at 1680 cm^{-1} (C=O stretching) along with a medium intensity band at 2700 cm^{-1} (C—H stretching for aldehyde) the compound under study must be an aldehyde.
3. For distinguishing inter and intramolecular hydrogen bonding, the spectra of the sample should be scanned at two different concentrations.
4. It may be remembered that a particular functional group which exihibits many characteristic absorption bands can normally be identified more readily than a functional group which shows only one characteristic absorption band. Thus, ketones, where only C=O stretching vibrational absorptional band is seen are less easily identified than esters which display C=O and C—O stretching vibrational absorption bands.

Similarly amides which display C=O stretching, N—H stretching and N—H bending are more readily identified than esters.

5. For easy identification of functional groups present in a compound, the IR spectral region (4000 cm^{-1} – 650 cm^{-1}) may be divided into several portions as mentioned below.
   (i) 3600 cm^{-1} – 3200 cm^{-1}: The presence of bands in this region suggests the presence of O—H and N—H functionalities. From the position, intensity and breadth of the bands the nature of association (free, inter and intramolecular) may be ascertained.
   (ii) 3200 cm^{-1} – 3000 cm^{-1}: Absorption due to ≡C—H and Ar—H stretching occur in this region.
   (iii) 3000 cm^{-1} – 2500 cm^{-1}: The absorptions due to C—H stretching of methyl and methylene groups are observed in this region. A very broad band in the region 3000 cm^{-1} – 2500 cm^{-1} is most characteristic of carboxylic acids. For aldehydes weak bands around 2710 cm^{-1} and 2820 cm^{-1} are observed.
   (iv) 1800 cm^{-1} – 1660 cm^{-1}: In this region strong bands for carbonylic functionalities like keto, anhydride esters, amides due to C=O stretching occur.
   (v) 1600 cm^{-1} – 1000 cm^{-1}: This region is very important for identification of nitro and ether groups.
   (vi) Below 1000 cm^{-1}: Substituted aromatic compounds display strong bands in this portion.
   (vii) Last but not the least is the finnger-print region (1500 cm^{-1} – 650 cm^{-1}). The identical compounds have exact correspondence in this region.

Finally it must be remembered that it is not possible to identify a particular compound only from infrared spectral measurement. All other available data, both spectroscopic (UV, NMR, mass, etc.) and chemical should be in accord with any structure suggested.

27. Discuss some important applications of infrared spectroscopic technique.

➜ Infrared spectroscopy is one of the most important techniques at the hands of chemists particularly organic chemists. It has made the task of organic chemists comparatively simpler. Some of its remarkable applications are discussed in sequel.

(a) *Identification of an organic compounds*: The simplest and most definitive application of IR spectra is in the confirmation of identity. For a compound obtained from two different sources, say natural and synthetic, should give identical IR spectra, i.e., two spectra must be superimposable. Every band should coincide with its counterpart not only in the position but also in intensity, if both spectra are recorded under identical conditions using the same spectrophotometer. Even if they are taken in different instruments, the band positions must coincide though their intensities may be slightly different. This method of establishment of identity is much better than other methods like mixed melting point determination. The finger-print region (1500 cm^{-1} – 650 cm^{-1}) is most supreme in this regard. The different compounds containing the same functional group may have similar absorptions above 1500 cm^{-1} but their spectra differ significantly in the finger-print region.

(b) *Identification of functional group in organic compounds*: The presence or absence of absorption at certain wavelength regions in the IR spectrum can be interpreted in terms of the presence or absence of certain functionalities in the compound. For example, the oxygen in a compound containing carbon, oxygen and hydrogen only can be present as O—H, C═O (simple carbonyl, ester, carboxylic acid or anhydride) or as C—O—C (ether). The presence or absence of absorption band in the carbonyl region (1750 cm^{-1} – 1650 cm^{-1}) or hydroxyl region (3600 cm^{-1} – 3200 cm^{-1}) can serve to ascertain some of these possibilites. It is, thus, possible to establish a co-relation between IR absorptions and the functional groups.

(c) *Structure deduction*: Elucidation of complete structure of even moderately complex compound is difficult only by IR spectrum because it furnishes little information about the carbon skeleton of a compound. However, in some cases IR can precisely be used to establish the correct structure. Following examples may be cited in this context.

(i) Both α– and γ– pyrones exhibit resonance, the ionic form is aromatic and hence stable. There are two basic sites—the two oxygen atoms in both the pyrones. Obviously, the negatively charged oxygen in the side chain is more basic. So greater the resonance, greater is the basicity. The presence of conjugated $\nu_{C=O}$ band of γ-pyrone at 1664 cm^{-1} shows that it is more basic while in α-pyrone $\nu_{C=O}$ (1727 cm^{-1}) suggests that it is non-conjugated carbonyl.

Thus,

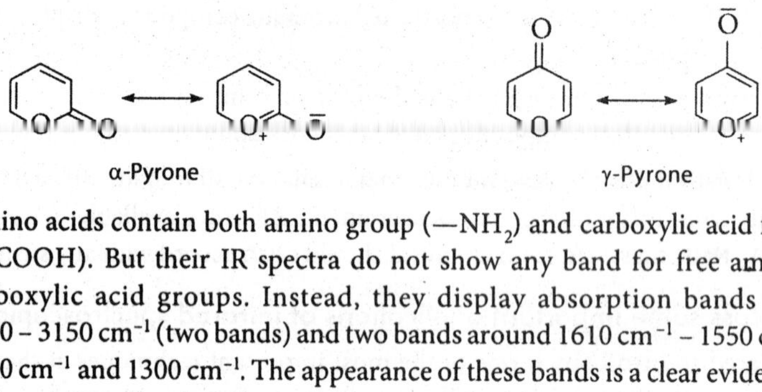

α-Pyrone γ-Pyrone

(ii) Amino acids contain both amino group (—NH_2) and carboxylic acid function (—COOH). But their IR spectra do not show any band for free amino and carboxylic acid groups. Instead, they display absorption bands around 3350 – 3150 cm^{-1} (two bands) and two bands around 1610 cm^{-1} – 1550 cm^{-1} and 1430 cm^{-1} and 1300 cm^{-1}. The appearance of these bands is a clear evidence that both amino and carboxylic groups exit as salts. This suggests that amino acids exist as zwitterion (A).

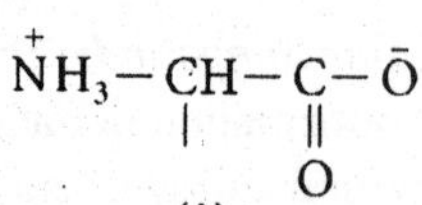

(A)
Zwitterion

Similar is the case with sulphanilic acid which exhibits bands for $\overset{+}{N}H_3$ and SO_3^- groups. So the structure of sulphanilic acid may be expressed as (B).

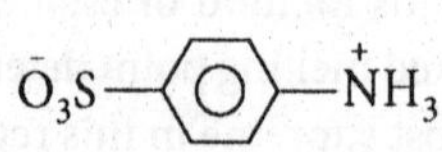

(B)

(iii) The structure of carbosytril has been settled by the help of IR spectrum. It has been suggested that carbostyril may exist in two tautomeric forms (C) and (D).

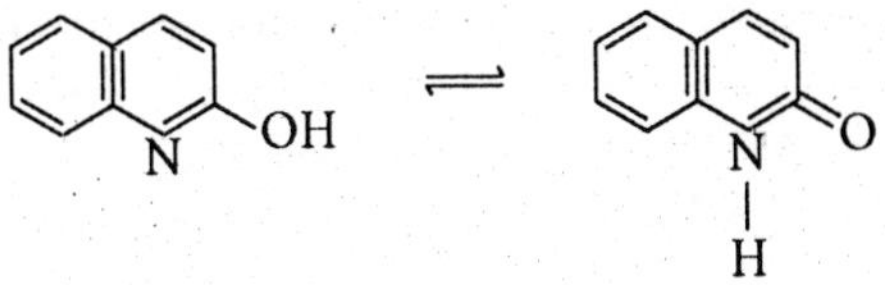

(C) (D)

The IR spectrum reveals that carbostyril displays a strong band around 1660 cm^{-1} (amide) rather than free O—H absorption band near 3600 cm^{-1} in dilute solution. Hence carbostyril predominantly exists in amide form (D).

(d) *Verification of purity of a sample*: Infrared spectroscopic technique may be utilised for the deterction of impurities in a sample. However, sometimes results may be confusing because the IR bands may not always be very intense and hence small amount of impurity present may not be detected. But when fairly intense band is obtained, impurity can be detected. Thus, alcoholic samples frequently are contaminated with corresponding ketonic compounds. In such situation ketonic impurities give intense C=O absorption band. When high amount of impurity is present, blurred and indistinct spectra are obtained.

(e) *Study of a chemical reaction*: Chemical reactions can successfully be monitered by infrared spectroscopic technique. Thus, the reduction of ketonic compound to alcohol can be studied, whether the reduction proceeds in the desired direction can be verified by recording IR spectrum in the midway of the reaction. Appearance of O—H absorption band along with C=O absorption band suggests that the reaction is occurring in the desired direction.

Isomerisation of verbenone to chrysantheneone has been successfully monitered by this technique. Irradiation of verbenone in cyclohexanone solvent readily isomerises.

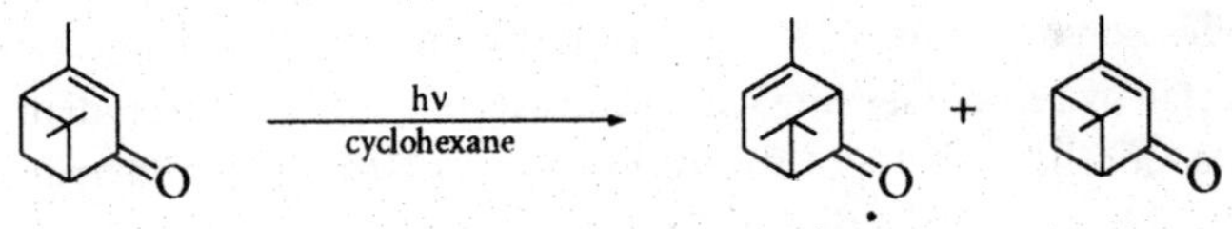

Verbenone ν_{max} (CCl_4) 1685 cm^{-1} — Chrysanthenone ν_{max} (CCl_4) 1787 cm^{-1} — Verbenone ν_{max} (CCl_4) 1685 cm^{-1}

(f) *Study of geometrical isomerism*: This technique is highly useful in ascertaining *cis* and *trans* isomerism. Normally trans isomers are, in general, symmetrical and no appreciable change in dipolemoment occurs. But in the case of cis-isomers dipolemoment change occurs.

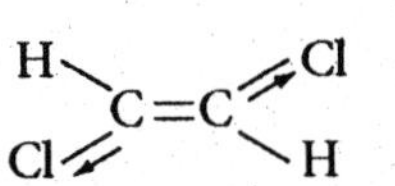

Trans
Dipolemoment = 0

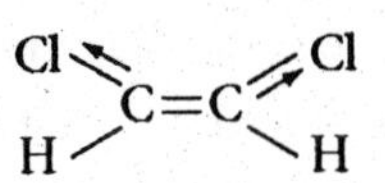

cis
Dipolemoment = 1.85D

(g) *Study of Keto-Enol tautomerism*: Whether a particluar compound exists in the keto form or as an equilibrium mixture of keto-enol forms can be studied. In the case of mixture, along with carbonyl stretching frequency a broad O—H stretching and a C=C stretching frequencies are also observed as in the case of ethyl aceto acetate.

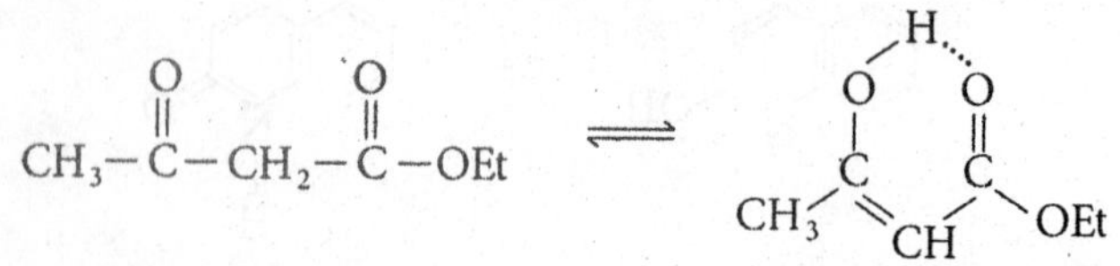

Two C=O groups
ν_{max} (CCl_4) 1733, 1710 (S) cm^{-1}

one C=O, one bonded
OH and one C=C
ν_{max} (CCl_4) 3300 cm^{-1}, 1650 cm^{-1} and 1600 cm^{-1}

(h) *Conformational analysis*: Axial and equatorial substituents in cyclohexane ring absorb differently in infrared region. Equatorial bond is less sterically hindered than axial bond. Hence equatorial substituent normally absorbs at higher frequency than does the axial substituent.

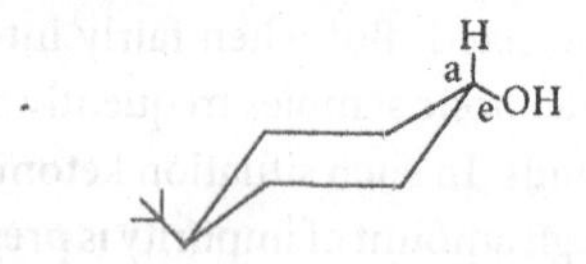

ν_{max} C—O ≈ 1060 cm^{-1}

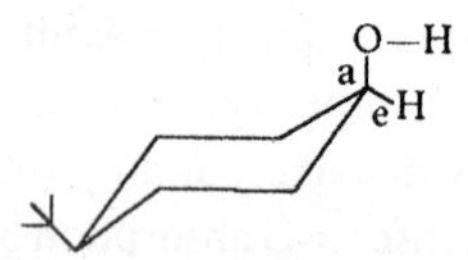

ν_{max} C—O ~ 950 cm^{-1}

(i) *Application in other fields*: Infrared spectroscopy has also been applied in other fields of importance as metioned below.

(i) *Determination of alcohol level in the breath and in the blood*: The diffusion of blood alcohol through the lung occurs into the blow of breath. This is subjected to infrared study in the O—H stretching region near 3400 cm^{-1}. The appearance of any absorption band in this region confirms the presence of alcohol in the breath and hence in the blood.

(ii) *Detection of automobile pollutants*: The exhaust from poorly maintained automobiles contains carbon monoxide and hydrocarbons. These two pollutants are detected by remote sensing which consists of a source of IR radiation on the side of the road for the intensity of absorption of CO (2160 cm^{-1}) and hydrocarbons (3000 cm^{-1} and 2800 cm^{-1}). On the roadside at the site of remote sensing, a composite machine consisting of a detector, calibrator, video is kept and a computer records the number plate along with the percentage of pollutants. On this basis, the owners of the vehicle may be penalised.

28. How would you distinguish the following pairs from IR spectral data.

(i) Phenol and cyclohexanol

(ii) I-hexene and I-hexyne

(iii) Ethylbenzene and Orthoxylene

(iv) Ethanol and dimethyl ether

(v) CH_3—CH_2—CHO and CH_3—$COCH_3$

(vi) Maleic and fumeric acid

(vii) Ethanol and ethylamine

(viii) cis and trans–2–butene

(ix) 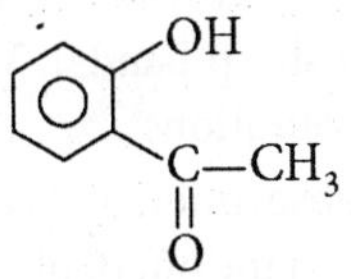 and 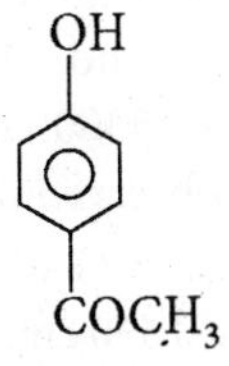

(x) Axial and equatorial O–H group

(xi) Phenyl acetate and methyl benzoate

(xii) *Trans* and *cis*-cinnamic acid.

(i) Phenol is aromatic while cyclohexanol is aliphatic in nature. So both of them show different spectral absorption bands. In phenol free O—H stretching appears near 3600 cm^{-1}. In addition strong C—O stretching gives bands around 1200 and 1410 cm^{-1} – 1300 cm^{-1}. But cyclohexanol shows sharp band in the region 3700 cm^{-1} – 3500 cm^{-1} for O—H stretching and a strong band at 1100 cm^{-1} for C—O stretching.

(ii) The spectrum of I-hexene exhibits C—H stretching at 3095 cm^{-1} together with C=C stretching absorption at 1640 cm^{-1} (strong). On the other hand, the spectrum of I-hexyne shows C—H stretching at 3300 cm^{-1} along with weak C≡C stretching at 1640 cm^{-1}.

(iii) The structures of ethyl benzene and orthoxylene may be written as

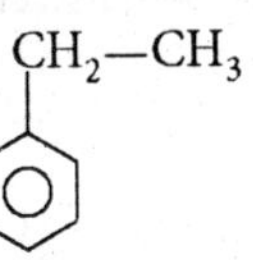

Ethyl benzene

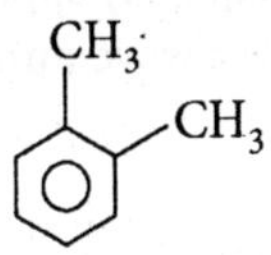

o-Xylene

Ethyl benzene is monosubstituted benzene while *o*-xylene is a disubstituted benzene derivative. Ethyl benzene displays bands at 710 cm^{-1} — 690 cm^{-1} (s) and 770 cm^{-1} – 730 cm^{-1} (s) while *o*-Xylene exhibits bands at 840 cm^{-1} – 800 cm^{-1} (m) and 770 cm^{-1} – 735 cm^{-1} (s).

(iv) In ethanol strong band in the region 3700 cm^{-1} – 3500 cm^{-1} for O—H stretching along with strong C—O stretching band at 1350 – 1260 cm^{-1} will be observed but in dimethyl ether band in the region 3700 cm^{-1} – 3500 cm^{-1} would be missing but a C—O—C stretching band may be displayed in the region 1150 cm^{-1} – 1070 cm^{-1} (s) along with C—H stretching band in the region 2850 cm^{-1} – 2810 cm^{-1} (m).

(v) The C=O stretching for propionaldehyde will appear at 1740 cm^{-1} while that of acetone will appear at 1715 cm^{-1}. Further, aldehyde will show two C—H stretching (symmetric and asymemtric) absroption bands one at 2720 cm^{-1} and the other at 2820 cm^{-1} which are absent in acetone.

(vi) Maleic acid is *cis*-isomer while *fumeric* acid is *trans*-isomer. We know that *trans*-isomer is a centrosymmetric molecule as a result of which it is IR-inactive. But *cis*-isomer is IR-active and it will show C=C stretching absorption bands.

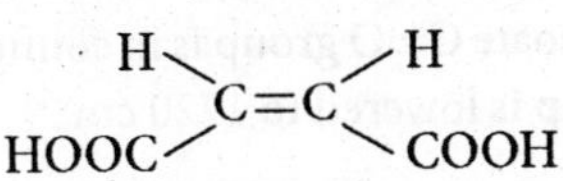

Maleic acid

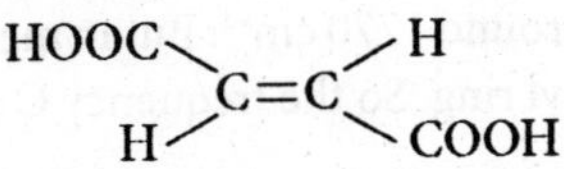

Fumeric acid

(vii) Ethanol is a primary alcohol and will display bands in regions 3700 cm^{-1} – 3500 cm^{-1} (s). 1350 cm^{-1}– 1260 cm^{-1} (s) and ≈ 1050 cm^{-1} (s). But ethylamine will exihibit absorption bands at 3500 cm^{-1} – 3300 cm^{-1} (m) (two sharp bands). Weak bands around 1440 cm^{-1} may also be observed due to C—N vibrations.

(viii) Both *cis* and *trans*-butenes show =C—H stretching absorption at 3020 cm^{-1} and C=C stretching absorption near 1670 cm^{-1}. They differ in the out of plane bending, *cis*-2-butene absorbing at 670 cm^{-1} while *trans*-isomer absorbs strongly around 970 cm^{-1}.

(ix) Ortho hydroxy acetophenone will form intramolecular H-bond as shown below.

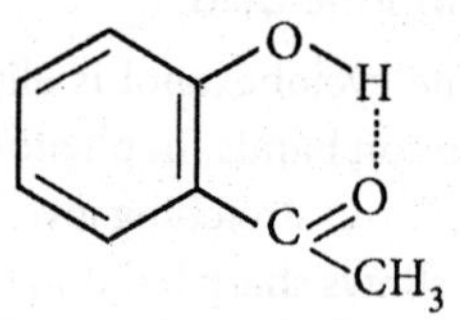

So stretching frequencies of phenolic O—H and C=O will be shifted from their normal positions. The C=O stretching occurs around 1634 cm^{-1} and for O—H bands appear at 3300 cm^{-1} and for C—O stretching at ~ 1200 and 1410 cm^{-1} – 1300 cm^{-1}. But in the case *p*-hydroxy acetone no such H-bonding occurs but due to conjugation frequencies are shifted to some extent. In this compound C=O stretching frequency appears at about 1680 cm^{-1} and O—H stretching frequency is shifted to lower wave number.

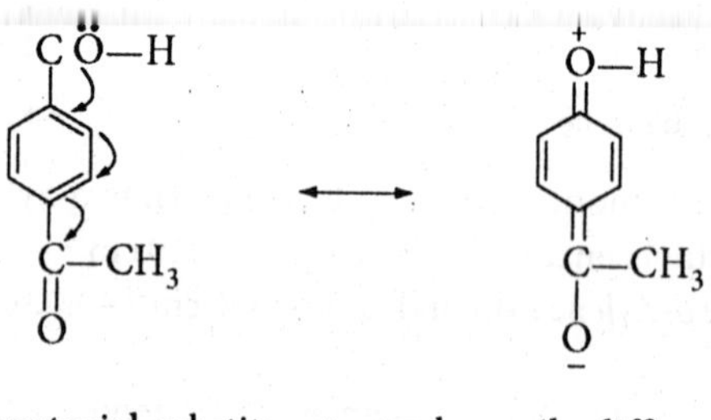

(x) The axial and equatorial substituents can be easily differentiated by IR spectrum. The frequency of equatorial group is higher than that of axial group. This is due to the fact that equatorial bond is less sterically hindered than axial bond.

(xi) The structures of phenyl acetate and methyl benzoate may be written as below.

Phenyl acetate Methyl benzoate

In phenyl acetate, electron withdrawing phenyl group is attached to the single bonded (C—O) oxygen and hence C=O frequency is raised to some extent and appears around 1770 cm^{-1}. But in methylbenzoate C=O group is in configuration with phenyl ring. So the frequency C=O group is lowered to 1720 cm^{-1}.

(xii) The structures of *trans*- and *cis*-cinnamic acid may be expressed as

trans-Cinnamic acid *cis*-Cinnamic acid

cis-cinnamic acid absorbs at a higher wave number than trans- isomers. This is due to the fact that repulsive action of carboxylic carbonyl goes out of the plane of the double bond. As a result conjugation is diminished and hence C=O absorption occurs at higher wave number.

29. Explain the following:

(a) Propene and propyne show C—C multiple bond stretching bands in IR spectrum while ethylene and ethyne do not show such bands.

(b) cis 1,2-Dichloroethylene is IR active but its trans isomer is not.

(c) 2-*tert*-Butylphenol displays IR absorption bands at 3600 cm^{-1} and 3645 cm^{-1} but 2,6-di*tert*-butylphenol shows only one band at 3645 cm^{-1}.

(d) Concentrated solution of both ethanol and 1,2-glycol in carbon tetrachloride exhibit broadband at 3350 cm^{-1} for O—H stretching. But on dilution with CCl_4 though there is no shifting of band for O—H stretching, IR spectrum of ethanol shows an additional band at 3600 cm^{-1}.

(e) Aldehydes display their carbonyl stretching frequency at slighly higher wave number than ketones.

(f) The carbonyl stretching frequency of the carboxylate anion differs from that of parent carboxylic acid.

(g) An aliphatic unsaturated aldehyde containing unconjugated double bond shows C=O and C=C stretching frequencies respectively at 1735 cm^{-1} and 1650 cm^{-1} but crotonaldehyde exhibits corresponding bands respectively at 1700 cm^{-1} and 1630 cm^{-1}.

(h) Carbonyl stretching frequency in l-acetyl cyclohexene slightly differs from that of l-acetyl-2-methyl cyclohexene.

(i) α, β-unsaturated esters absorb at a lower wave number as compared to its saturated counterpart.

(j) 2,4,6-cycloheptatrienone has exceptionally low carbonyl frequency.

➜ (a) Propene and propyne are IR active as they are not symmetrical.

$CH_3—CH=CH_2$ $CH_3—C\equiv C—H$

Propene Propyne

So these two compounds absorb IR radiations and hence exhibit band for carbon-carbond multiple bonds. On the other hand, ethylene and ethyne are symmetrical and so they are IR inactive and they do not display such bands.

$CH_2=CH_2$ $H—C\equiv C—H$

Ethylene Ethyne

(b) *cis*-1,2-Dicholroethylene is not symmetrical and hence it is IR active but *trans*-isomer is symmetrical, and so IR inactive.

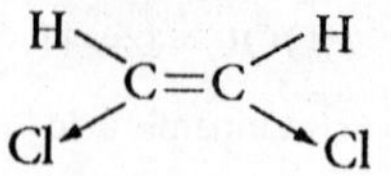

cis-1, 2-Dichloroethylene

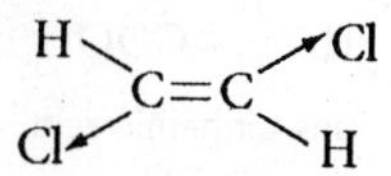

trans-1,2-Dichloroethylene

(c) The structures of 2-tert-butylphenol and 2,6-*tert*-dibutyl phenol may be written as below:

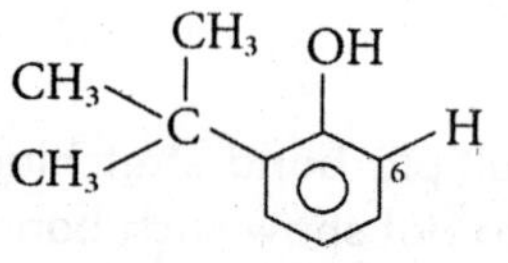

(a)
2-*tert*-Butylphenol

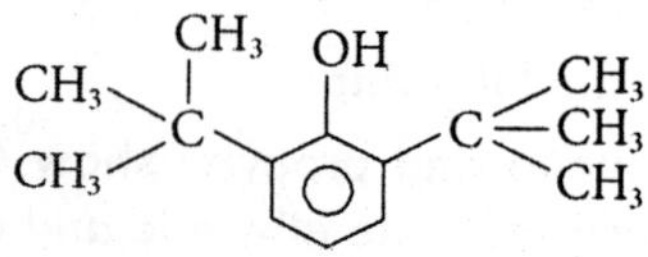

(b)
2, 6-tert-Butylphenyl

From the structures it is evident that in (a) position C_6 is free and O–H at position I may form H–bond with C_6–H. So (a) show bands for both free (36450 cm^{-1}) and bonded O–H (3600 cm^{-1}). But in (b) as there is no H-atom at C_6-position no hydrogen bond can occur. Hence (b) exhibits only band for free O–H group.

(d) Ethanol in solution forms intermolecular hydrogen bond but 1,2-glycol is intramolecularly H-bonded molecule. As shown below:

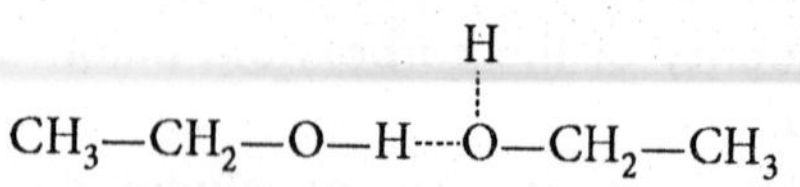

Ethanol
(Intermolecular H-bond)

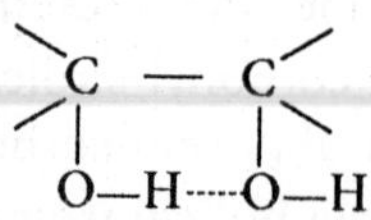

1,2-Glycol
(Intramolecular H-bond)

We know that intermolecular H-bond is concentration dependent but intramolecular H-bond is indifferent towards concentration. So on dilution, ethanol exhibits two bands but 1,2-glycol only one.

(e) We know that frequency of a functional group depends upon inductive effect. Since in ketones more number of electron releasing groups are present, frequency is shifted to lower wave number.

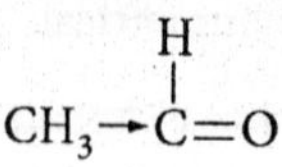

Aldehyde
(1730 cm^{-1})

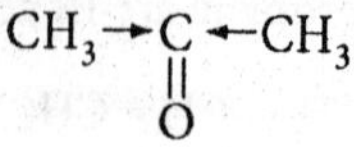

Ketone
(1715 cm^{-1})

(f) Carboxylate anion is a resonating species and the resonating forms are

These two forms are virtually symmetrical and so the two bands—one for C=O and other for C—O are of the same strength. But the parent carboxylic group has two nonsymmetrical resonating forms

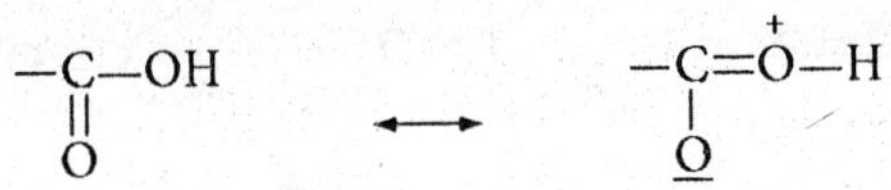

For this reason carbonyl stretching frequencies of both are different.

(g) Crotonaldehyde is a conjugated carbonyl compound and exists in resonance hybrid as shown below:

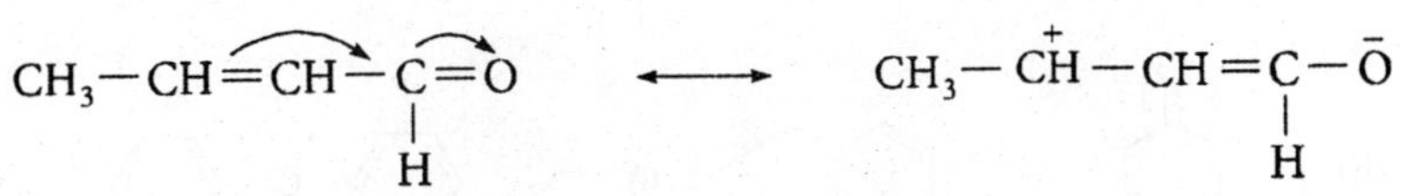

As a result of such resonance, frequencies of both aldehyde carbonyl and unsaturation are shifted to longer wavelength. But aliphatic aldehyde having isolated double bond does not show such resonance and hence there is no question of shifting of frequencies.

(h) I-acetyl cyclohexene and I-acetyl-2-methyl cyclohexene may be structurally written as below:

(a) I-Acetyl cyclohexene (b) I-Acetyl-2-methyl cyclohexene

Compound (b) is sterically hindered. As a result planarity of the double bond is to some extent reduced. So perfect resonance does not occur in it. But in (a) normal resonance occurs as there is no steric hindrance. So the frequency of carbonyl group in (b) is higher than in (a).

(i) αβ-unsaturated ester exists in resonance as

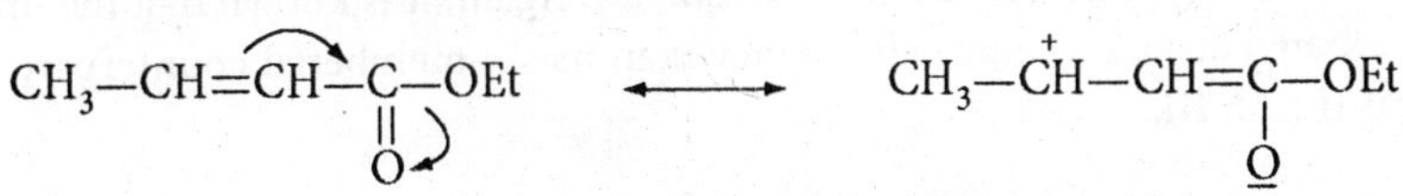

Due to this resonance αβ-unsaturated esters absorb at lower wave number than its saturated counterpart.

(j) 2,4,6-cycloheptatrienone exists in resonance hybrid as shown below:

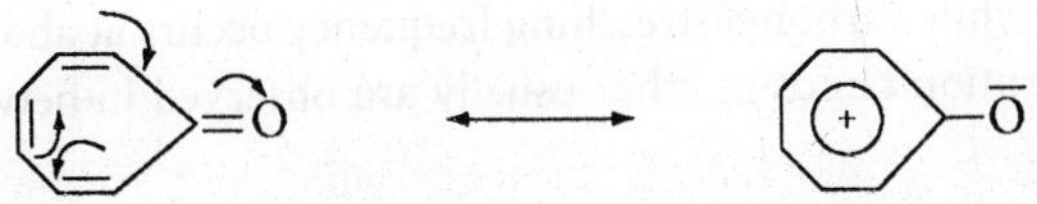

Due to this effect carbonyl stretching frequency is abnormally lowered.

30. Arrange the following carbonyl compounds in order of increasing carbonyl frequency.

(a) 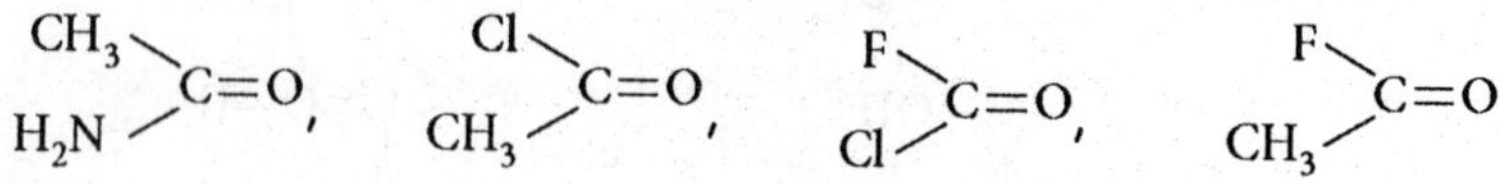

I II III IV

→ (a) Due to different inductive effects, these carbonyl compounds absorb IR radiation at different regions. The highest inductive effect is in III and lowest in (I). So the order is III > IV > II > I.

(b) 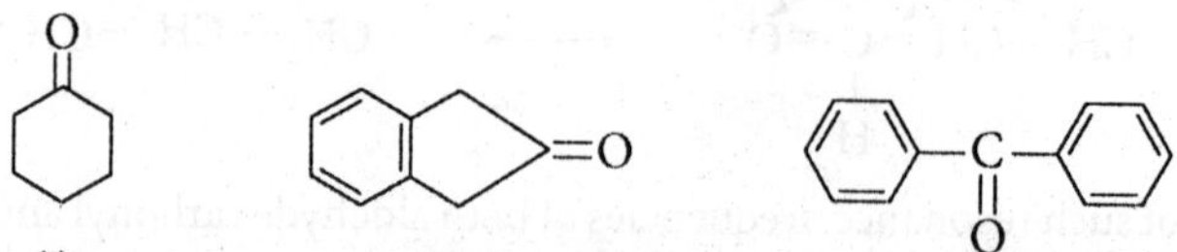

(I) (II) (III)

II is a non-conjugated five membered cyclic ketone, hence it will absorb at higher frequency than other two. (I) is six-membered cyclic ketone, but III is perfectly conjugated. So the order is III > I > II.

(c) 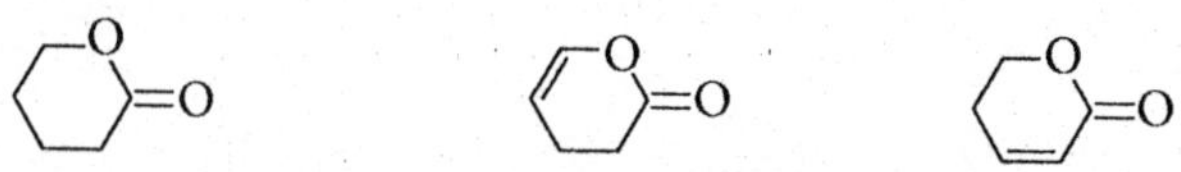

(I) (II) (III)

All three are lactones. In III lactone carbonyl is conjugation with the double bond. So its frequency will be reduced much than (I) which is its saturated counterpart. But in (II) lactone oxygen is in conjugation with the carbon-carbon double. This will increase the carbonyl ferquency. Hence the order II > I > III.

(d) 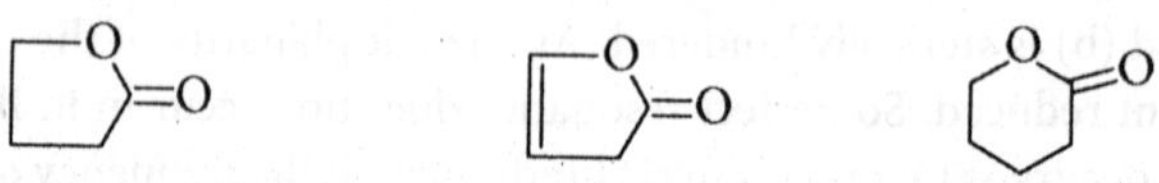

(I) (II) (III)

It is known that double bond in conjugation with the singly bonded oxygen atom in lactones or esters increases the C=O frequency. Again, it is known that five-membered ring carbonyl absorb at higher frequency than its six-membered counterpart. Thus the order is II > I > III.

31. An oxygen atom in an organic compound may be present either as hydroxyl, ether or as carbonyl function. How would you ascertain them by IR spectroscopy?

→ These functional groups absorbs IR radiation at different frequencies. Hence they show absorption bands at different regions. Normally hydroxyl group absorb at about 3600 cm^{-1} – 3400 cm^{-1}, while carbonyl stretching frequency occurs at about 1700 cm^{-1} – 1650 cm^{-1} and the absorption bands of ether usually are observed in between 1250 cm^{-1} and 1000 cm^{-1}.

32. An organic compound supposed to have any one of the following structures (a), (b) and (c). Its IR spectrum displays absorption bands at 1680 cm^{-1} and 3300 cm^{-1}. Suggest the most probable structure.

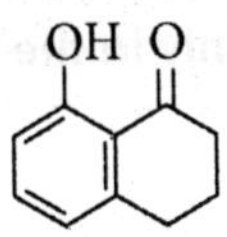

(a)

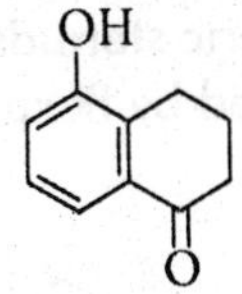

(b)

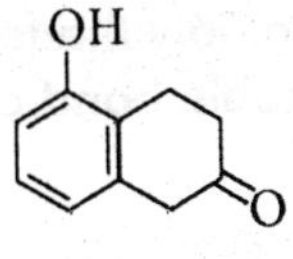

(c)

The band at 3300 cm^{-1} clearly suggests that OH group in the compound is H–bonded while the carbonyl stretching frequency at 1680 cm^{-1} reveals that it is conjugated in nature. H–bond formation takes peace in between OH and >C=O groups. Considering all these evidences, the most probable structure of the compound is (a).

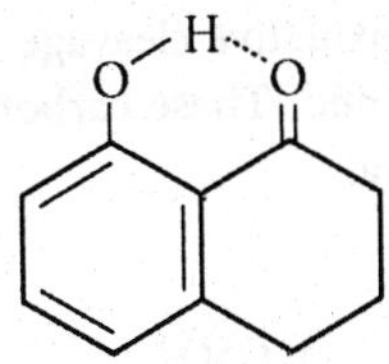

(a)

33. 1, 2-Dimethyl Cyclohexanol when treated with concentrated H_2SO_4, forms three unsaturated products. Suggest these products. How will you identify them from IR spectroscopy?

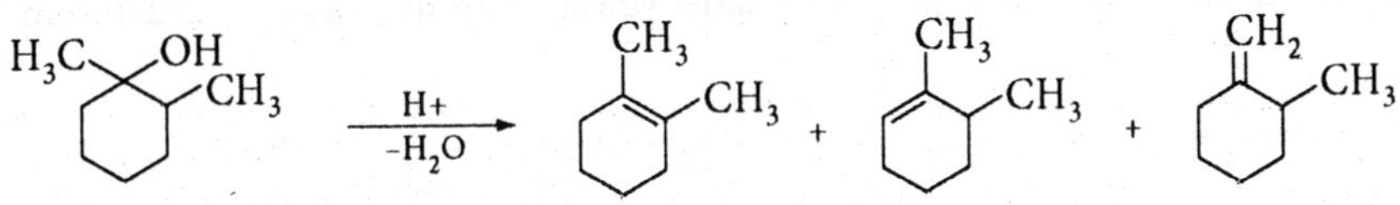

1,2-Dimethyl Cyclohexanol (I) (II) (III)

(I), (II) and (III) are the probable products. They can be easily identified by IR spectroscopy. (I) is symmetrical and so is IR inactive and no IR absorption band will appear in the IR spectrum. (III) has exocyclic methylene group. It will show absorption band around 895 cm^{-1} - 885 cm^{-1} while (II) is a trisubstituted alkene and absorbs around 840 cm^{-1} – 790 cm^{-1}.

34. An organic compound having molecular formula, C_2H_5NO, shows in its IR spectrum an absorption band at 1680 cm^{-1}. When reduced with $LiAlH_4$ it forms C_2H_7N whose IR spectrum lacks the peak at 1680 cm^{-1} instead it exihibits a band at 3300 cm^{-1}. Suggest the probable structure.

The absorption band at 1680 cm^{-1} is most probably due to amide linkage ($-\underset{\underset{O}{\|}}{C}-NH_2$). It is known that amide on reduction with $LiAIH_4$ generates amine, i.e.,

$$R-\underset{\underset{O}{\|}}{C}-NH_2 \xrightarrow{LiAlH_4} R-CH_2-NH_2$$

This reduced compound does not have any amide linkage but it has amino group (—NH—). So it will display —N—H stretching frequency at 3300 cm^{-1} instead of $-\underset{\underset{O}{\|}}{C}-NH$ stretching frequency.

35. Explain how the following isomeric steroids which differ only in the position of the double bond can be identified by IR spectra.

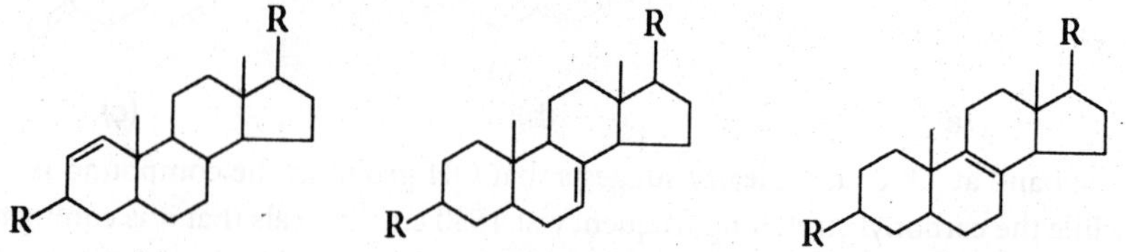

(a) (b) (c)

For identification, these steroids are hydroxylated by OSO_4 when glycols are formed. These glycols are subjected to oxidative cleavage by HIO_4 when carbonyl compounds (aldehydes or ketones) are formed. These carbonyls (aldehydes and ketones) can be easily distinguished by IR spectra.

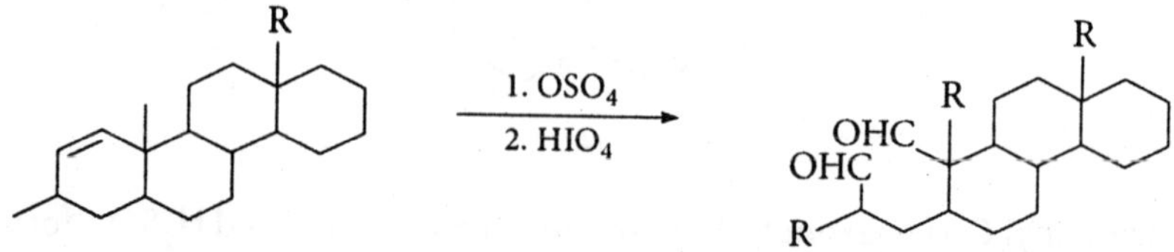

(a) (a′)

(a′) in its IR spectrum shows $\nu_{C=0}$ for aldehyde at 1730 cm^{-1} and ν_{C-H} at 2700 cm^{-1}.

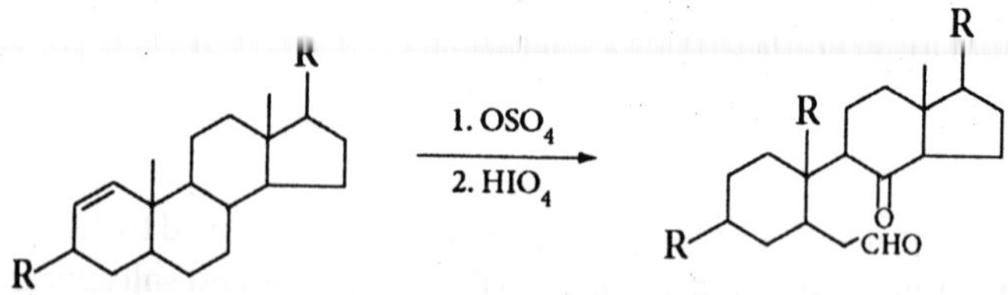

(b) (b′)

The IR spectrum of (b′) displays $\nu_{C=0}$ 1725 and ν_{C-H} 2700 cm^{-1} for aldehydic group and $\nu_{C=0}$ 1705 cm^{-1} for keto carbonyl.

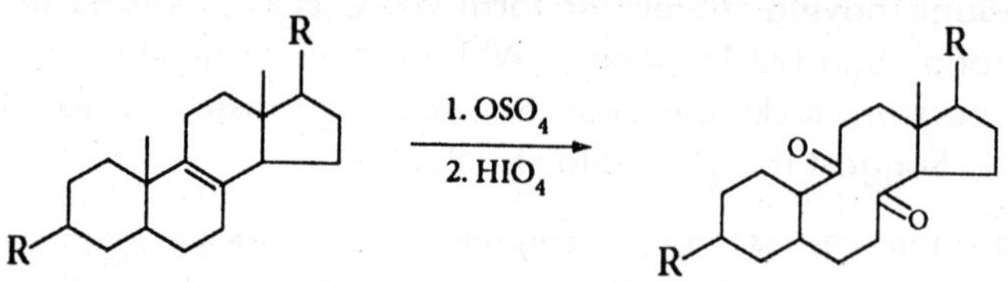

(c) (c′)

(c′) has only keto carbonyl and hence exihibits only one absorption band at 1705 cm^{-1} for keto carbonyl.

Thus the above thrice isomeric steroids can identified.

36. An organic compound, C_8H_{18} records in its IR spectrum the following bands at 2960, 2860, 1460, 1395, 1385, 1375, 1365 cm^{-1} and several bands in the fingerprint region. Suggest the probable structure.

➜ The molecular formula, C_8H_{18} of the compound conforms to the general formula C_nH_{2n+2}. So it is alkane in nature, presence of several bands in the fingerprint region is an indication of a branched alkane. The bands at 2960 cm^{-1}, 1860 cm^{-1} are suggestive of the presence of CH_3 groups. Peaks at 1395 cm^{-1}, 1385 cm^{-1} and 1365 cm^{-1} reveals the presence of $C(CH_3)_3$ while bands at 1375 cm^{-1} to 1365 cm^{-1} may be ascribed to $(CH_3)_2CH$—grouping. The peak at 1460 cm^{-1} is due to —CH_2 group. Thus the probable structure of the compound is

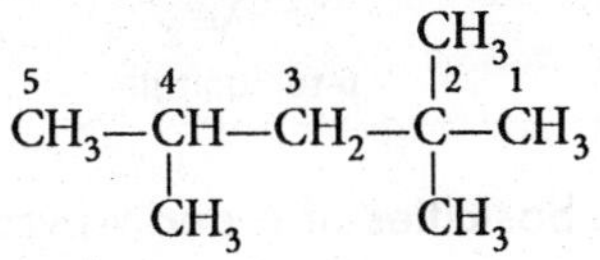

2, 2, 4-trimethyl pentane

37. An organic compound, C_3H_9N (A) displays the following bands in IR spectrum: 3012 (m), 3423 (s), 3236 (m) and 1615 cm^{-1} (m). When the compound is treated with nitrous acid compound (B) is formed which shows strong peak at 3430 cm^{-1}. Deduce the structures of (A) and (B) and explain the reaction involved.

➜ The two bands at 3423 and 3236 cm^{-1} in the IR spectrum of (A) are due to asymmetrical and symmetrical N—H stretching.So the compound contains an —NH_2 grouping in its molecule. The bands at 3012 cm^{-1} and 1615 cm^{-1} may be assigned respectively to C—H stretching and N—H bending. Thus the most probable structure of (A) is CH_3—CH_2—CH_2—NH_2. When this is treated with HNO_2, alcohol, CH_3—CH_2—CH_2—OH is formed which exhibits band at 3430 cm^{-1}.

Thus,

$$\underset{\substack{(A)\\ n\text{-Propyl amine}}}{CH_3{-}CH_2{-}CH_2{-}NH_2} \xrightarrow{HNO_2} \underset{\substack{(B)\\ n\text{-Propyl alcohol}}}{CH_3{-}CH_2{-}CH_2{-}OH}$$

38. An organic compound, C_8H_6 decolourises bromine in carbon tetra chloride and gives a white precipitate with ammoniacal silver nitrate solution. Its IR spectrum displays the following bands.

3300 (sh), 3085–3040 (w), 2110 (w), 760 (w), 690 (w), 650 (br) and 610 (w) cm^{-1}. Suggest the probable structure.

➜ Decolourisation of bromine is indicative of the unsaturated nature of the compound while formation of precipitation by ammoniacal silver nitrate suggests the presence of teminal acetylenic moiety.

The sharp band at 3300 cm^{-1} is due to C—H stretching vibration while weak band at 3085 cm^{-1} - 3040 cm^{-1} reveals aromatic C—H stretching vibration. The weak band at 2110 cm^{-1} may be assigned to $C\equiv C$ stretching vibrations. The presence of weak bands at 690 cm^{-1} and 760 cm^{-1} suggests the presence of monosubstituted phenyl ring. The broad and weak absorption bands at 650 cm^{-1} and 610 cm^{-1} are due to acetylenic C—H bending vibrations. So the probable structure of the compound is Ph—C≡C—H (Phenyl acetylene).

39. An organic compound, C_8H_7N exhibits following IR absorption bands: 3030 (w), 2920 (w), 2217 (s), 1607 (m), 1508 (m) and 817 (s) cm^{-1}. Deduce its structure.

➡ The extremely low content of hydrogen in the compound indicates that it is probably aromatics in nature which is confirmed by the presence of bands at 3030 cm^{-1}, 1607 cm^{-1} and 1508 cm^{-1}. The strong bands at 2217 cm^{-1} is ascribed to nitrile group (—CN). The strong band at 817 cm^{-1} may be Ar—H bending vibrations. All these information suggest that the compound is *p*-toluonitrile, i.e.,

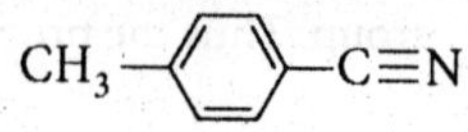

p-toluonitrile

40. Determine the relative basicities of α and γ-pyrones from their IR spectral data as mentioned below:

α, γ-Dimethyl-α-pyrone: 1727 (s), 1706 (s), 1640 (m) and 1570 (m).
α-α¹-Dimethyl-γ-pyrone: 1664 (s), 1612 (s), 1592 (m).

➡ Both the pyrone derivaties exhibit resonance as shown:

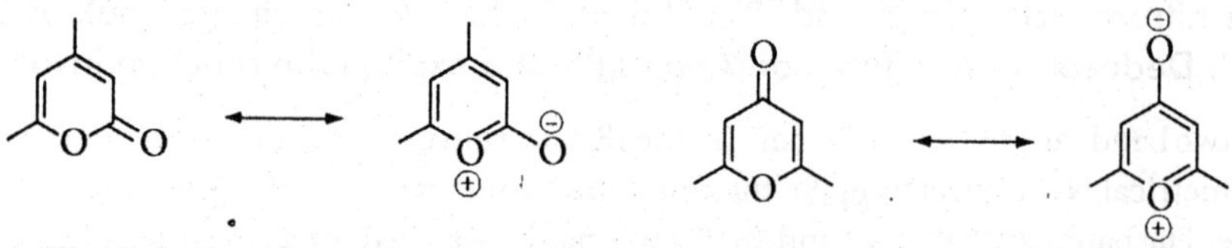

α,γ-Dimethyl-pyrone α,α'-Dimethyl-γ-pyrone

The different IR bands of α-pyrone may be assigned as 1727 and 1706 (C=O), 1640, 1570 cm^{-1} (C=C), and those of γ-pyrone may be assigned as 1664 cm^{-1} (conjugated C=O) 1612 and 1592 (C=C). There are two basic sites—two oxygen atoms in both the pyrones. However, negatively charged oxygen is more basic. So greater the resonance, more is the basicity. The presence of the conjugated C=O band of γ-pyrone at 1664 cm^{-1} shows that it is more basic than α-pyrone in which both $\nu_{C=0}$ correspond to non-conjugated C=0 vibrations.

41. Which of the following compounds (A) or (B) shows low C=O stretching frequency?

➡

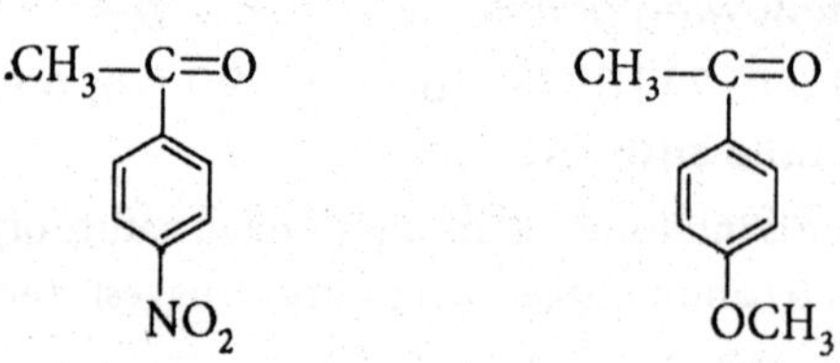

(A) (B)

—OCH_3 group is electron releasing. So it will decrease C=O stretching frequency while —NO_2 group is electron withdrawing and it will try to shift C=O stretching frequency in the higher direction i.e.

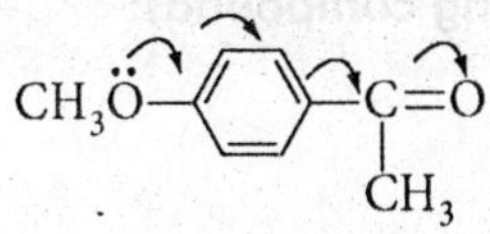

(B)

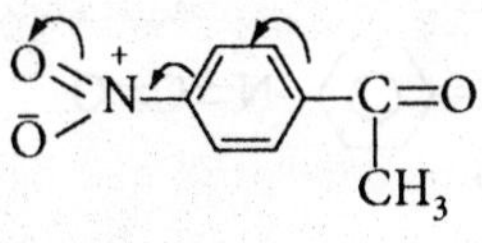

(A)

Thus compound (B) shows low C=O frequency.

42. Predict the frequency shift of aldehydic carbonyl absorption in the following compounds:

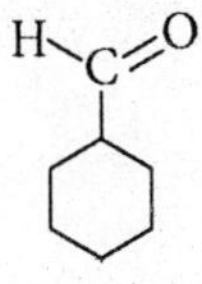

(A)

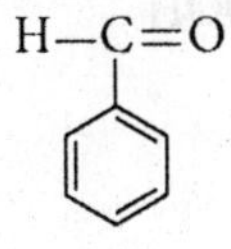

(B)

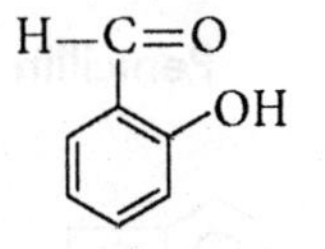

(C)

(A) is a saturated aldehyde. So it exhibits IR band around 1730 cm^{-1} but the compound (B) is aromatic. So aldehydic C=O stretching vibration will shift to lower frequency around 1680 cm^{-1}. But in (C) there is an intramolecular H-bonding like

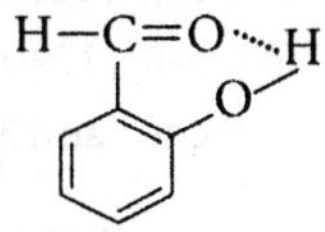

Hence aldehydic C=O absorption is further shifted to lower frequency around 1650 cm^{-1} – 1660 cm^{-1}.

43. Comment on the O—H and C=O stretching vibrations for the following pairs.

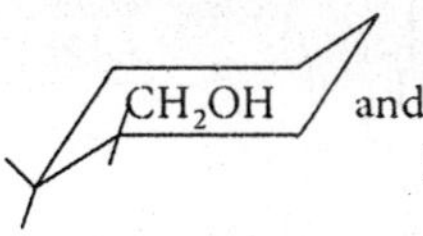

and

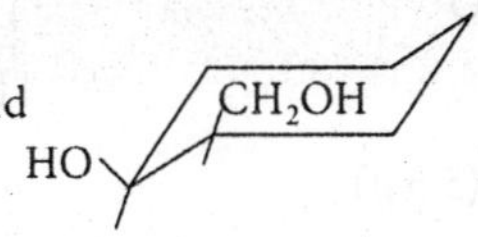

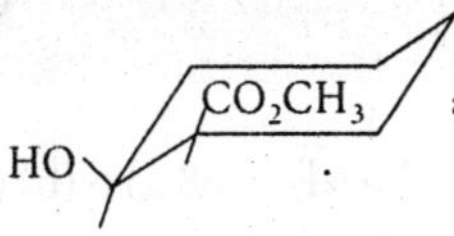

and

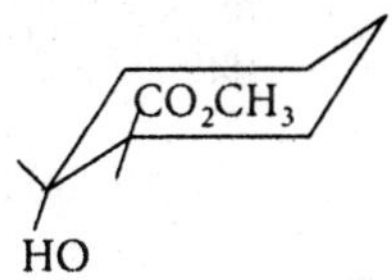

(A)	(B)	(C)	(D)
3641 cm^{-1}	3629 cm^{-1}, 3550 cm	3548, 1709 cm^{-1}	3635, 1723 cm^{-1}

In the pair (A) and (B) only one member (B) forms intramolecular H-bond as

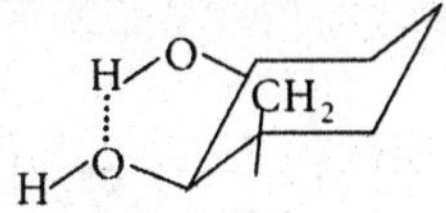

(B)

Similarly, in the pair (C) & (D) only (C) can form intramolecular H-bond as shown below.

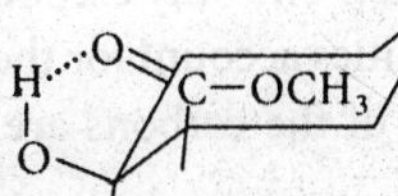

(C)

Due to these H-bond formation the pairs differ in their absorption bands.

44. Indicate the expected IR peaks for the following compounds:

(a) 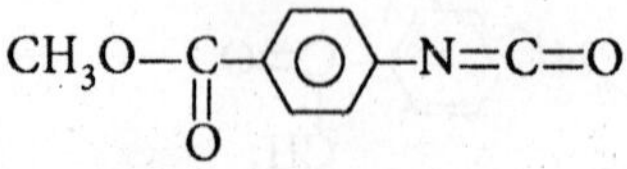

(b) 4-Hydroxy-3-penten-2-one

(c) 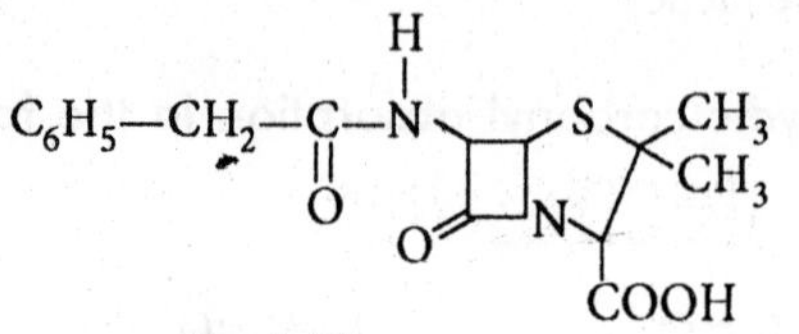

Penicillin

(d) 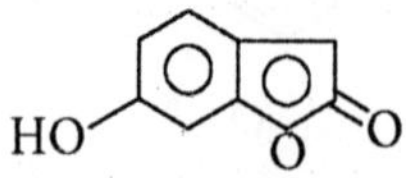

➡ (a) 1750 cm^{-1} and 2100 cm^{-1}.
(b) 1600 cm^{-1}, 1650 cm^{-1}, and 3500 cm^{-1}.
(c) 1690 cm^{-1}, 1740 cm^{-1}, 1750 cm^{-1} and 3300 cm^{-1}
(d) 1770 cm^{-1} and 3500 cm^{-1}.

45. Arrange the following structure-groups in order of increasing wavelength of carbonyl absorption in IR spectrum.

➡ (a) 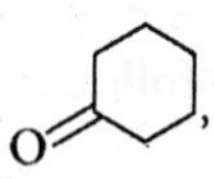, (b) 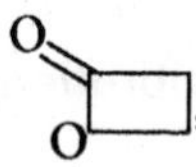, (c) 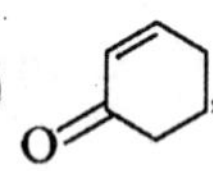, (d) 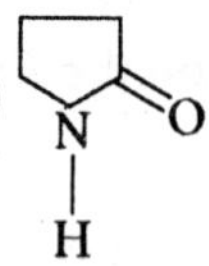

(c) (5.95 μ) > (a) (5.85 μ) > (d) (5.68 μ) > (b) (5.5 μ)

46. Deduce the structures that fit the following spectral facts.
(a) C_4H_8O shows no bands above 1500 cm^{-1} except one at 2860 cm^{-1}.
(b) $C_5H_3NO_2$ exhibits bands at 1725 cm^{-1}, 2210 cm^{-1} and 2280 cm^{-1}.

➡ (a) Must be a saturated ether like

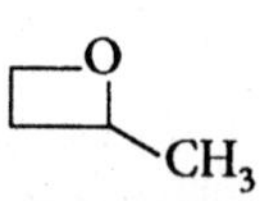 or 

(b) Two tripple bond bands (compound with only one nitrogen) with only one nitrogen must be due to C≡C and C≡N. The only carbonyl could be ketone or ester and not in a ring. These three functionalities account for the compound. Hence there is no ring or more unsaturation and four of the carbons are placed, an ester acounts for the fifth carbon and both oxygens. The unique solution is $N\equiv C-C\equiv C-COOCH_3$.

CHAPTER 3

ULTRAVIOLET SPECTROSCOPY

1. What do you mean by ultraviolet absorption spectroscopy? What is its absorption range? Why is it called electronic spectroscopy?

➻ Ultraviolet absorption spectroscopy deals with the measurement of energy absorbed when electrons are promoted to higher energy levels.

The wavelength range of this technique is 200–800 nm of which 200–400 nm is ultraviolet region and 400–800 nm region is known as visible region. The corresponding energy range of this region is 35–150 kcal/mole. By the absorption of this amount of energy electronic transitions of the molecule occur. For this reason this spectroscopic technique is also termed as *electronic spectroscopy.* The transition is indicated by an arrow as shown below:

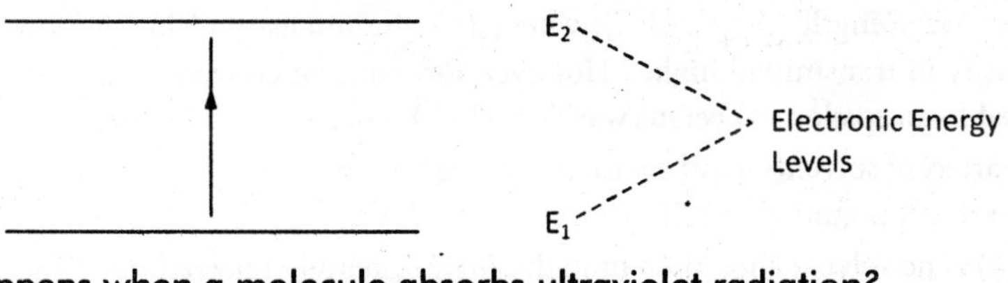

2. What happens when a molecule absorbs ultraviolet radiation?

➻ When a molecule absorbs electromagnetic radiation from ultraviolet region, electronic transition, i.e., promotion of electrons from low energy ground state orbital to higher energy molecular orbital takes place.

3. What is an ultraviolet spectrum?

➻ Ultraviolet spectrum is nothing but a plot of wavelength of light absorbed versus intensity of absorption. The wavelength at which maximum amount of energy required for a particular transition is absorbed appears in the spectrum as a peak. A typical UV-spectrum may be shown as below:

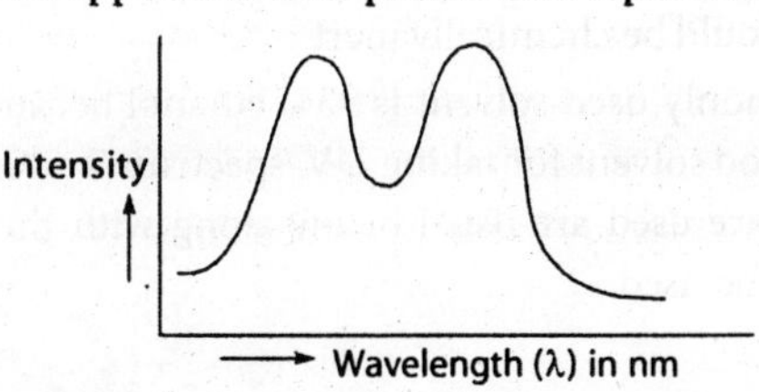

4. How would you measures the UV-spectrum?

➜ The ultraviolet or visible spectrum is usually taken in solution. Normally very dilute solution is used. Generally, 1 mg of a compound with molecular weight of 100–200 is accurately weighted and dissolved in an appropriate solvent and the volume is made up to to 100 ml. A portion of this solution is taken in a cell made of quartz or silica. The cell is made in such a way that the beam of light passes through a 1 cm thickness of solution. A matched cell containing pure solvent is also prepared and each cell is placed in the appropriate place in the spectrometer. Both the cells are mounted in a special holder in the spectrometer in such a way that two equal beams of UV-light are passed-one through the sample cell and other through the solvent cell. The most suitable source of light in the UV-region (200–400 nm) is the hydrogen discharge lamp. A tungsten filament lamp is usually used for the visible region (400–800 nm). The spectrometer electronically subtracts the absorption of the solvent in the reference beam (cell containing only solvent) from the absorption of the solution in the sample beam. In this way the effects owing to the absorption of light by the solvent is minimised. The spectrum is automatically plotted on most machines as absorbance (A) versus wavelength ($A = \log I_0/I$), where I_0 is the intensity of incident light and I is the intensity of transmitted hight). However, for working convenience, these are often converted to an epsilon (ε) versus wavelength (λ) plot as shown in the figure above:

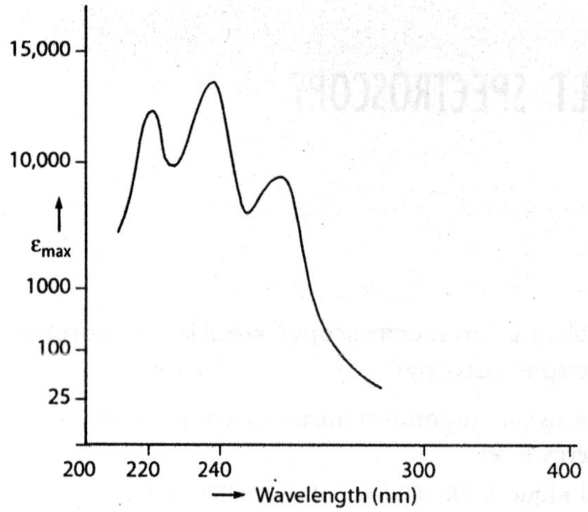

A variety of solvents may be used for taking UV-spectrum. But the choice of an appropriate solvent is guided by following criteria:

(i) The solvent should be of high purity, generally referred to as 'spectrograde'. Care should be taken to keep lint and dust from contaminating the final solution.

(ii) A good solvent should be transparent over the range of wavelength. Generally, solvent which do not contain conjugated systems are most suitable for running UV-spectrum.

(iii) Another criterion is the polarity of the solvent. A non-polar solvent does not form hydrogen-bond while a polar solvent can form hydrogen bond. So nature of the spectrum may be different in these two types of solvent.

(iv) The solvent should be good enough for dissolving most of the organic compounds.

(v) Solvent should be chemically inert.

The most commonly used solvent is 95% ethanol because it fulfils most of the criteria required for a good solvent for taking UV-spectrum and it is comparatively cheap. Other solvents, which are used are listed below along with the minimum wavelengths from which they may be used.

Solvent	Minimum wavelength (nm)
Ethanol	200
Methanol	200
Ether	215
Benzene	280
Acetone	330
Hexane	195
Carbon tetrachloride	265
Cyclohexane	200
Chloroform	245

Most spectrometers are automatic recording double beam instruments and consist of the following components:

(i) Radiation source covering the required wavelength range
(ii) Sample and reference cells
(iii) Monochromator for selecting a narrow band of wavelengths
(iv) Amplifier
(v) Recording device.

A schematic diagram of a UV-spectrometer (UV-spectrophotometer) is given below.

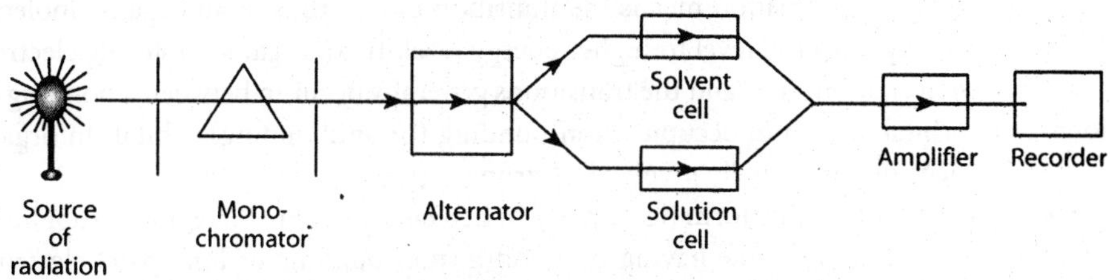

Double-beam UV-Spectrophotometer

5. Explain why glass cells are not used in UV measurement?

→ Glass absorbs UV-light strongly from 300 nm down and is not particularly useful for organic work.

6. Explain why ethanol is a good solvent for UV-measurement but not for IR.

→ See answer against Question No. 4 and add the following:
Since ethanol contains a hydroxyl group, its IR spectrum will show bands for O—H, C—O stretching and O—H bending and so complications may arise.

7. Ultraviolet absorption bands are generally broad compared to infrared. Explain.

→ It is known that a molecule in any electronic state has a number of vibrational and rotational energy levels. Therefore, when an electronic transition occurs vibrational and rotational transitions also occur because electronic transition requires more energy than the requirement of vibrational and rotational transitions. In other words, the electronic spectrum of a molecule consists of a number of vibrational and rotational lines in addition to electronic lines and will look like a 'band of spectral lines'. This does actually happen when the spectrum of a compound is taken in gas phase. But, more often, UV-spectrum is taken in solutions. In solutions, molecules are not free to rotate or vibrate due to solvent-solute interaction and so different spectral lines merge to give a broad absorption band when spectra are taken in solution.

8. Give a short account of different regions of ultraviolate spectrum.

➜ The whole region of ultraviolet spectrum is divided broadly into following three regions.

1. *Far ultraviolet region*: The wavelength region below 200 nm is known as far ultraviolet or vacuum ultraviolet region. It is usually obscured by atmospheric absorptions and it prevents the hazardus, i.e., high energy ultraviolet radiation in the sunlight from striking the surface of the earth. This region is difficult to measure and is of little diagnostic value.
2. *Near or quartz ultraviolet region*: The wavelength range 200–400 nm is the region normally meant by ultraviolet region. The atmosphere is quite transparent in this region. This region is of greatest importance and is usually studied for colourless compounds.
3. *Visible region*: The region from 400–800 nm is termed as visible region. Coloured compounds absorb in this region.

It may be mentioned that though 200–800 nm range includes both UV and visible radiations it is customary to refer this whole range as UV-range.

9. What do you mean by electronic transitions? Discuss various types of electronic transitions that are possible in an organic compound and indicate them by energy diagram.

➜ The term electronic transition means the transition of electrons of an organic molecule from lower energy ground level to higher energy level. In an organic molecule electrons resides in molecular orbitals and the transitions generally occur in between a bonding (or lone-pair) orbital and an unoccupied non-bonding (or antibonding) orbital. In organic molecules electrons are usually involved as strong σ (sigma) bonds, weak π bonds or are present in the non-bonding form (lone-pair). When energy from UV region is supplied to an arbitrary organic molecule having σ, π and *n* (non-bonding or lone-pair) electrons, electrons will absorb energy and will be promoted to their excited states, i.e., to their anti-bonding orbitals. The anti-bonding orbitals are indicated by the asterix (*). Thus during excitation σ-electrons will be promoted to σ* and π-electrons to π*. But as *n*-electrons having no bonding moelcular orbital of their own, they are promoted to the anti-bonding orbitals of σ- and π-molecular orbitals. Thus, in an arbitrary organic molecule having σ, π and *n* electrons will have four types of transitions such as

$$\text{(i) } \sigma \rightarrow \sigma^*, \text{ (ii) } \pi \rightarrow \pi^*, \text{ (iii) } n \rightarrow \sigma^*, \text{ (iv) } n \rightarrow \pi^*$$

The energy levels of these transitions may be indicated by the following diagram:

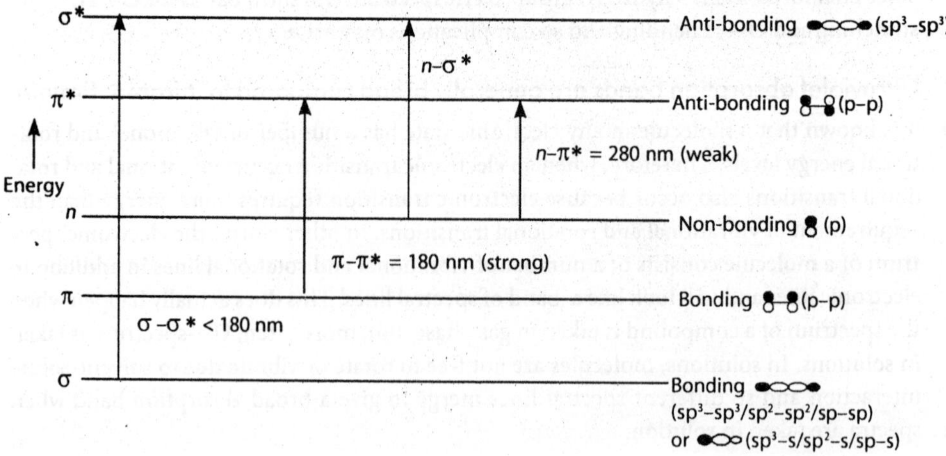

From the diagram it is evident that different electronic transitions require differemt amount of energies for their transitions. Out of these four types of transitions σ-σ^* require much more energy than available in usual ultraviolet region. So normally for any organic molecule, σ-σ^* transition does not occur. So only three types of electronic transitions occur in this technique as metioned below:

(i) $\pi \rightarrow \pi^*$

(ii) $n \rightarrow \pi^*$

(iii) $n \rightarrow \sigma^*$

Among these three types of electronic transitions $n \rightarrow \pi^*$ require minimum amount of energy while for n-σ^* transitions maximum amount of energy is rerquired. But $\pi \rightarrow \pi^*$ transitions occur in the energy range in between the above two transitions. In this connection it may be mentioned that a particular transition occurs only at a particular wavelength at which maximum amount of energy required for that transition is available. This wavelength at which maximum amount of energy required for a particular transition is available is indicated as λ_{max}. Thus from UV spectrum one or more λ_{max} values are obtained. These λ_{max} values reveal the nature of the electrons present in the molecule.

10. In a molecule containing σ, π and n electrons indicate the types of possible electronic transitions. Which transition involves longer wavelength radiation? Explain diagrammatically.

➜ See answer against Question No. 9.

11. Antibonding orbitals are expressed by the symbols $\sigma*$ and $\pi*$, but not by $n*$. Explain.

➜ Since n electrons are not engaged in any bonding molecular orbitals there are no corresponding anti-bonding (n^*) orbitals. n Electrons, therefore, on excitation, are promoted to anti-bonding molecular orbitals of σ- and π- bonds.

12. (a) What information can be obtained from the UV-spectral measurements? (b) What do you mean by chromophore and auxochrome?

➜ (a) From UV-spectral measurements we can get some λ_{max} values. From these λ_{max} values we can have an idea regarding the nature of electrons present in the molecule under consideration. It may be remembered that the nature of the electrons means the functional nature of the electrons, i.e., how the electrons function in the molecule. It is known that electrons function in the molecule through functional group. Thus from the UV-spectral studies nature of the functional group may be ascertained. However, it must be remembered that all the functional groups generally occuring in organic compound cannot be determined by UV-spectral study, only certain functional groups can be determined. Such functional groups are known as *chromophores*. So from UV-spectroscopic measurement nature of chromophore present in the molecule can be determined.

(b) *Chromophores*: Originally, the term *chromophore* was used to mean the system responsible for imparting colour to a compound and was derived from Greek word *chromophoros* meaning colour carrier. The term has now been retained within an extended interpretation to indicate any isolated functional group not in conjugation with any other group which exhibits absorption of a characteristic nature in the ultraviolet or visible region. The simplest examples of chromophores are >C=O, C=C and NO_2. Simple compounds containing chromophoric systems are ethylene (CH_2=CH_2), acetone (CH_3—CO—CH_3) and benzene (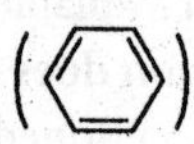). Each chromophore has some characteristic properties as mentioned below:

(i) A particular chromophore present in different compounds will absorb UV-radiation at very nearly the same wavelength.

(ii) A particular chromophore present in different compounds though may absorb at more or less same wavelength but the intensities of absorption in different compounds may or may not be the same.

The intensity of ultraviolet ray absorbed is indicated by ε (epsilon)—the molar extinction coefficient. Important examples of organic chromophores are tabulated below:

Some simple organic chromophores

Chromophores		Wavelength (λ_{max}) nm	Intensity (ε_{max})
C=O	aldehyde	223	150
	ketone	279	15
C=O		175	14000
C≡C		195	2100
$-NO_2$		270	17
—C≡N		165	5
C=C–C=C			215 20000
C=C–C=O			220 10000
		315	30
C=C–C≡O			220 7500
		230	7500
Benzene		184	60000
		204	9400
		255	204

It may be mentioned that in organic chemistry conjugated chromophores are of greatest importance.

Auxochrome: Auxochrome is an atom or a group of atoms which itself is not a chromophore but which enhances the wavelength (λ_{max}) and intensifies the absorption of UV-radiation by the chromophore. For examples —OH, —OR, $-NH_2$, $-NR_2$ (each containing lone pair), CH_3, etc. do not itself absorb in UV-region but when attached to any chromophore shift the λ_{max}. However, this definition of auxochrome has been modified in the light of modern theory. The effect of auxochrome is due to their ability to extend the conjugation of a chromophore by sharing of the non-bonded electrons and thus auxochrome becomes a part of a new extended chromophore and their action only differs in degree from the effect of extending the conjugation by adding a chromophore to another chromophore. Some common examples of auxochromes are as below.

 is chromophore; it has λ_{max} at 250 nm. But when the auxochrome —OH is attached

to it, i.e., 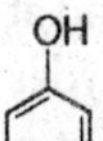 λ_{max} shifts to 280 nm. Similarly, auxochrome —NH_2 when attached, λ_{max} of benzene shifts to 270 nm.

13. (a) Define chromophore. (b) Classify the chromophores according to the type of electronic transitions giving examples.

(a) Chromophore is an isolated functional group which exhibits absorption of a characteristic nature in the ultraviolet or visible region. Common examples of chromophore are the groups such as >C=O, C=C, —NO_2, etc.

(b) According to electronic transitions, chromophores may be classified under following headings.

(i) *Chromophores involving $n \rightarrow \sigma^*$ transition*: These chromophores contain single bonds with one or more atoms having lone pair of electrons, i.e., saturated halides, alcohols, ether, aldehydes, ketones, amines, etc., These transitions require less energy and generally absorb above 150 nm and around 200 nm. For example, H_2O, CH_3OH, CH_3NH_2 and CH_3I absorb respectively at 167, 183, 213 and 255 nm. This type of chromophores are sensitive to hydrogen bonding. Thus alcohols form hydrogen bonds with solvent molecules due to the participations of non-bonded electrons on hetroatom and so require more energy for absorption, i.e. λ_{max} shift to lower wavelength.

(ii) *Chromophores showing $\pi \rightarrow \pi^*$ transition*: Chromophores giving rise to $\pi \rightarrow \pi^*$ transitions are those which contain electrons in π-molecular orbitals. Thus, these transitions are observed in compounds with unsaturated centres, i.e., simple alkene, aromatic, carbonyl, etc. This transition requires lesser energy than $n \rightarrow \sigma^*$ transition. Compounds containing isolated multiple bonds, i.e. ,$CH_2=CH_2$, $CH_2=CH-CH_2-CH_2-CH=CH_2$, etc. show λ_{max} above 150 nm and around 200 nm. For example, $CH_2=CH_2$ absorbs at 170 nm and saturated ketone shows λ_{max} at 155 nm.

In molecules containing conjugated bonds there are a number of π molecular orbitals and the π-orbitals of separate alkene systems combine to give new π-orbitals. In such systems transition between the highest energy containing molecular orbital occupied by electrons (HOMO) (bonding M.O) and lowest energy containing molecular orbital (anti-bonding) unoccupied by electrons (LUMO) occur. Such transition is of very low energy (higher wavelength) than simple $\pi \rightarrow \pi^*$ transition of unconjugated alkene. The wavelength of this type transition increases with the increase in the length of conjugated chain. Thus, while ethylene absorbs at 170 nm, butadiene has λ_{max} = 217 nm. If the chain extends to ten or more carbon atoms the absorption passes into visibile region. β-carotene (vitamin A) having eleven conjugated double bonds absorbs visible light. Conjugated ketones show both an $n \rightarrow \pi^*$ and a $\pi \rightarrow \pi^*$ transitions and are of lower energy, i.e., at higher wavelengths than those of isolated system.

(iii) *Chromophores displaying $n \rightarrow \pi^*$ transition*: Chromophores giving rise to $n \rightarrow \pi^*$ transition are those which contain electrons in both π and *n* molecular orbitals. In this transition, an electron of non-bonding orbital on a heteroatom is excited to anti-bonding orbitals. This transition requires least amount of energy than other.

transitions, and so this transition gives rise to an absorption band at longer wavelength. The most common example is >C=O group in aldehydes and ketones. They undergo both $\pi \rightarrow \pi^*$ and $n \rightarrow \pi^*$ transitions, the latter at a longer wavelength with low intensity than $\pi \rightarrow \pi^*$. Formaldehyde, for example, shows two absorption bands, a strong band at 170 nm (due to $\pi \rightarrow \pi^*$) and a weak band at 280 nm (due to $n \rightarrow \pi^*$) as shown in the figure below>

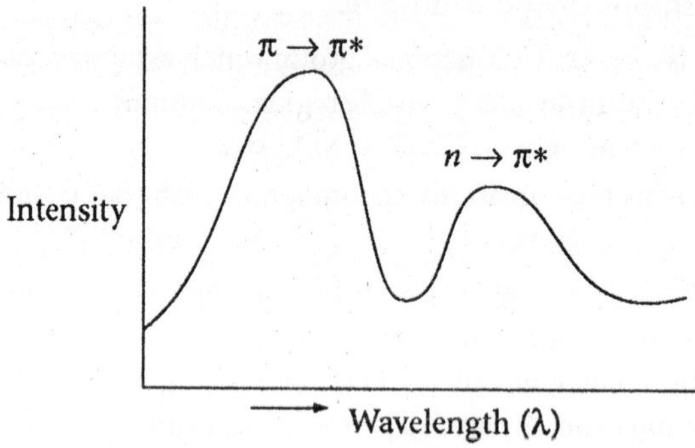

UV-Spectrum of Formaldehyde

It may be mentioned that if n electrons are present in conjugated system the increase in wavelength is much more pronounced. Thus, while the chromophore >C=O absorbs at 170 nm ($\pi \rightarrow \pi^*$) and 280 nm ($n \rightarrow \pi^*$), >C=C—C=O absorbs at 240 ($\pi \rightarrow \pi^*$) and 280 nm (n $\rightarrow \pi^*$).

14. State the selection rule governing electronic transition.

➜ In many chromophores there may be certain electronic transitions, which would at first sight appear to be possible, are not generally observed in the ultraviolet spectrum. Consequently certain restrictions operate during the transitions. These restrictions are known as *selection rules.* A few generalisations of the selection rules may be stated as below:

(i) The transitions which involve a change in the spin quantum number of an electron during transition are forbidden (i.e., not allowed).

(ii) The second generalisation is related to the multiplicity of states. According to this generalisation singlet-triplet transitions are forbidden.

(iii) The two molecular orbitals—ground state and excited state, should not have the same symmetry for allowed transition. If these two molecular orbitals possess same symmetry, transition is forbidden.

In general, the allowedness or forbiddenness of any transition may be ascertained from the values of molar extinction coefficient (ε, epsilon). It has been observed that the transitions with epsilon values more than 10^4 are usually termed as allowed transitions. Such transitions generally originate due to $\pi \rightarrow \pi^*$ transitions. For examples, the absorption at 217 nm (ε 2100) of butadiene is a case of *allowed transition.* On the other hand, for the forbidden transitions, the values of epsilon are generally below 10^4. This transition arises due to excitation of one electron from the lone pair present on the heteroatom to an antibonding π^* orbital ($n \rightarrow \pi^*$). For example, the transition of non-bonding electron to π^* orbital near 300 nm in the case of carbonyl compounds with ε value between 10–100 is the result of *forbidden transition.* A simple illustration may be cited in this context. Benzophenone, C_6H_5—CO—C_6H_5 displays two types of transitions as below:

(a) $\pi \rightarrow \pi^*$, ε 20,000 (allowed)
(b) $n \rightarrow \pi^*$, ε 180 (forbidden).

15. What do you understand by singlet and triplet state of molecule? Which state has the lower energy and why?

➜ In a typical organic molecule there are even number of electrons. These electrons are usually paired in the ground state. When radiation with correct energy is supplied, absorption occurs without the inversion of spin of electrons and so the electron spin remains paired in the excited state. Such excited state is normally termed as *excited singlet state* and is indicated by S_1. But in many cases, excitation of electrons takes place wih the inversion of spin. This new excited state with two unpaired electrons is designated as excited *triplet state* expressed by T_1. It may be mentioned that direct excitation of the ground state (S_0) to a triplet state is not possible and so T_1 is invariably occurs through S_1. These phenomenon for ethylene molecule can be shown as in the figures below. In this molecule absorption of radiation of the correct wavelength promotes one of π electrons to the anti-bonding π* orbital.

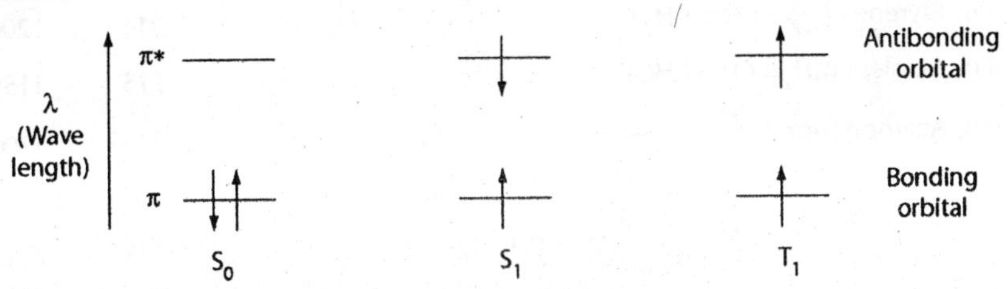

Singlet and triplet states of ethylene

The terms *singlet* and *triplet* originate from the fact that singlet states do not split in a magnetic field but triplet states have three possible energy states in an applied magnetic field. A triplet is more stable and is of lower energy than the corresponding excited singlet state because the two electrons in this state are not in close proximity to each other in space when their spins are parallel and so less inter-electronic repulsion is expected between the unpaired electrons than between paired electrons.

16. The intensity of $\pi \rightarrow \pi^*$ transitions is 10 to 100 times stronger than $n \rightarrow \pi^*$ transitions. Explain.

➜ $n \rightarrow \pi^*$ transition is always less intense than $\pi \rightarrow \pi^*$ transition because the electrons in the *n*-orbital are situated perpendicular to the plane of π-bond and hence to the plane of the π* orbital and so the probability of the jumping of an electron from *n* orbital to π* orbital is very low and in fact, zero according to the symmetry selection rule. However, vibrations of atoms bring about a partial overlap between perpendicular planes and so $n \rightarrow \pi^*$ transitions occur only to a limited extent. On the other hand, the intensity of $\pi \rightarrow \pi^*$ transition is very strong because the π- and π*- orbitals are situated in the same plane. As a result jumping of an electron from π to π* orbital is highly probable.

17. What do you mean by absorption band? Designate different types of absorption bands observed in ultraviolet spectrum.

➜ Normally in ultraviolet spectroscopy excitation of electrons is accompanied by vibrational and rotational change; as a result broad absorption band rather than sharp peak appears in the UV-spectrum. Several absorption bands have been recognised in the UV-spectrum. These absorption bands may be designated either by using electronic transitions or by the letter designation as illustrated below:

(i) *K-bands:* Band due to π—π^* transition, in fact, is the *K*-band. However, *K*-band originates from the German word Konjugierte and the band due to $\pi \rightarrow \pi^*$ transition in a compound with conjugated π-system is usually referred to as *K*-band. Such band is usually very intense with ε_{max} is of the order 10^4. Aromatic compounds, as they have π-bands are expected to exhibit *K*-band but such aromatic system as a whole behaves as a chromophore, *K*-band in aromatic compounds will be observed only when they are in conjugation with other chromophore. Some examples of compounds in which *K*-band appears are given below:

Compounds showing *K*-band

Compound	Transition	λ_{max} (nm)	ε_{max}
(i) Butadiene ($CH_2=CH—CH=CH_2$)	$\pi \rightarrow \pi^*$	217	21000
(ii) Styrene (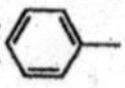—$CH=CH_2$)	$\pi \rightarrow \pi^*$	214	12000
(iii) Acrolein ($CH_2=CH—CHO$)	$\pi \rightarrow \pi^*$	215	11500
(iv) Acetophenone (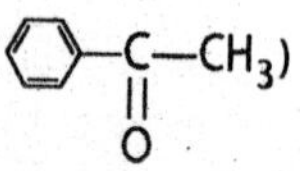)	$\pi \rightarrow \pi^*$	240	13000
(v) Pentadiene 1,3 ($CH_2=CH—CH=CH—CH_3$)	$\pi \rightarrow \pi^*$	223	22500
(vi) 2,3-Dimethylebutadiene ($CH_2=C(CH_3)—C(CH_3)=CH_2$)	$\pi \rightarrow \pi^*$	225	21500

It may be mentioned that *K*-band is due to an allowed transition and in compounds containing electrons other than π, bands due to other type of transitions may appear along with *K*-band.

(ii) *R-bands:* This band arises due to $n \rightarrow \pi^*$ transition of single chromophoric group with at least one lone pair of electrons on hetero-atom. These bands are also termed as forbidden bands as they originate from *forbidden* transition. The intensities of these bands are generally very low with ε_{max} below 100. But as non-bonded electrons are loosely held, they require less energy for excitation and hence these bands appear at higher wavelength. The name of this band has been designed from the German word *Radikal.* A few examples of compounds in which these bands occur are cited below:

Compounds showing R-band

Compound	Transition	λ_{max} (nm)	ε_{max}
(i) CH_3CHO	$n \rightarrow \pi^*$	290	10
(ii) $CH_3—C(=O)—CH_3$	$n \rightarrow \pi^*$	280	15
(iii) $CH_3—CH=CH—CHO$	$n \rightarrow \pi^*$	320	15

Contd.

Compound	Transition	X_{max}(nm)	ε_{max}
(iv) $CH_2=CH-CHO$	$n \rightarrow \pi^*$	315	15
(v) 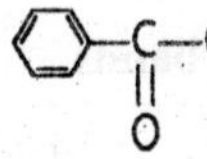	$n \bullet p^*$	320	50

It may be mentioned that the *R*-band is the most characteristic for carbonyl group (Aldehydes or Ketones).

(iii) *B-Bands:* B-Bands are also known as *benzenoid bands* and are observed in aromatic compounds—both homo- and hetrocyclic. These bands originate from $\pi \rightarrow \pi^*$ transitions in aromatic compounds. It may be mentioned that along with electronic transitions vibrational transitions also occur. As a result of such vibration, aromatic ring may lose planarity to some extent. So $\pi \rightarrow \pi^*$ transitions in aromatic systems take place in a different way than $\pi \rightarrow \pi^*$ transitions in non-aromatic compounds. Generally, benzenoid bands in aromatic compounds appear in the range of 230–270 nm. These bands are shifted to longer wavelength when any other chromophore is attached to the aromatic ring. These are low intense bands and are characteristic of aromatic compounds. However, these bands are susceptible to solvent effects and are diminished or frequently destroyed in polar solvents. In the case of substituted aromatic compounds, these bands may not always be observed. Some aromatic compounds displaying these bands are mentioned below.

Aromatic Compounds showing B-band

Compound	Transition	λ_{max}(nm)	ε_{max}
(i) Benzene 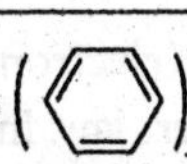	$\pi \rightarrow \pi^*$	256	210
(ii) Toluene 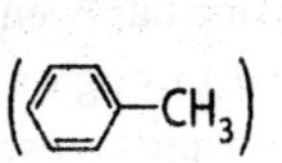	$\pi \rightarrow \pi^*$	260	170
(iii) Styrene 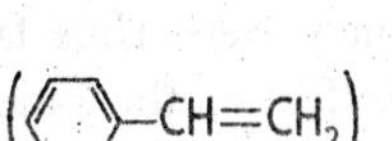	$\pi \rightarrow \pi^*$	282	450
(iv) Benzoic Acid 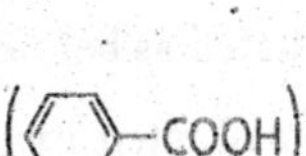	$\pi \rightarrow \pi^*$	272	810
(v) Nitro Benzene 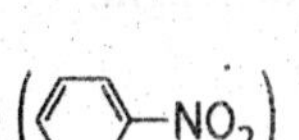	$\pi \rightarrow \pi^*$	282	1000
(vi) Naphthalene 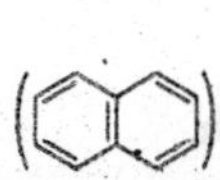	$\pi \rightarrow \pi^*$	310	290

(iv) *E-Bands:* This type of bands arise from the electronic transitions of aromatic system, i.e., due to the closed circuit of π electrons in aromatic systems. It is termed as ethylenic bands and are characteristic of aromatic systems like *B*-bands. *E*-bands are further characterised as E_1 and E_2 bands. E_1 and E_2 bands of benzene occur near 180 and 200 nm respectively and their ε_{max} varies between 2000 and 14000. In general, E_1 appears at lower wavelength and is usually more intense than E_2 band for the same compound. The table below shows E_1 and E_2 bands of a few aromatic compounds.

Aromatic Compounds exhibitings E_1 and E_2 bands

Compound	E_1-band		E_2-band	
	λ_{max} (nm)	ε_{max}	λ_{max} (nm)	ε_{max}
Benzene 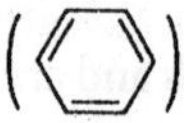	180	60,000	200	8,000
Pyrrole 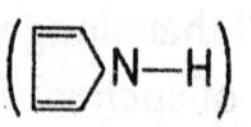	230	10,800	290	750
Quinoline 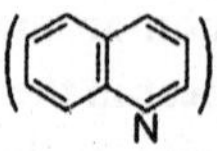	288	40,000	272	3,000
Naphthalene 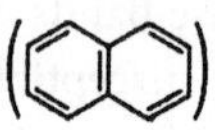	220	1,30,000	280	9,500
Anthracene 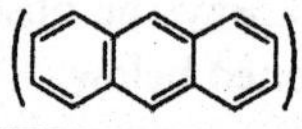	255	1,80,000	375	9,000

18. State absorption laws of light. Discuss how the intensity of electronic transition can be determined from these laws.

➜ The intensity of electronic transition or the intensity of light (electromagnetic radiation) absorbed by the molecule is determined by using Lambert-Beers law. In fact, this law is the combination of two laws—one due to Lambert and other one has been proposed by Beer. It is obvious that when a light beam (radiation) is passed through a sample of a given compound, the ratio of intensities of the incident light (I_0) and transmitted light (I) will depend upon how many molecules the light beam encounters. Thus, factors like concentration, temperature and the length have a marked effect on this ratio. Lambert's law is related to path length while Beer's law is concerned with the concentration of the sample through which light passes. These two laws may be stated as below:

Lambert's law: This law states that the rate of change of intensity of radiation with space (path length) is proportional to the intensity of incident light.

The above law can be mathematically expressed as

$$\frac{dI_0}{dx} \propto I_0, \qquad \text{(i)}$$

where I_0 is intensity of incident light, x is the path length.

Beer's law: Beer stated his law as—the rate of intensity of radiation with space (path length) is proportional to the concentration of the sample through which radiation passes. The above statement can be expressed mathematically as

$$-\frac{dI_0}{dx} \propto C, \qquad \text{(ii)}$$

where C is molar concentration.

Combining the expressions (i) and (ii), we get

$$\frac{dI_0}{dx} \propto I_0 C, \qquad \text{(iii)}$$

$$\text{or,} \; -\frac{dI_0}{dx} = \varepsilon I_0 C,$$

where ε (epsilon) is a proportionality constant known as *molar extinction coefficient* or *molar absorptivity*

$$\text{or,} \; -\frac{dI_0}{I_0} = \varepsilon C \cdot dx$$

$$\text{or,} \; -\int_{I_0}^{I} \frac{dI_0}{I_0} = \varepsilon C \int_{x=0}^{x=x} dx .$$

Here I is the intensity of transmitted light.

$$\therefore \; \log \frac{I_0}{I} = \varepsilon C x . \qquad \text{(iv)}$$

N.B. Negative (–) sign indicates the decrease in intensity.

The term $\log \frac{I_0}{I}$ is known as *absorbance* of the sample and is denoted by *A*, formerly absorbance was called optical desnsity (OD). Thus, equation (iv) may be written as

$$A = \varepsilon C x \qquad \text{(v)}$$

$$\text{or,} \; \varepsilon = \frac{A}{C.x} . \qquad \text{(vi)}$$

From the equation (vi) it is thus evident that epsilon (ε) is directly related to the intensity of light absorbed by the molecule. Hence, from the value of ε intensity of electronic transition can be determined.

19. What do you mean by molar extinction coefficient? What information can be obtained from it?

➻ The molar extinction coefficient or molar absorptivity, ε, is a constant for a particular compound at a given wavelength and is most commonly expressed as ε_{max} at an absorption band maximum (λ_{max}). In general, its value is very high. That is why, in most of the cases it is expressed as its logarithm, i.e. as log, ε_{max}. It is related as

$$\log \frac{I_0}{I} = \varepsilon C x ,$$

where I_0 is the intensity of incident light, I is the intensity of transmitted light, C is the molar concentration and x is the path length through which light passes. For convenience the size of the cell in which the solution of the sample taken is one cubic centimetre. So the expression becomes

$$\log\frac{I_0}{I} = \varepsilon C$$

$$\text{or,}\quad \varepsilon = \frac{C}{\log\frac{I_0}{I}}$$

The term $\log\frac{I_0}{I}$ is known as *absorbance* and is denoted by A. Thus $\varepsilon = \frac{C}{A}$. A is recorded directly by the instrument. So from the values of A and molar concentration of the sample taken value of ε can be determined.

Further, epsilon is related to the transition probability as

$$\varepsilon = 0.87 \times 10^{20} \times P \times a,$$

where P is the probability of transition and a is the target area which means the size of chromophore. It may be mentioned that size of the chromophore depends upon delocalisation of electrons. More the delocalisation, more is the size of the chromophore. Generally delocalisation occurs through conjugation and resonance. Thus from the value of epsilon, following informations may be obtained:

(i) Intensity of electronic transitions

(ii) Probability of electronic transitions. It has been observed that if the value of epsilon is greater than 1000, the transition is allowed, while if its value becomes less than 1000 the transition is forbidden.

(iii) Size of the chromophore. It has been found that when the value of ε_{max} is of the order 10^5 the length of the chromophore becomes 10 Å or 10^{-7} cm.

20. Define and explain the following: (i) Absorbance and (ii) Transmittance.

(i) *Absorbance*: Absorbance is a measure of the absorption of radiation by the sample. It is the logarithm of the ratio of the intensities of the incident light and the transmitted light. It is usually denoted by A. Thus, $A = \log\frac{I_0}{I}$, where I_0 is the intensity of incident light and I is the intensity of transmitted light. It is known as Optical Density (OD). It is a dimensionless quantity and its value can be calculated from the molar extinction coefficeient (ε), molar concentration (C) and the path length (x) of the material through which light passes according to the following equation:

$$A = \log\frac{I_0}{I} = \varepsilon C x.$$

It is directly recorded by the instrument

(ii) *Transmittance*: The transmittance for a substance is the ratio of intensity of transmitted light and incident light and is given by the relation:

$$T = \frac{I}{I_0},$$

where I is the intensity of transmitted light and I_0 is the intensity of incident light and T stands for transmittance. It is purely a number. From the above relation it is evident that the intensity of any ultraviolet absorption band can be expressed in terms of transmittance.

21. Define the following terms:

(i) Bathochromic Shift
(ii) Hypsochromic Shift
(iii) Hyperchromic Shift
(iv) Hypochromic Shift.

What structural features may produce such shifts in an organic compound?

(i) *Bathochromic Shift*: Shift of an absorption band to longer wavelength is termed as *bathochromic shift* or *red shift*. It is caused by the change of solvent or by the presence of an auxochrome. Thus, $\pi \rightarrow \pi^*$ transition is shifted to longer wavelength by increasing the polarity and introducing an auxochrome. Thus

(i) $CH_2{=}CH_2$, λ_{max}^{hexane} 170 nm, $\lambda_{max}^{ethanol}$ 175 nm

(ii) , λ_{max} 250 nm but in 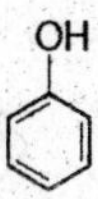 λ_{max} is 280 nm.

(ii) *Hypsochromic Shift*: When a shift of λ_{max} towards shorter wavelength occurs, the shift is designated as *hypsochromic* or *blue shift*. This may be caused by a change in the polarity of the solvent or due to the structural change. The structural change involves the change in chromophore or auxochrome. Thus,

Acetone λ_{max}^{hexane} 279 nm $\lambda_{max}^{H_2O}$ 264 nm

H—C=O (with H on C) λ_{max} 293 nm; H—C(=O)NH₂ λ_{max} 205 nm

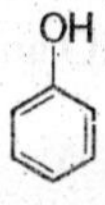 λ_{max} 280 nm

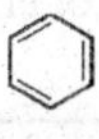 λ_{max} 250 nm

(iii) *Hyperchromic Shift*: It is an effect due to which intensity of absorption maximum is increased, i.e., ε_{max} increases. Like bathochromic and hypsochromic shift, this shift is produced by the change in structure or medium.

Thus,

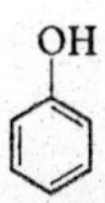 $\lambda_{max}^{ethanol}$ 280 nm ε_{max} 1450

 λ_{max}^{NaOH} 287 nm ε_{max} 2600

$\lambda_{max}^{ethanol}$ 257 nm ε_{max} 2750

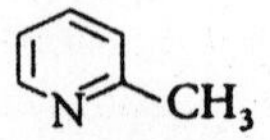

$\lambda_{max}^{ethanol}$ 262 nm ε_{max} 3560

(iv) *Hypochromic Shift*: It is reverse of hyperchromic shift. It may be defined as a phenomenon due to which intensity of absorption maximum is decreased. This type of shifts is generally observd by the introduction of a group which distorts the geometry of the molecule. Thus,

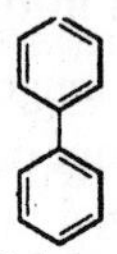

Biphenyl

λ_{max} 250 nm, ε_{max} 19,000

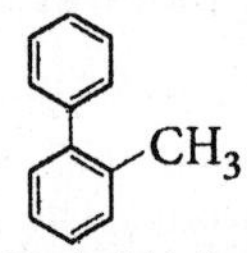

2-Methyl biphenyl

λ_{max} 237 nm, ε_{max} 10,250

However, hypochromic shift is also observed by changing the medium as in the case of benzoic acid.

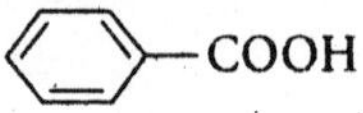

Benzoic acid

$\lambda_{max}^{ethanol}$ 273 nm, ε_{max} 970

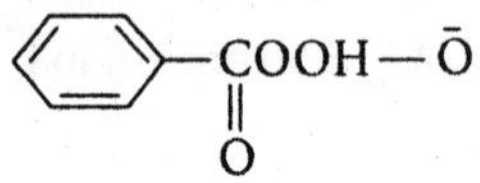

Benzoatic

λ_{max}^{NaOH} 268 nm, ε_{max} 500

22. Discuss the factors which affect the position of ultraviolet bands.

➜ The following factors have been found to affect the position of UV-bands.

(a) *Conjugation*: When two or more chromophoric groups are in conjugation in a molecule, the position of absorption band is shifted to higher wavelength (Bathochromic shift). It is known that in electronic transition there is a promotion of an electron from HOMO (highest energy containing molecular orbital occupied by paired electrons) to LUMO (lowest energy containing molecular orbital). So it is obvious that the smooth transition will occur when difference of energy between HOMO and LUMO is less.

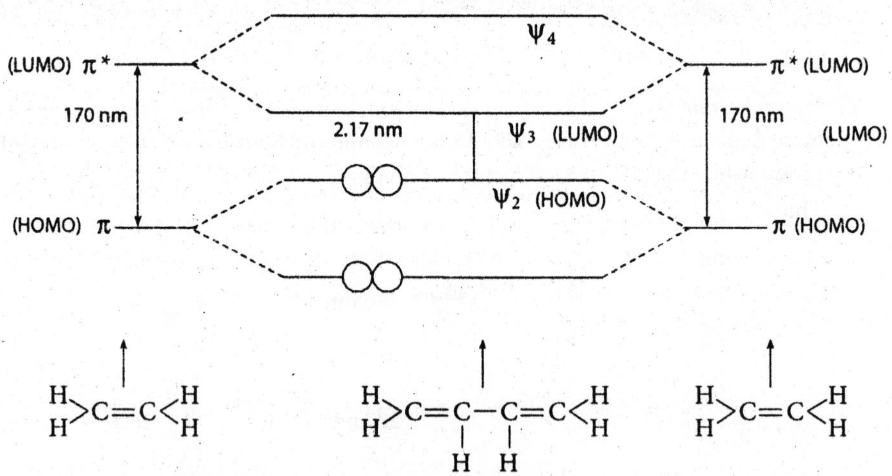

Due to conjugation delocalisation of electrons in the molecule occurs. As a result, a new set of molecular orbitals (both bonding and anti-bonding) are formed. Thus, it lowers the energy gap between HOMO and LUMO and hence bathochromic shift occurs. This situation may be illustrated by the above diagram taking ethylene and butadiene as examples.

It may further be mentioned that greater the number of conjugation, the lower will be the energy gap between HOMO and LUMO and λ_{max} will be more shifted to higher wavelength. Sufficient numbr of conjugation shifts the absorption to wavelength that reach the visible region. Thus, lucopene which is responsible for the red colour of tomatoes has eleven conjugated double bonds. This type of bathochromic shift of absorption maxima is also observed in other conjugated systems of chromophoric groups as tabulated below:

Some common conjugated chromophores

Chromophore	Example	λ_{max}	ε_{max}	Solvent
$>C=C-C=C<$	Butadiene ($CH_2=CH-CH=CH_2$)	217	21000	Hexane
$>C=C-C=O$	Crotanaldehyde ($CH_3-CH=CH-C=O$)	228	1800	Hexane
$>C=C-COOH$	*cis*-Crotonic acid ($CH_3-CH=CH-COOH$)	205	13500	Ethanol
$>C=C-NO_2$	Nitropropene ($CH_3-CH=CH-NO_2$)	230	9500	Ethanol
$>C(=O)-C=O$	Glyoxal (CHO—CHO)	195 280	35 03	Hexane
$>C(=O)-C=O$	$CH_2=CH-OCH_3$ (MethylVinyl Ether)	460 190	04 10,000	Ethanol

It may be mentioned that due to greater distance between the ends of the chromophore conjugated diene, in general, gives maxima of higher intensity than the diene with isolated double bonds.

(b) *Effect of Steric hindrance:* UV spectroscopy is very sensitive to the distortion of the chromophore. Due to distortion planarity of π-bond system may be affected, as a result shifting of absorption band occurs. Depending upon the nature of distortion, bathochromic (red) or hypsochromic (blue) shifts may occur. Thus verbenene exhibits λ_{max} at 245 nm but usual calculation shows a value of 227 nm.

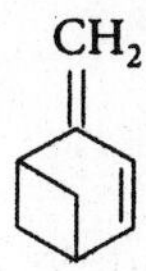

Exhibited λ_{max} 245 nm, calculated value of λ_{max} 227 nm

Verbenene

Similar is the situation with diene (*A*) which is expected to show λ_{max} at 273 nm but due to the loss of planarity by the distortion (i.e., loss of conjugation) shows λ_{max} at 220 nm.

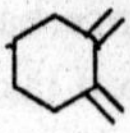

Exhibited λ_{max} 220 nm, expected λ_{max} 273 nm

(A)

Biophenyl system is also affected by steric hindrance.
Thus,

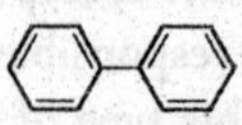

Biphenyl
(Planar)
λ_{max} 250 nm

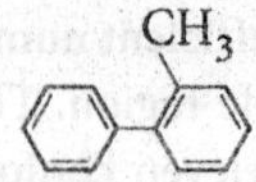

2-Methyl biphenyl
(Planarity is decreased)
λ_{max} 237 nm

(c) *Effect due to geometrical isomerism*: In the case of geometrical isomerism shifting of absorption maxima is also observed. *Trans*-isomer normally absorbs at higher wavelength than *cis*-isomer. *Cis*-isomer due to steric effect is non-planar whereas trans-isomer is free from steric hindrance. So effective overlap of molecular orbitals takes place in *trans*. Thus,

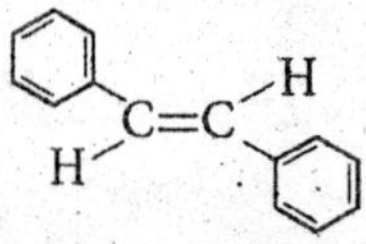

Trans-Stilbene
λ_{max} 295 nm

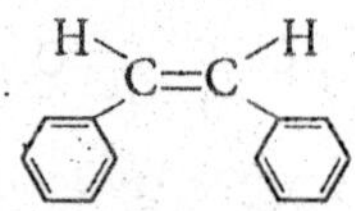

Cis-Stilbene
λ_{max} 278 nm

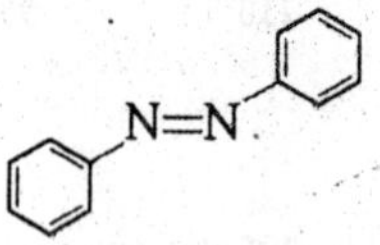

Trans-azobenzene
λ_{max}
$\pi \rightarrow \pi^*$ 320 nm
$n \rightarrow \pi^*$ 443 nm

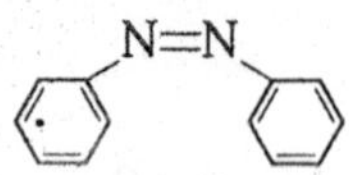

Cis-azobenzene
λ_{max}
$\pi \rightarrow \pi^*$ 281 nm
$n \rightarrow \pi^*$ 433 nm

(d) *S-cis (Cisoid) and S-trans (Transoid) Conformation*: Dienes can exist in S-*cis* and S-*trans* forms due to rotation about their central single bond. Acyclic dienes, in general, exist in preferred *transoid* conformation. But cyclic diene systems may exist either in *cisoid or* in *transoid form.* The cisoid cyclic dienes are known as *homoannular* dienes while *transoid* cyclic dienes are termed as *heteroannular dienes*. The heteroannular diene (S-*trans*) are spectrally related to butadiene. Thus,

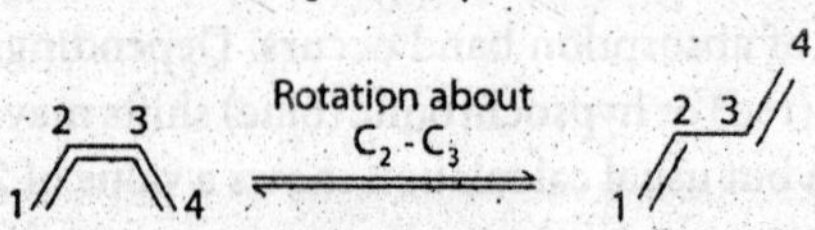

S-cis *S-trans*

1,3-Butadiene

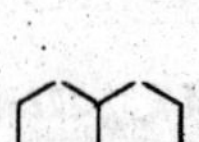

Heteroannular diene
(S-*trans*), λ_{max} 235 nm

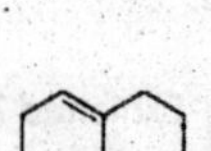

Homoannular diene
(S-*cis*), λ_{max} 273 nm

In general, *cis*-conjugated dienes or polyenes absorb at longer wavelength and have lower intensity than their trans-counterparts. It is not entirely clear why the S-cis conformation leads to longer wavelength absorption than does the *S-trans* conformation. However, it has been argued that since π-electrons in *cis*-isomer are more crowded, more repulsion between them occurs. This leads to the instability of the ground state energy level. But this interelectronic repulsion is reduced when an electron is promoted to anti-bonding (π*) molecular orbital. As a result, instability becomes less in excited state. So λ_{max} is higher and ε_{max} is lower in *cis*-conjugated-enes. The lower value of ε_{max} is due to shorter distance between the ends of the chromophore. As homoannular diene is S-*cis* ene, it absorbs at higher wavelength with lower intensity than heteroannular diene which is S-*trans* ene.

(e) *Exocyclic Double Bond*: When an exocyclic double bond becomes the part of the conjugated system, the shift of absorption maxima occurs toward longer wavelength. The presence of double bond outside the ring distorts the chromophore. As a result, the ground state energy level of the chromophore becomes unstable but promotion of one electron to anti-bonding (π*) molecular orbital makes excited state less unstable. So absorption maxima shifts to higher wavelength. It has been observed that a bathochromic shift of λ_{max} by 5 nm generally occurs. Thus, λ_{max} of β-phellandrene will have λ_{max} as calculated below:

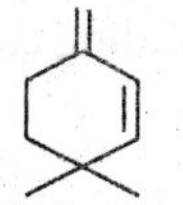

β-phellandrene

217 nm (Base value)
+ 5 nm (one exocylic double bond)
+ 5 x 2 nm (two ring residues)
232 nm

(f) *Substituents and Ring Residues*: It has been observed that when substituent or ring residue is attached to a conjugated chromophore absorption maxima is shifted to higher wavelength. It has been suggested that the inductive and resonance effects offered by the substituents are responsible for the red shift of λ_{max}. The magnitude of the shift depends upon the nature of substituents. For example, alkyl group and an amino group shift λ_{max} by 5 nm and 50 nm respectively due to small inductive effect of the alkyl group and the strong resonance effect of the amino group. Ring residue, i.e., the part of a carbocyclic ring is considered to be equivalent to alkyl substituent. The following example (A) may be cited in this connection. (A) is a hetro-annular diene with one exocyclic double bond and two ring residues. Hence, it will have λ_{max} as calculated below.

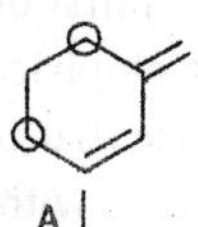

A

Ringresidue

217 nm (Base value
+ 5 nm (1 exo cylic double bond)
+ 5 x 2 nm (two ring residues)
232 nm

(g) *Solvent Effects*: Polarity of the solvents used has profound effect on the intensity and position of absorption maxima. By changing the polarity of the solvents, absorption maxima and its intensity are changed. It has been found that the increase in polarity of the solvent generally shifts bands due to $n \rightarrow \pi^*$ and $n \rightarrow \sigma^*$ transitions to shorter wavelength and bands caused by $\pi \rightarrow \pi^*$ transition to longer wavelength. These effects may be explained in terms of solute-solvent interaction as stated below.

In $\pi \rightarrow \pi^*$ transitions, the excited states are more polar than the ground state. So, if a polar solvent is used, it will interact more with the excited state of the molecule than the ground state. As a result, the excited state is stabilised (its energy is lowered) more compared to the ground state. With a non-polar solvent the interaction will be poor as indicated in the figure below.

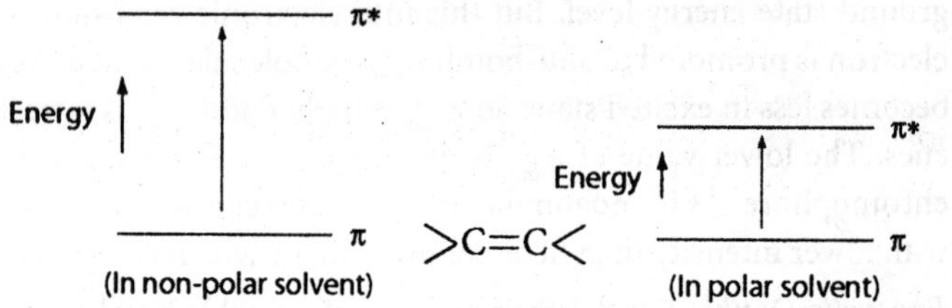

Therefore, the energy required for $\pi \rightarrow \pi^*$ in a polar solvent will be less (wavelength will be longer) than in a non-polar solvent.

Molecules with lone pair (non-bonding) electrons are able to interact with polar-protic solvent to form hydrogen bonds and this interaction is greater in the ground state than in excited state. So the ground state is stabilised more than the excited state. As a result, $n \rightarrow \pi^*$ and $n \rightarrow \sigma^*$ transitions will require more energy than in non-polar solvents. So the absorption maxima will be shifted to lower wavelength (blue shift). Greater the hydrogen bonding ability of the solvent, more will be the shift.

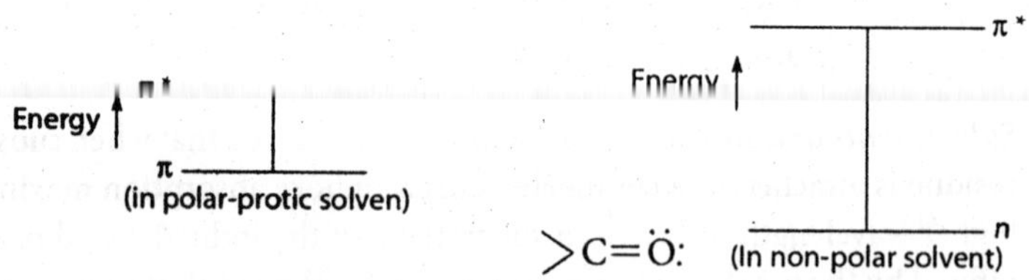

It may be mentioned that conjugated ene and aromatic hydrocarbons experience less solvent shift than conjugated enone. This is because π^* molecular orbital in enone is more stabilished than in ene.

23. Describe the Woodward-Fieser rules for calculating absorption maxima in diene and enones with examples.

➜ The absorption of ultraviolet radiation by conjugated enes and enones is, in fact, influenced by the structure in a surprisingly regular way, was first observed by Nobel laureate RB Woodward in 1941 and he proposed two separate empirical rules for predicting absorption maxima of these two conjugated systems. Since that time these rules have been modified by Fieser and by Scott as a result of experience with a large number of compounds of these two categories. To use these rules, it is simply a matter of choosing the correct base value and summing the contributions made by each substituent feature. These two rules may be summarised as below.

1. ***Rules for Diene Absorption***

Base value assigned to parent open chain or heteroannular diene	217 nm
Base value assigned to parent homoannular diene	253 nm

Contd.

Increment for

(a) each alkyl substituent or ring residue 5 nm

(b) each exocyclic double bond 5 nm

(c) each double bond extending the conjugation 30 nm

(d) auxochrome –O–Acyl 0 nm

–O–Alkyl 6 nm

–Cl, –Br 5 nm

λ_{max} (Calc) Total:

Examples cited in this connection are as under:

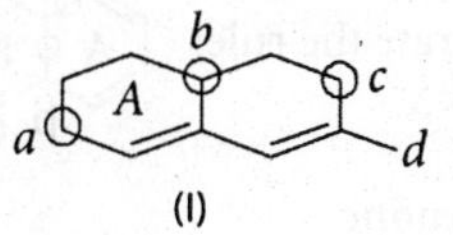

(I)

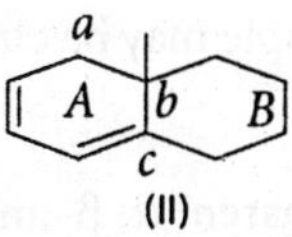

(II)

(I) is heteroannular diene		(II) is homoannular diene	
Base value	217 nm	Base value	253 nm
Increment for		Increment for	
(a) three ring residues (a, b, c) 3×5	15 nm	(a) three ring residues (a, b, c) 3×5	15 nm
(b) one alkyl substituent (d)	5 nm		
(c) double bond in ring (A) is exocyclic	5 nm	(b) double bond in ring (A) is exocyclic 1 × 5	5 nm
Total :	247 nm	Total :	273 nm

Compound (I) has been found to have λ_{max} value at 237 nm where (II) exhibits λ_{max} at 275 nm. It has been observed that in most of the cases calculated and observed values of λ_{max} match within ± 5 nm.

2. *Rules for conjugated Enone absorption*

Base value assigned to parent α, β-unsaturated six-memberd ring or acyclic enone 215 nm

Base value assigned to parent α, β-unsaturated five-membered ring enone 202 nm

Increment for

(a) each double bond extending conjugation 30 nm

(b) each alkyl group or ring residue at α 10 nm

at β 12 nm

at γ and higher 18 nm

(c) auxochrome

(i) OH at α 35 nm

at β 30 nm

at δ 50 nm

(ii) OAC at α, β 6 nm

(iii) OMe at α 35 nm

β 30 nm

γ 17 nm

Contd.

(iv) –Cl at α	15 nm
at β	12 nm
(v) –Br at α	25 nm
β	30 nm
(d) each axocyclic double bond	5 nm
(e) each homoannular diene component	39 nm
Calcl λ_{max} Total:	

Following example may be cited to illustrate the rule: 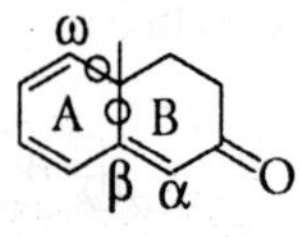

Base value for parent α, β-unsaturated enone	215 nm
Increment for (i) one β-substituents (O) 1× 12	12 nm
(ii) two double bonds extending the conjugation 2 × 30	60 nm
(iii) one ω-substituent (O) 1 × 18	18 nm
(iv) one exocyclic double bond in ring (B)	5 nm
(v) one homoannular diene component 1 × 39	39 nm
Total :	349 nm

The observed value of λ_{max} is 348 nm. In this case also calculated and experimental values match within ±5 nm.

24. Describe some important applications of ultraviolet spectroscopy.

➡ The technique of UV-spectroscopy has been proved to be a versatile tool at the hands of chemists and has a wide range of applications. Some of the important applications in brief are described below.

(a) *Detection of Chromophores (Functional groups)*: The most important application of this technique is the detection of certain chromophoric groups in a molecule, like C=C, C=O, C=N, extent of conjugations, etc. Each chromophore has its characteristic absorption. Again, absence of a band at a particular wavelength may be considered as an evidence for the absence of a particular functional group in the compound.

(b) *Structure Determination*: Although UV technique alone is seldom enough for complete structural elucidation of unknown compound, however, by comparison of UV-spectrum of an unknown compound with the spectra of model compounds it is possible in some cases to identify a compound. Following examples may be cited in this context.

Unknown Compound	**Model Compound**
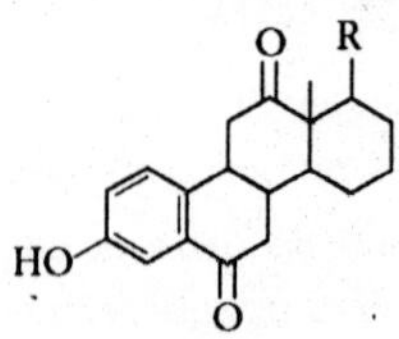	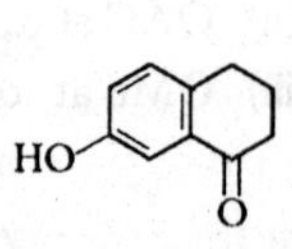

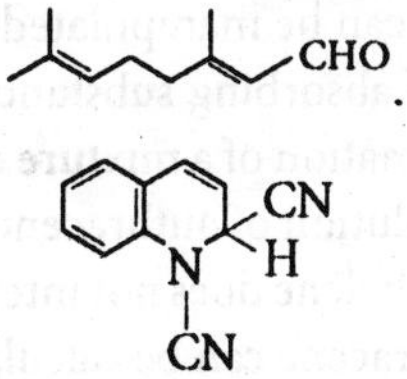

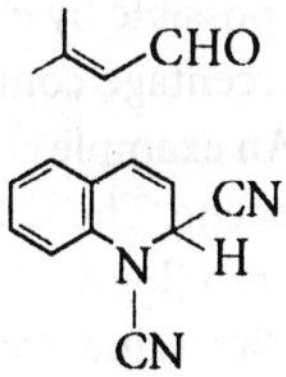

Another interesting example in this regard is the determination of the position of the group in azelain, a flavonol glycoside. This glycoside has the following part structure.

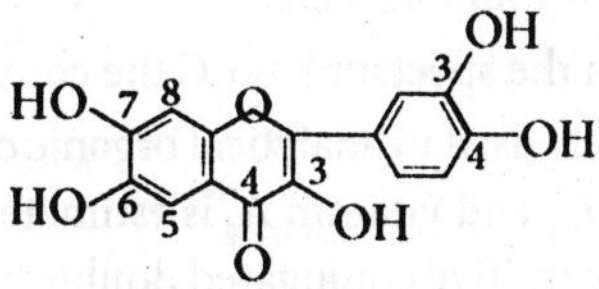

(A)

Aglycone (A) + CH_3 (as —OCH_3) + Sugar
(Azelain)

Flavonols in general show two strong UV bands, one at 250 nm and other at 370 nm. Now treatment of azelain with mild base (sodium acetate) shows a red shift of 19 nm in λ_{max} 250 nm. This shows 7—OH is free as shown below.

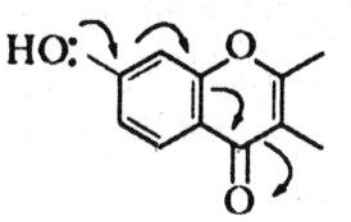

Again, the aglycone (A) can be readily degraded by NaOEt. This suggests sugar moiety is attached to 3—OH. Again, azelain forms a complex with boric acid suggesting the presence of two free hydroxyls at two consecutive positions. Hence 3—OH + 4—OH are free. Consequently —OCH_3 group is present at the position 5.

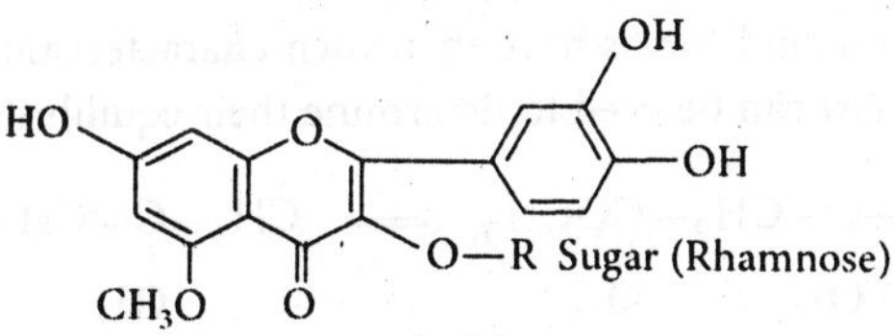

Azelain

(c) *Distinction between Isomeric Compounds*: By using UV-spectral data, isomeric compounds can be distinguished. Thus the compounds (I) and (II) can be easily distinguished.

$$(CH_3)_2{-}C{=}CH{-}\underset{\underset{O}{\|}}{C}{-}CH_3 \qquad\qquad CH_2{=}\underset{\underset{CH_3}{|}}{C}{-}CH_2{-}\underset{\underset{O}{\|}}{C}{-}CH_3$$

(I) (II)

Compound (I) is a conjugated enone while (II) is not. The forbidden $n \to \pi^*$ band for carbonyl group in (I) will appear at longer wavelength compared to that for (II).

(d) *Quatitative Analysis*: Using Lambert-Beer's law the absorbance data can be used to determine the concentration of an absorbing material in a very dilute solution which

is not possible by chemical means. The law can be manipulated to determine even the percentage composition in a mixture of absorbing substances present in solution. An example cited in this case is the estimation of a mixture of naphthalene and anthracene. The UV-spectrum in ethanol solution of authracene shows absorption at 375 nm (log ε = 3.87); absorption of naphthalene does not interfere in this region. From this value, concentration of pure anthracene can be calculated and hence the proportion of anthracene and naphthalene in the mixture. Equation to be used is

$$A \text{ (absorbance)} = \varepsilon . C . l$$

l is path length of the sample for convenience of calculation, sample is always taken *I* cc cell. Hence, value of $l = 1$ and $A = \varepsilon . C$

A and ε are obtained from the spectrum. So *C* the concentration can be calculated. This technique is extensively used in analytical organic chemistry. In natural fats and oils % content of vitamin A_1 and vitamin A_2 is estimated conveniently by measurement of intensities at 325 nm (five conjugated double bonds vitamin A_1) and at 351 nm (six conjugated double bonds, vitamin A_2) respectively. Comparison of the intensities with those of solutions of two vitamins of known cocentration enables to determine the desired concentration

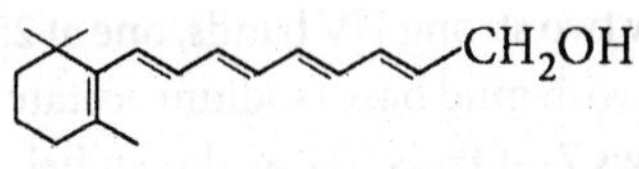

Vit A_1 (λ_{max} = 325 nm)

Vit A_2 (λ_{max} = 351 nm)

(e) *Study of Equilibria*: This spectroscopy is used to determine equilibrium constants in (i) acid-base dissociation (ii) complex formation, (iii) tautomeric equilibria. For example, let us consider a keto-enol tantomerism.

The keto and enol forms have their own characteristic absorption and hence Lamber-Beer's law can be used to determine their equilibrium concentration.

$$CH_3-\underset{\underset{O}{\|}}{C}-CH_2-\underset{\underset{O}{\|}}{C}OCH_3 \rightleftharpoons CH_3-\underset{\underset{OH}{|}}{C}=CH-\underset{\underset{O}{\|}}{C}OCH_3$$

λ_{max} = 275 nm (ε = 20) λ_{max} = 245 nm (ε = 18000)

(f) *Determination of the Configuration of Geometrical Isomers*: It has been established that in the case of geometrical isomersim *trans* isomer displays longer wavelength absorption with higher intensity than its *cis* isomer. Thus

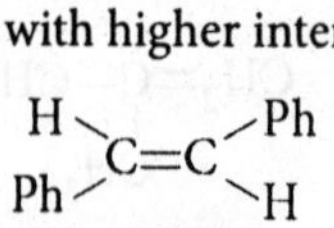

trans-Stilbene

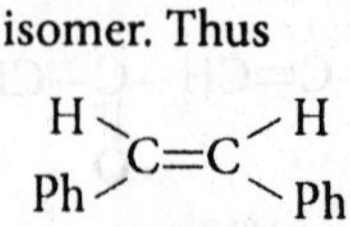

cis-Stilbene

λ_{max} = 275 nm (ε = 20) λ_{max} = 245 nm (ε = 18000)

(g) *Study of Charge Transfer Complexes*: In this types of complexes a donar donates a pair of electrons to a π-orbital of a double bond or an aromatic system. Such complexes generally show a spectrum which is not the sum of the spectra of the parent individual molecules. These molecules are normally coloured and show a peak in the visible region. Formation of picrate between picric acid and aromatic hydrocarbon is an example of this type of complexes.

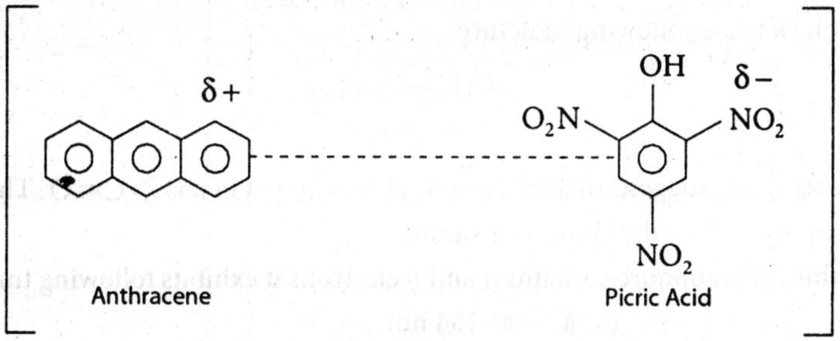

Picrate of Anthracene

(h) *Use in Sunscreen Lotion*: The principle of UV-spectroscopic technique is used in the preparation of sunscreen lotion. Substituted benzenes absorb between 250–290 nm with considerable ε_{max} value. This property has been applied in the preparation of sunscreen lotion to protect the skin from potential cancer-causing UV-radiation. Commonly used aromatic compound is para-amino benzoic acid (PABA). It has a λ_{max} at 289 nm with ε_{max} value around 19,000.

(i) *Geometry of Molecules in Excited States*: UV spectra have been used to predict the geometrical changes in molecules on electronic excitation. Thus, ethylene and formaldehyde in ground state are planar but on excitation the spectral data reveal that they are bent. Such information are very basic to a detailed understanding of systems undergoing photochemical reactions.

25. Explain the various types of electronic transitions possible in formaldehyde and sketch the spectrum indicating the transitions.

➡ The structure of formaldehyde is

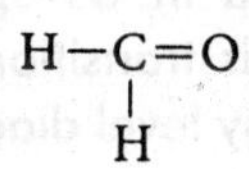

From this structure it is seen that its chromophore $>C=\ddot{O}:$ has two types of electrons—*n* and π. So it undergoes $\pi \rightarrow \pi^*$, $n \rightarrow \pi^*$ and $n \rightarrow \sigma^*$ transitions. The $\pi \rightarrow \pi^*$ transition gives rise to a strong band at 170 nm and two weak bands at 190 nm and 280 nm appears in the spectrum as shown in the spectrum.

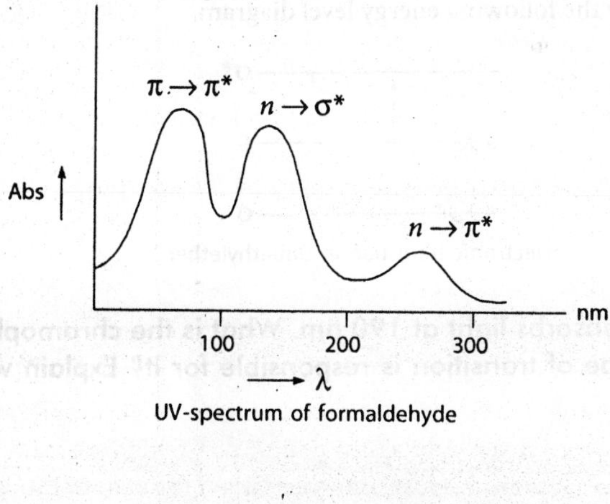

UV-spectrum of formaldehyde

26. Acetone absorbs at 154, 187 and 280 nm. What chromophore is responsible for each of these absorptions? What type of transition causes each absorption? Sketch the spectrum and indicate the transitions.

➻ Acetone has the following structure.

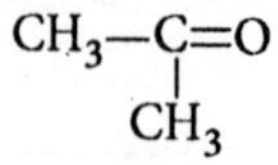

The structure suggests that the chromophore present in it is $>C=\ddot{O}:$. This chromophore is responsible for the above transitions.

As this chromophore contains π and n electrons it exhibits following transitions.

(i) $\pi \rightarrow \pi^*$ 154 nm

(ii) $n \rightarrow \sigma^*$ 187 nm

(iii) $n \rightarrow \pi^*$ 280 nm.

The spectrum may be drawn as under.

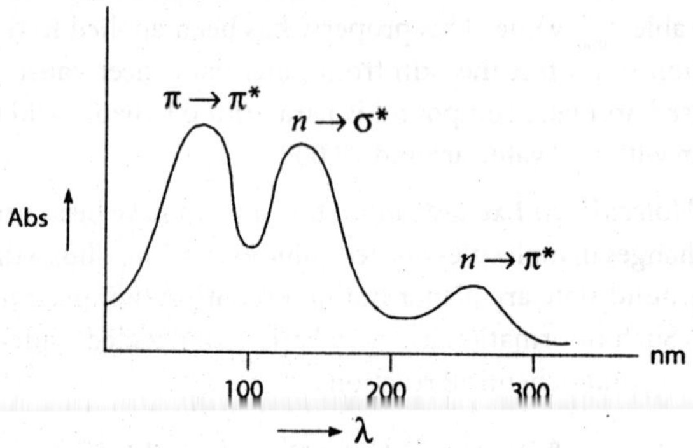

UV-spectrum of Acetone

27. Dimethyl ether displays a band in UV-spectrum at 185 nm. What is the chromophore responsible for this transition? What kind of transition is this? Illustrate your answer with energy level diagram.

➻ The structural formula of dimethyl ether stands as $CH_3-\ddot{\underset{\cdot\cdot}{O}}-CH_3$. It has only one chromophore $-\ddot{\underset{\cdot\cdot}{O}}-$.

It will show only $n \rightarrow \sigma^*$ transition. So the band at 185 nm is due to this transition. This may be shown by the following energy level diagram.

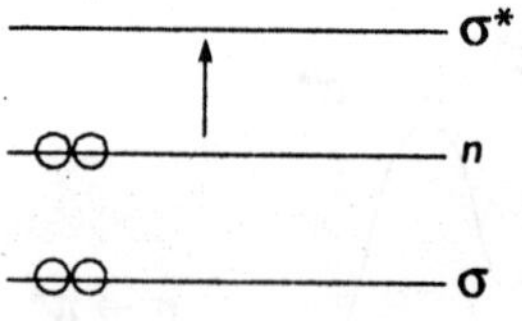

Electronic transition in Dimethylether

28. Cyclopentene absorbs light at 190 nm. What is the chromophore responsible for it? What type of transition is responsible for it? Explain with energy level diagram.

➜ Cyclopentene, 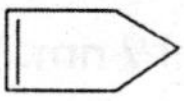 contains a double bond. So the chromphore is >C=C<. The corresponding transition is $\pi \rightarrow \pi^*$. The energy level diagram is given below.

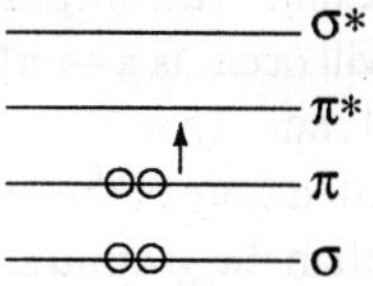

29. What type of transitions are possible for the following compounds.
(a) $CH_2{=}CH{-}O{-}CH_3$, (b) $CH_2{=}CH{-}CH_2{-}CH_2{-}O{-}CH_3$

➜ Compound (a) is a conjugated ether and so it will exhibit (i) $n \rightarrow \sigma^*$, (ii) $n \rightarrow \pi^*$, and $\pi \rightarrow \pi^*$.

The compound (b) contains two chromophores but they are not conjugated. Hence transitions expected are

(i) $\pi \rightarrow \pi^*$ (due to isolated C=C bond)

(ii) $n \rightarrow \sigma^*$ (due to isolated lone point)

30. Which of the following two compounds will absorb at higher wavelength and why?
(a) $CH_2{=}CH{-}CH_2{-}CH{=}CH_2$ (b) $CH_2{=}CH{-}CH{=}CH_2$

➜ Though (a) and (b) contains two π-bonds but in (b) these two π bonds are conjugated while in (a) they are isolated. Hence, (b) will absorb at higher wavelength.

31. An organic compound may be either (a) or (b). Its UV-spectrum is as below. Identify the compound. (a) $CH_2{=}CH{-}OH$, (b) $CH_3{-}CH{=}O$.

➜ The scrutiny of the spectrum reveals that the compound shows two bands at 190 nm and 230 nm. The band at 190 nm may be assigned to $\pi \rightarrow \pi^*$ while the band at 230 nm is due to $n \rightarrow \pi^*$. These absorptions are typical of >C=O chromophoric compound. So the the compound is $CH_3{-}CHO$ (Acetaldehyde).

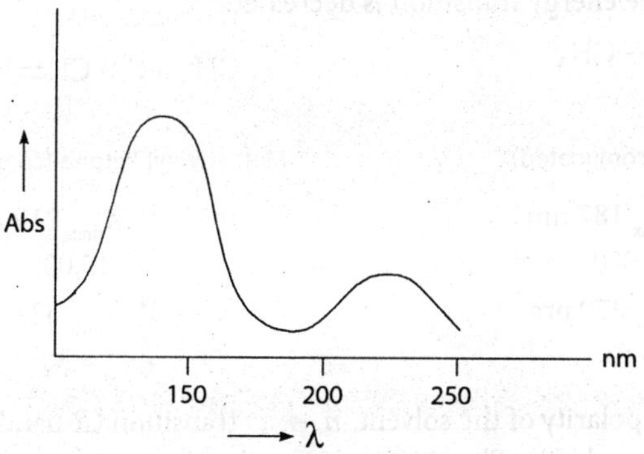

32. An organic compound in hexane exhibits λ_{max} at 305 nm and in ethanol shows λ_{max} at 307 nm. What should be the nature of transition?

➜ Ethanol is a polar and hexane is a non-polar solvent. The shift of λ_{max} to longer wavelength in more polar solvent is the characteristic of $\pi \rightarrow \pi^*$, becasuse the polar solvent stabilises the π^* state more than the π state.

33. A conjugated diene in hexane solution shows λ_{max} at 219 nm. What will happen if the solvent is changed to ethanol?

Since ethanol is a polar solvent and hexane is a non-polar solvent, λ_{max} will be changed. In conjugated diene the transition that will occur is $\pi \rightarrow \pi^*$. In $\pi \rightarrow \pi^*$ transitions, the excited states are more polar than the ground state. Therefore, when the polar solvent (ethanol) is used it will interact more with the excited state of the molecule than the ground state. As a result, excited state is stabilised more than the ground state and so the energy of the excited state will be lowered. Therefore, $\pi \rightarrow \pi^*$ will require less energy and λ_{max} will increase.

34. Describe different absorption bands displayed by saturated aldehydes and ketones.

Saturated aldehydes and ketones contain >C=O chromophore in their molecules. This chromophore will exhibit three absorption bands.

(i) $\pi \rightarrow \pi^*$ transition absorbs strongly near 170 nm (*K*-band)

(ii) An $n \rightarrow \sigma^*$ transition absorbs near 190 nm

(iii) An $n \rightarrow \pi^*$ transition absorbs at 270–300 nm (*R*-band)

The *R*-band is weak ($\varepsilon_{max} < 30$) and results from the forbidden transition of *n* electron to π^* orbital. The most intense band for these compounds is *K*-band due to $\pi \rightarrow \pi^*$.

35. What type of transition are observed in α, β-unsaturated carbonyl compounds? How absorption maximum and intensity are shifted when carbonyl group is not conjugated? Discuss the effect of solvent polarity on *R*-band.

In αβ-unsaturated carbonyl compounds two major type of transitions occur.
(i) $\pi \rightarrow \pi^*$ and (ii) $n \rightarrow \pi^*$.

Due to conjugation between ethelenic bond and carbonyl group, the energy gap between the ground state and excited state are lowered. As a result, absorptions due to these transitions occur at longer wavelengths (bathochromic shift). Thus, in methyl vinyl ketone both these transitions occur at longer wavelengths with more intensities than in the non-conjugated ketone, acetone. The excited states are more stabilised than the ground state, so magnitude of the energy transition is decreased.

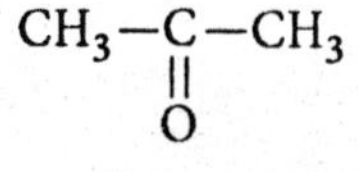

Acetone (non-conjugated)

$\pi \rightarrow \pi^*$, λ_{max} 187 nm

ε_{max} 10,000

$n \rightarrow \pi^*$, λ_{max} 279 nm

$\varepsilon_{max} \approx 10$

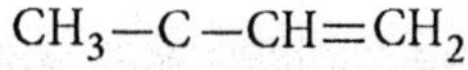

Methyl vinyl ketone (conjugated)

$\pi \rightarrow \pi^*$, λ_{max} 219 nm

ε_{max} 15,000

$n \rightarrow \pi^*$, λ_{max} 324 nm

$\varepsilon_{max} \approx 30$

By increasing the polarity of the solvent, $n \rightarrow \pi^*$ transition (*R*-band) undergoes a hypsochromic shift (blue shift). The shift is due to the change in the stability of the ground electronic states. $n \rightarrow \pi$ transition (*R*-band) experiences a blue shift as the degree of hydrogen bonding between the carbonyl oxygen and the solvent increases by increasing the polarity of the solvent or in other words, the energy of *n* electrons is strongly lowered by hydrogen bonding, i.e., *n* electrons are stabilised. So more energy is required to excite *n* electrons and hence λ_{max} is shifted to lower wavelength (blue shift).

36. How will you distinguish between *K*-band and *R*-band in carbonyl group of ketones? Do you expect any shift of the positions of absorption maxima and intensity by using a more polar solvent?

↠ The *K*-band of carbonyl group (>C=O) in the UV-spectrum arises due to $\pi \rightarrow \pi^*$ while *R*-band originates from $n \rightarrow \pi^*$ electronic transition. These two electronic excitations require different amount of energy and hence, the bands due to these transitions will appear at different wavelengths. In general, $\pi \rightarrow \pi^*$ transition occurs at about 170 nm as a strong band (K-band)(ε_{max} > 10,000) but $n \rightarrow \pi^*$ transition takes place at higher wavelength, near 270-300 nm as an weak band ($\varepsilon < 30$). Thus from the positions and intensities, these two bands (*K*-band and R-band) can be easily differentiated.

In $\pi \rightarrow \pi^*$ transition, π^* molecular orbital is more polar than π M.O. and so π^* will be stabilised by polar solvent. As a result transition will require less amount of energy. Consequently, λ_{max} of *K*-band will be shifted to higher wavelength with increased intensity. But in $n \rightarrow \pi^*$, *n* electrons containing molecular orbital is polar and so can readily form H-band with polar-prolic solvent. So in this transition ground state is stabilised. As a result more energy is required for this transition and so λ_{max} of *R*-band ($n \rightarrow \pi^*$) will be shifted to lower wavelength (blue shift) with the increase in intesnsity.

37. Discuss in short the behaviour of benzene and its derivatives in ultraviolet radiation.

↠ The behaviour of benzene and its derivatives in ultraviolet radiation has been extensively studied. It has been observed that the parent compound, benzene, displays four UV-bands at 184 nm (ε_{max} 60,000), 204 nm (ε_{max} 8,500), 254 nm (ε_{max} 250) and 340 nm (very low). The band of very low intensity at 340 nm is of little importance and is not observed under usual conditions of recording the spectrum. The intense band at 184 nm known as E_1 band is a result of an allowed transition but is not accessible in UV spectrometer. The weaker bands near 200 nm, often called *K*-band or E_2-band and bands near 250 nm known as *B*-band result respectively from forbidden transitions in the highly symmetrical benzene molecule.

The band around 250 nm (*B*-band) is sometimes called benzenoid band and is easily identifiable. It has almost the same intensity (ε = 250-300) and similar structure in benzene and its simple derivatives. This band is characterised by a large degree of fine structure in non-polar solvents but in polar solvents due to solute-solvent interactions fine structure is reduced. The presence of fine structure as shown in the figure is indicative of the simple aromatic molecules.

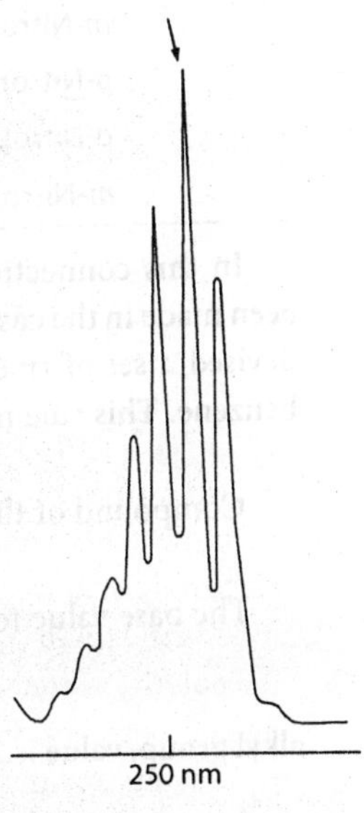

When benzene is substituted by alkyl groups, absorption are slightly shifted to longer wavelength but the fine structure remains intact. The bathochromic shift is due to hyperconjugation between the alkyl group and π system of the ring. The second alkyl group is most effective in producing bathochromic shift if it is in the para-position, but when the second group is at ortho position there is blue shift. This is due to steric interaction which effectively reduces hyperconjugative effect. However, when groups containing non-bonding electron pair (–OH, –NH_2, etc.) are present in the benzene ring the bathochromic shift is very pronounced and the fine structure is seriously diminished or wholly eliminated because of n–π conjugation. It may be mentioned that wavelength and intensity of bands increase with the increase in the extent of chromophore. On adding more and more conjugation to the benzene ring, *K*-band at 204 nm moves to longer wavelength, faster than *B*-band. Thus, in the spectrum of benzoic acid, *K*-band appears at 230 nm and *B*-band at 273 nm while the spectrum of benzene derivative with longer chromophore like cinnamic acid shows the *K*-band at 273 nm but no *B*-band.

Absorption maxima for some common mono and disubstituted benzenes are listed below:

λ_{max} nm for Monosubstituted benzene (Ph-*R*) (in ethanol)

R	*K*-band	ε_{max}	*B*-band	ε_{max}
-H	204	8400	254	250
-OH	211	6200	270	1500
-Cl	210	80000	264	490
-OMe	217	6400	270	1480
$-NH_2$	230	8500	280	1430
-COOH	230	11500	273	980
$-NO_2$	264	7800	–	–
-CHO	250	11400	–	–
$-CH{=}CH_2$	248	14000	282	750
-Ph	251	18300	–	–
-CH=CH=Ph	295	29000	–	–
-CH=CH-COOH	273	21000	–	–

λ_{max} nm for some Disubstituted benzene (in ethanol)

Compound	K-band	ε_{max}
p-Mtroaniline	380	13500
o-Nitroanilin	283	5500
m-Nitroaminile	280	5000
p-Nitrophenol	318	10000
o-Nitrophenol	280	6500
m-Nitrophenol	274	6000

In this connection it may be mentioned that some quantitative assessments have been made in the case of acyl deriatives of benzene, and like Woodward-Fieser rule, Scott devised a set of rules for calculating the absorption maximum for acyl derivatives of benzene. This rule may be stated as below:

Compound of the type 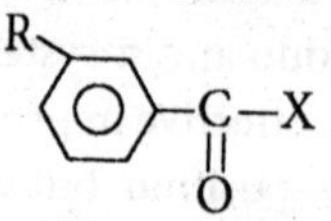

The base value for this type of compound is taken for 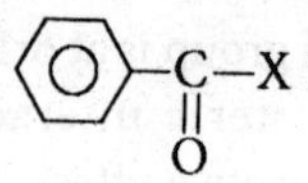 when X is any alkyl group, value is 246 nm, if X = O-alkyl, the value is 230 nm.

Increment for substituent (R) at	ortho	para	meta
Alkyl or ring residue	+3	+10	+3
OH, OR	+7	+25	+7
Cl	0	+10	0
Increment for substituent (R) at	**ortho**	**para**	**meta**
Br	+2	+15	+2
NH_2	+13	+58	+13
NHAC	+20	+45	+20
NR_2	+11	+85	+20

Following examples may be cited in this connection to illustrate the rule.

(a) Calculate λ_{max} for para-chloroacetophenon in ethanol.
The structure of *p*-chloroacetophenone may be written as

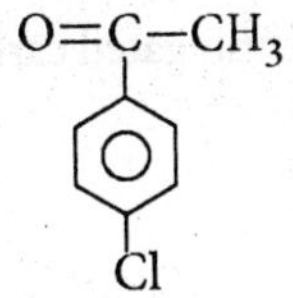

Base value	246 nm
Cl substitution at para position	+10 nm
Calc λ_{max}	256 nm. The observed value is 254 nm.

(b) Calculate the λ_{max} in ethanol of the following compound.

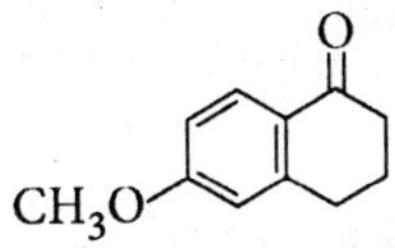

Base value Increment for	246 nm
(i) ortho ring residue	+3 nm
(ii) para methoxy	+25 nm
Calc λ_{max} (ethanol)	274 nm

The observed value is 276 nm in the same solvent.

38. State Scott rule to calculate λ_{max} value of acyl derivatives of aromatic hydrocarbon with suitable examples.

➜ See answer against question No. 37.

39. Mark the correct statement out of the following.

(a) Acetone absorbs at 280 nm, ε_{max} 16,000
(b) Ethylene absorbs at 165 nm, ε_{max} 10,000
(c) Butadiene absorbs at 217 nm, ε_{max} 21,000
(d) Benzene shows *K*-band at 254 nm, ε_{max} 60,000

- (a) is wrong, ε_{max} should be much low ≈ 50.
  (b) is correct.
  (c) is correct.
  (d) is wrong, *K*-band of benzene appears at about 200 nm, ε_{max} 8500.

40. Identify the type of transition in each of the following compounds.

(i) Ethanal = 290 nm, ε_{max} 17.
(ii) 1,3-Pentadiene = 224 nm, ε_{max} 22650
(iii) Crotonaldehyde = 214 nm, ε_{max} 15850
(iv) Styrene = 282 nm, ε_{max} 450

- (i) Ethanal (Acetaldehyde), CH_3—CHO has >C=O group as chromophore. The λ_{max} at 280 nm, ε 17 is due to $n \rightarrow \pi^*$ transition.
  (ii) 1,3-Pentadiene, CH_2=CH—CH=CH—CH_3, is a conjugated diene. So the λ_{max} at 224 nm, ε_{max} 22,500 is for $\pi \rightarrow \pi^*$.
  (iii) Crotonaldehyde, CH_3—CH=CH—CHO, is an α, β-conjugated carbonyl compound. The band at 214 nm, ε_{max} 15,850 originates from $\pi \rightarrow \pi^*$.

  CH = CH_2

  (iv) Styrene, 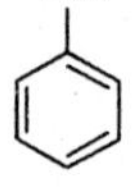 is a vinyl substituted benzene. The band at 282, nm ε_{max} 450 finds its origin from $\pi \rightarrow \pi^*$ (*B*-band). It is a forbidden transition.

41. An organic compound shows an absorption band at 235 nm. At this wavelength 20% of the incident light is transmitted by a 2 × 10⁻⁴ M solution of the sample in a cell having path length of 1 cm. Calculate molar extinction coefficient (ε).

- From Lambert-Beer law we know that

$$\log \frac{I_0}{I} = \varepsilon . c . l$$

In the problem, $l = 1$ cm, $c = 2 \times 10^{-4}$ M and $I = 20\% \, I_0$

i.e., $\log \dfrac{I_0}{0.2\, I_0} = \log 5 = 0.7$

$$\text{Thus, } \varepsilon = \frac{0.7}{2\times10^{-4}} = \frac{7}{2\times10^{-3}} = \frac{7\times10^{3}}{2} = 3.5 \times 10^3 .$$

42. An $\alpha\beta$-unsaturated ketone of mass 110 has an absorption band at 215 nm and ε = 10,000. A solution of this ketone showed absorbance *A* = 2.0 in a 1 cm cell. Calculate the concentration of ketone in this solution expressed in grams per litre.

- We know $A\left(\log \dfrac{I_0}{I}\right) = \varepsilon.\, c.\, l$

Here ε = 10,000, $l = 1$, $A = 2$.

$\therefore\ 2 = 10{,}000 \times 1 \times c$

or, $c = \dfrac{2}{10{,}000}$ mol / L = 2×10^{-4} mol / L.

We know that mass of the molecule is 110, So the concentration can be expressed as $2 \times 110 \times 10^{-4}$ g / L = 220×10^{-4} g / L = 2.2×10^{-2} g / L.

43. Biological oxidation-reduction involves the co-enzyme NAD (Nicotinamide-adenine-dinucleotide) and its reduction product NADH. NADH gives two strong UV bands at λ_{max} 260 nm (e = 15,000) and λ_{max} 240 nm (ε = 6220) while NAD gives only one band at λ_{max} 220 nm (ε = 18,000). A reaction mixture taken in a cell of 1.0 cm path length showed the following data.

λ_{max} 260 nm, absorbance = 1.200
λ_{max} 340 nm absorbance = 0.311
Estimate the relative amounts of NAD and NADH in the reaction mixture.

➜ The λ_{max} at 340 nm is due to only NADH.

ε_{340} = molar absorptivity of NADH at 340 nm = 6220

$$= \frac{\text{Absorbance}}{l \times c} \quad \left(\text{Relation } \varepsilon = \frac{A}{l \times c},\ l = \text{path length, } c = \text{conc., } A = \text{absorbance}\right)$$

$$= \frac{0.311}{1 \times c_{NADH}} \quad \text{or, } 6220 = \frac{0.311}{c_{NADH}},\ c_{NADH} = \frac{0.311}{6200} = 0.00005 \text{ g-mole/l}$$

This concentraton of NADH gives the absorbance (A′) of λ_{max} 260 nm band.

$$\varepsilon'_{260} = \frac{A'}{1 \times 0.00005} \quad \text{or} \quad 15000 = \frac{A'}{0.00005}$$

$$\text{or, } A' = 15000 \times 0.00005 = 0.75\cdot$$

The remaining absorbance (A″) observed in λ_{max} 260 nm is due to NAD.

∴ A″ = absorbance by NAD at λ_{max} 260 nm

$$A'' = 1.2 - 0.75 = 0.45$$

$$\varepsilon''260\text{nm} = \frac{A''}{c' \times l} = \frac{0.45}{c'\text{mol/litre}} \quad (l = 2)$$

$$\text{or, } \varepsilon'_{NAD} = \frac{0.45}{18{,}000} \quad (\varepsilon'' = 18{,}000)$$

$$= 0.000025 \text{ mol/litre}$$

The ratio of molar concentration of $\dfrac{NADH}{NAD} = \dfrac{c}{c'} = \dfrac{0.00005}{0.000025} = \dfrac{2}{1}$.

44. Two organic compounds (*A*) and (*B*) display their *K*-bands at λ_{max} 296 nm (ε_{max} 10,700) and λ_{max} 281 nm (ε = 20,800). Which band belongs to which compound?

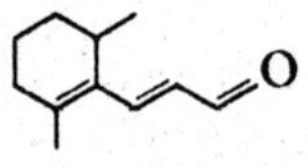

(A)

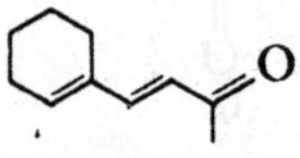

(B)

➻ The structure with λ_{max} 296 nm has nearly half of the intensity (ε_{max} = 10,700) of *K*-band than the other. This decrease in intensity is due to steric hindrance by methyl groups on the conjugated system. The scrutiny of these structures reveals that the compound with decreased intensity of the *K*-band is (A).

45. An organic compound (*A*), $C_{11}H_{16}O$ on reduction with hydrogen in presence of palladium forms (*B*) $C_{11}H_{18}O$. The compound (*B*) has been found to have the following structure.

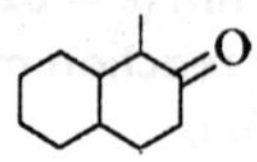

(B)

The compound (A) displays strong λ_{max} at 225 nm which absorption was not present in (*B*).

➻ Since during hydrogenation two hydrogen atoms have been added, compound (*A*) contains an unsaturation and a keto group which remained intact during hydrogenation. From λ_{max} value two alternative structures (I) ot (II) may be proposed for the compound (*A*).

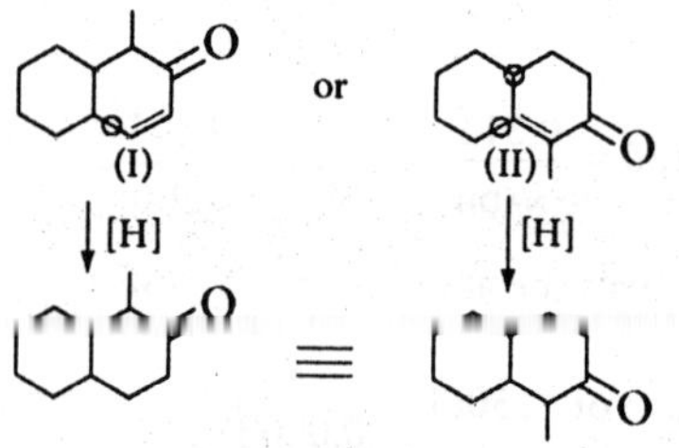

Now, the desired reduction product has λ_{max} at 225 nm. Following Woodward-Fisser rule, λ_{max} [*A* (I) and (II)] may be calculated as:

For structure (I)		For structure (II)	
Base value	215 nm	Base value	215 nm
β-Substitution	12 nm	α-substitution	10 nm
		Two β-substitutions	24 nm
		One exocyclic double bond	5 nm
Calc λ_{max}	227 nm	Calc λ_{max}	254 nm

Since the compound (*A*) shows λ_{max} at 225 nm, the structure of (*A*) is (I).

46. Which of the following compounds would be expected to absorb light of longest and shortest wavelengths?

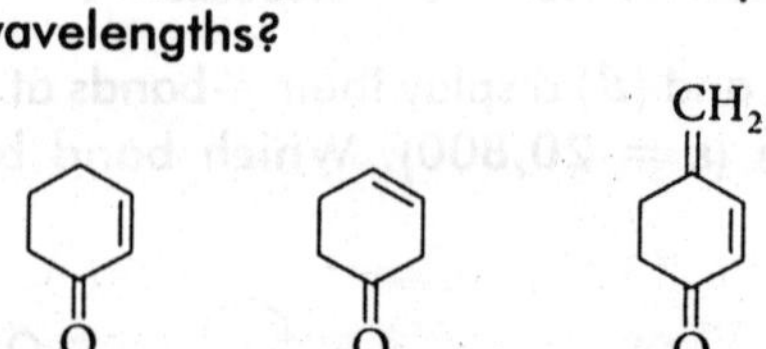

(I) (II) (III)

➻ Compounds (I) and (III) are conjugated enone but the compound (II) is non-conjugated ketone.

Hence, compound (II) will absorb *shortest* wavelength light.

Out of (I) and (III), (III) will absorb the light of longest wavelength because it has one double bond extending the conjugation more than (I). Moreover, (III) has an exocyclic double bond.

47. Acetone on boiling with barium hydroxide furnished two compounds (*A*) and (*B*). How with the help of UV-spectroscopy (*A*) and (*B*) can be characterised?

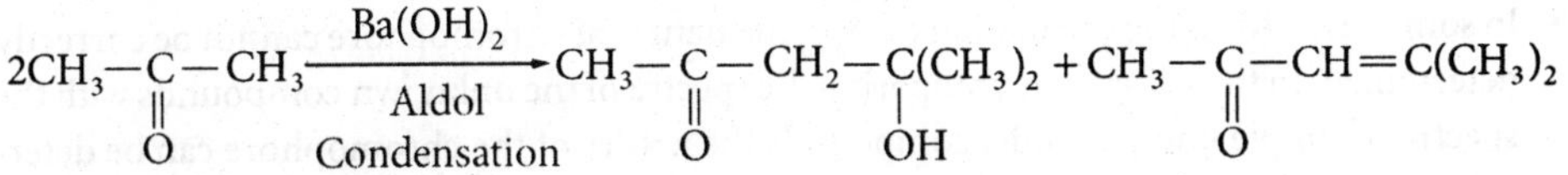

Acetone (*A*) (*B*)

➜ Compound (A) is a saturated carbonyl compound and so it will show an weak absorption at 275-295 nm region due to $n \rightarrow \pi^*$ and a strong absorption around 170 nm due to $\pi \rightarrow \pi^*$ generally not observable in UV-spectroscopy. But the compound (*B*) is an α, β-unsaturated ketone with two β-substituents. So it will have a characteristic λ_{max} at about 239 nm. Hence, they can be identified.

48. How many bands are expected to appear in the UV-spectrum of aniline? What will happen if the spectrum is taken in dilute hydrochloric acid solution?

➜ The structure of aniline may be written as below.

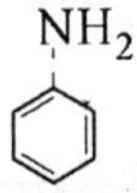

It is a benzene derivative with a substitution ($-NH_2$) having lone pair of electrons. Since the lone pair is in conjugation with the benzene ring, resonance will occur.

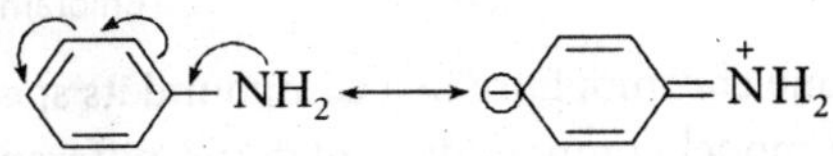

As it is a benzene derivatives following bands are expected.

(i) *K*-band (ii) *B*-band and (iii) a strong band due to allowed $\pi \rightarrow \pi^*$ transition. But as this band appears below 200 nm, it is not accessible in the UV-spectrometer. So in practice only two bands (i) and (ii) are expected.

But due to $n \rightarrow \pi$ conjugation as shown above these absorption bands are shifted to longer wavelength compared to those of benzene.

When the spectrum is taken in dilute hydrochloric acid $n \rightarrow \pi$ conjugation is removed i.e.,

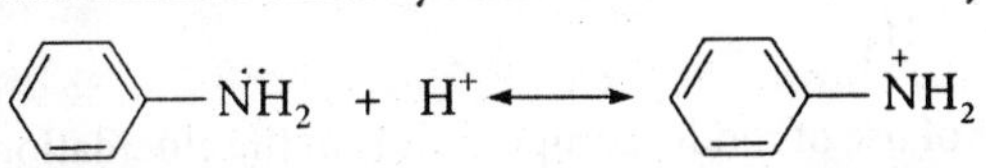

Thus, the main band (*K*-band) is shifted to normal wavelength of benzene, i.e., in acid solution spectrum will be comparable with that of benzene.

49. Aniline absorbs at 230 nm (ε 8000), however, in acid solution the main absorption band is seen at 230 nm (ε 7500) and is comparable with that of benzene.

➜ See answer against Question No. 48.

50. Which out of benzene (colourless) or quinone (yellow) has easily promoted electrons?

➻ Quinone has more easily promoted electrons. It is a non-aromatic enone having non-bonded electrons. In this compound π electrons are in conjugation with *n*-electrons. As a result electrons are readily available for promotion.

51. Discuss in short how the nature of chromophore in a compound can be ascertained with the help of model compounds.

➻ In some cases, from the calculation of λ_{max} the nature of chromophore cannot be correctly determined. In those cases, by comparing the spectra of the unknown compounds with the spectra of simple known model compounds the nature of the chromophore can be determined. The deduction of identity is based on the fact that UV-spectrum of a given chromophore is independent of the non-chromophoric part of the molecule. The spectra of large number of model compounds are readily available for comparison purpose. Thus,

Model Compound

(i) 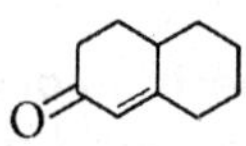 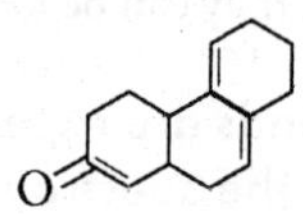

(ii) 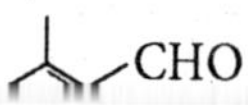 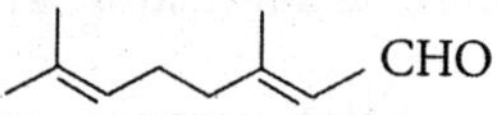

Citral

(iii) 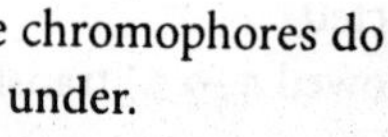 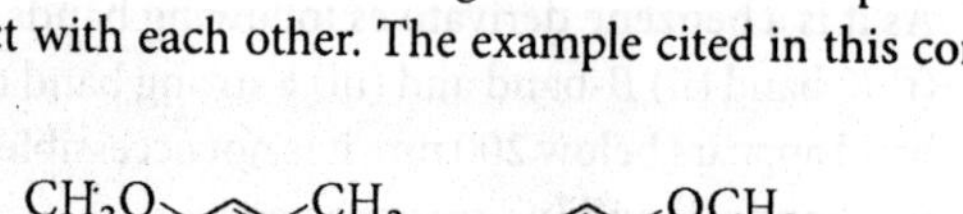

Chloramphenicol

Even when there is more than one chromophore in a compound its spectrum will be the sum total of the spectra of the model compounds containing different chromophores, provided the chromophores do not react with each other. The example cited in this connection is as under.

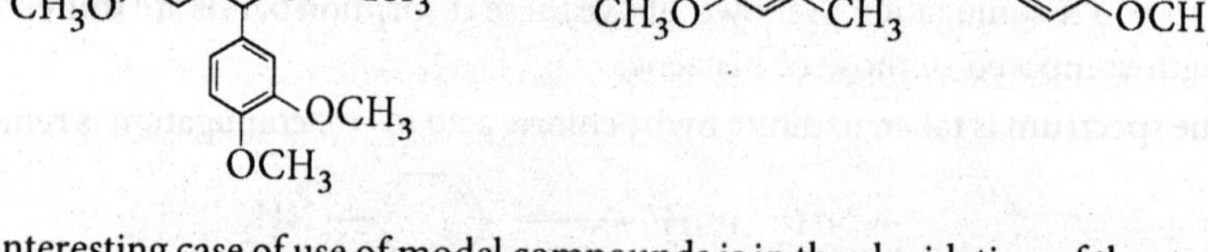

An interesting case of use of model compounds is in the elucidation of the stereochemistry, as in the case of strychnine. The spectrum of strychnine is compared with those of two model compounds–N–methyl–N–acetyl–O–toluidine and cis-hexahydrocarbazole. It has been found that the spectrum of N–methyl–N–acetyl–O–toluidine is entirely different than that of strychnine. So the dihydro-indole ring present in strychine is not planar. On the other hand, the spectrum of strychnine and that of *cis*-hexahydrocarbazole are found to be super-imposable. Hence, the desired ring junction in strychnine is *cis*.

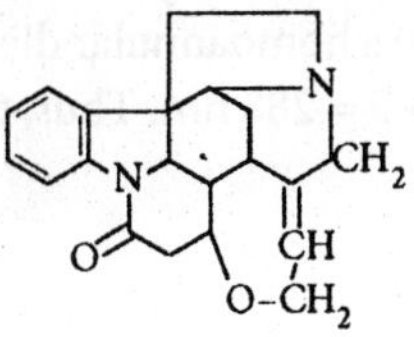

Styrine
λ_{max} (ε_{max}) 257 nm (16,000)
281 nm (4,100)
290 nm (3,400)

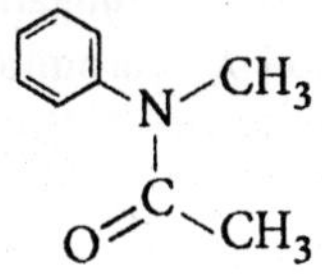

N-Methyl-N-acetyl-
o-toluidine
λ_{max} 270 nm (300)

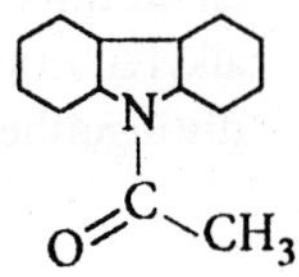

cis-Hexahydro carbazole
λ_{max} (ε_{max}) 257 nm (16,000)
281 nm (3,700)
290 nm (3,400)

52. How would you distinguish the following pair from UV-spectroscopy?

(a)  and 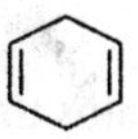

(b) 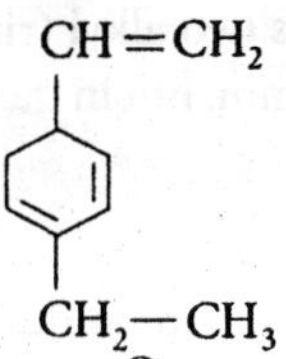 and 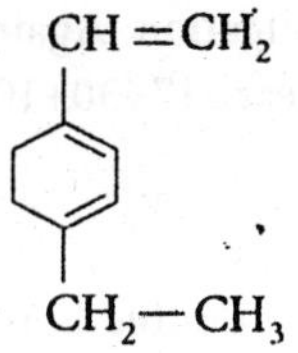

(c) 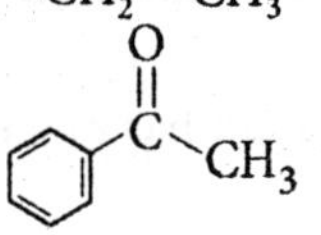 and 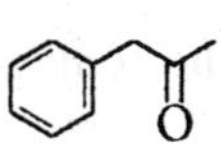

(d) 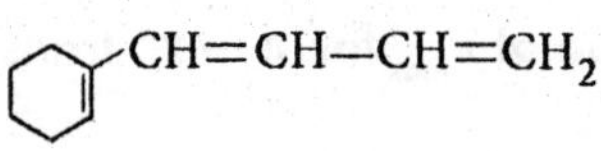 and 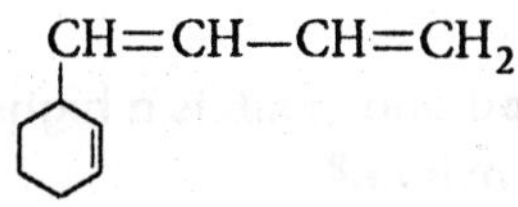

(a) In compound  both the double bonds are in conjugation and present in a ring. So it is a homoannular diene. Hence, it exhibits a λ_{max} around 253 nm. But in the other compound  two double bonds are isolated. So it will not show any characteristic λ_{max} observable in UV-region.

(b) The compound 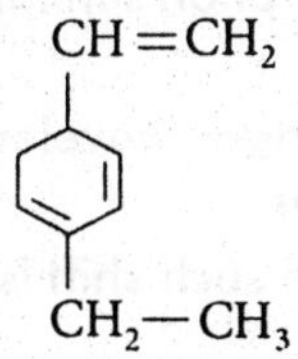 is a homoannular diene with one alkyl substituted.

The other double bond is not in conjugation with homodiene component. The compound will have λ_{max} around 253 + 5 = 258 nm. In other compound, 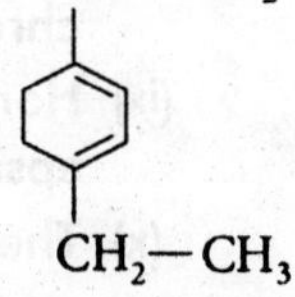

all the three double bonds are in conjugation. It is also a homoannular diene with one alkyl substituent. It will show λ_{max} around 253 + 30 + 5 = 288 nm. Thus, they may be distinguished.

(c) The compound 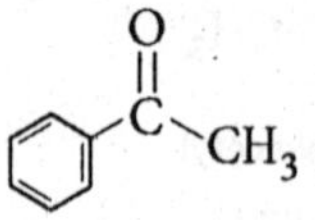 is an acyl benzene derivative. So it will display a chracteristic absorption at about 246 nm while in the compound 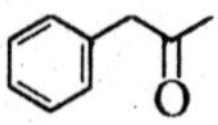 the acyl group is not directly attached to aromatic ring. So it will show sepctrum similar to pure benzene. Thus, they can be differentiated.

(d) The compound, 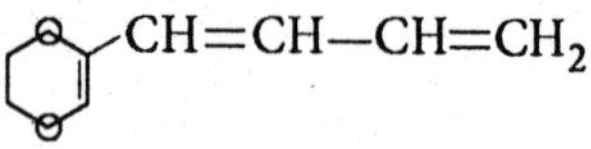 contains three double bonds and all these three double bonds are in conjugation. Again, it has two alkyl (ring residue) substistuents. So it will exhibit λ_{max} near 217+30+10 = 257 nm. But in the compound

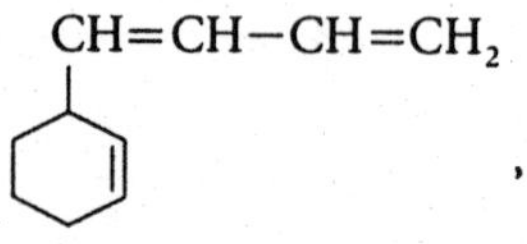

, out of three double bonds two are in conjugation (ring double bond is not in conjugation). The conjugated system has one substituent. Thus, it will have λ_{max} around 217 + 5 = 222 nm.

53. Explain the following:

(i) Why butadiene exhibits a higher value of λ_{max} for $\pi \rightarrow \pi^*$ transition than that of ethylene?

(ii) Why β-carotene absorbs in the visible range and what is its colour?

(iii) Why λ_{max} for (I) is lower than that of (II)?

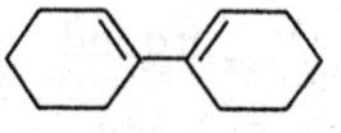

(I)

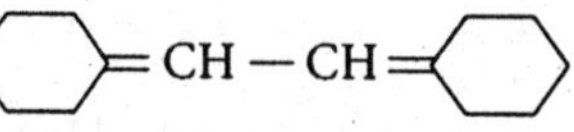

(II)

(iv) Biphenyl exhibits an intense absorption band at 252 nm (ε 19000) but its 2,2 e-dimethyl derivative shows absorption similar to *o*-Xylene (λ_{max} 262, ε_{max} 270).

(v) λ_{max} of the benzoid band shifts to longer wavelength as the number of linearly fused benzene rings increases.

(vi) *p*-Nitroaniline shows a red shift while such shift is not observed in ortho and *meta* nitroaniline.

(vii) A conjugated diene absorbs at a higher wavelength with higher value of extinction coefficient as compared to diene with isolated double bonds.

(viii) Why hydroxyl group is regarded as an auxochrome but not as a chromophore?

(ix) Homoannular diene absorbs at higher wavelength with lower value of epsilon than heteroannular diene.

(x) The intensity of $\pi \rightarrow \pi^*$ transition is 10 to 100 times stronger than $n \rightarrow \pi^*$.

(xi) Polar solvent usually shifts the $\pi \rightarrow \pi^*$ to longer wavelength and $n \rightarrow \pi^*$ transition to shorter wavelength.

(xii) The wavelenght as well as the extinction coefficient increases with the increase in conjugation in the compound.

(xiii) *K*-band in conjugated diene experiences a red shift if the polarity of the solvent is increased.

(xiv) *o*-Nitrophenol gives a UV-band at λ_{max} 350 nm when the spectrum is recorded in 0.1 NHCl solution, but the band is observed at λ_{max} 415 nm in 0.1 N NaOH solution.

(xv) Which out of benzene or quinone has more easily promoted electrons.

➜ (i) Butadiene is conjugated diene. So its HOMO is ψ_2 and LUMO is ψ_3. The energy difference between these two molecular orbitals is less than the energy difference for transition in ethylene. (Electronic transition occurs from HOMO to LUMO). So λ_{max} for butadiene is higher than that of ethylene .

(ii) β-Carotene has eleven conjugated double bonds. As a result it absorbs in the visible region of the spectrum, λ_{max} 452 nm, because the energy difference between HOMO and LUMO decreases with the increase in the number of conjugated double bonds. It absorbs orange colour.

(iii) λ_{max} for (I) is lower than that of (II). The reason is explained below:

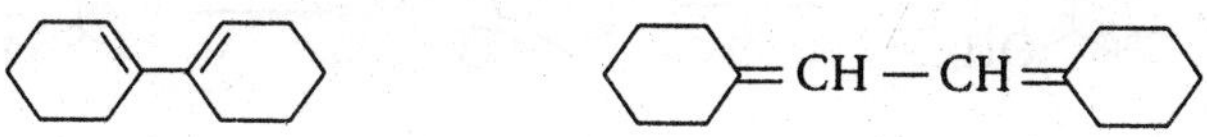

(I) (II)

In (I), there are four substituents (ring residues) while the compound (II) has also four ring residues. But in (II) both the double bonds are exocyclic. So (II) will have λ_{max} $(2 \times 5) = 10$ nm more than that of (I).

(iv) Simple biphenyl is planar, as a result conjugation occurs smoothly. Hence, it gives intense absorption band at 252 nm (*K*-band). But 2,2′-dimethyl biphenyl is sterically hindred and its planarity to some extent is less than simple biphenyl. As a result conjugation does not occur. So it may be looked upon as two molecules of *o*–Xylene and shows absorption similar to *o*–Xylene.

(v) It is known that with increase in the number of conjugated π-orbitals, the energy gap between HOMO and LUMO decreases. So with the increase in the number of linearly fused benzene λ_{max} shifts to longer wavelength.

(vi) In *p*-nitroaniline, there is extension of conjugation due to resonance. But in *meta* isomers extension of conjugation is not possible. In ortho isomer due to H-bond no such resonance is possible. Thus

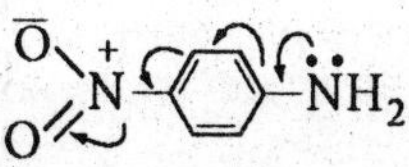

p-Nitroaniline

(vii) See answer against question no. 22(a).

(viii) An isolated hydroxyl group (—O–H) has no characteristic property of absorbing UV light but when associated with any chromophore it can change absorption property of the chromophore. Thus

CH_3OH 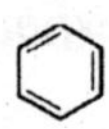 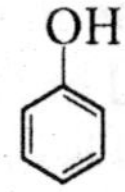

λ_{max} below UV-region λ_{max} 250 nm λ_{max} 280 nm

(ix) See answer against Question No. 22 (d).

(x) See answer against Question No. 16.

(xi) See answer against Question No. 22 (g).

(xii) See answer against Question No. 22 (a).

(xiii) *Hints*: *K*-band originates from $\pi \rightarrow \pi^*$ transition. This band is sensitive to solvent polarity. [See answer against Question No. 22(g)]

(xiv) *o*-Nitrophenol is acidic in nature and the presence of electron withdrawing nitro group at *ortho* position increases its acidity. Hence, in alkaline solution (0.1N NaOH) it readily forms phenolate anion. Due to readily available electrons resonance is extended as shown below.

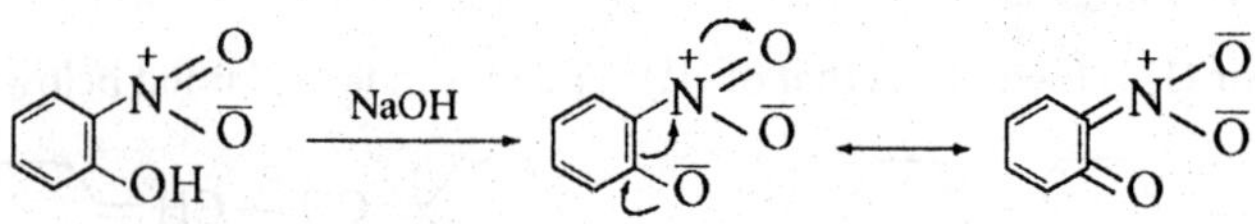

As a result λ_{max} is shifted to higher wavelength (415 nm).

(xv) Out of benzene (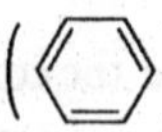) and quinone (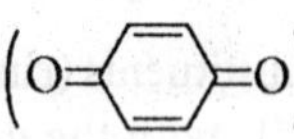), quinone has n electrons. So such electrons are readily promoted. Hence, quinone has easily promoted electrons.

54. Using Woodward-Fieser rules, calculate values of λ_{max} of the following compounds.

Compound	Answer		
(i) 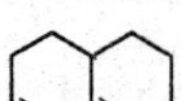	Base value (Heteroannular diene)	=	217 nm
	Ring residues 3 × 5	=	15 nm
	Exocyclic double bonds 1 × 5	=	5 nm
	Calc. λ_{max}		237 nm
(ii) 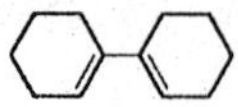	Base value (Heteroannular diene)	=	217 nm
	Ring residues 4 × 5	=	20 nm
	Calc. λ_{max}	=	237 nm

(iii) 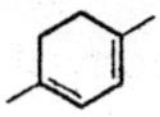

Base value (Homoannular diene) = 253 nm
Substituents 2 × 5 = 10 nm
Ring residues 2 × 5 = 10 nm

Calc. λ_{max} = 273 nm

(iv) 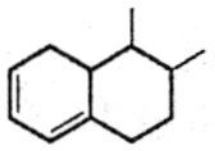

Base value (Homoannular diene) = 253 nm
Ring residues 3 × 5 = 15 nm
Exocyclic double bonds 1 × 5 = 5 nm

Calc. λ_{max} = 273 nm

(v) 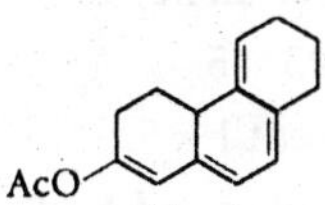

Base value (Homoannular diene) = 253 nm
Double bond extending the conjugation 2 × 30 = 60 nm
Ring residues 5 × 5 = 25 nm
–OAc substituent = 0 nm
Exocyclic double bonds 3 × 5 = 15 nm

Calc. λ_{max} = 353 nm

(vi) 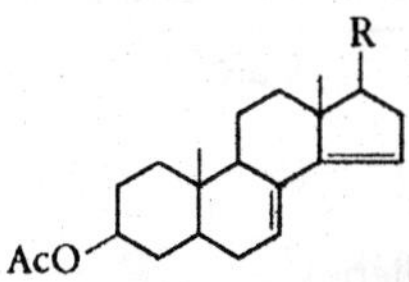

Base value (Heteroannular diene) = 217 nm
Ring residues 4 × 5 = 20 nm
Exocyclic double bonds 2 × 5 = 10 nm

Calc. λ_{max} = 247 nm

(vii) 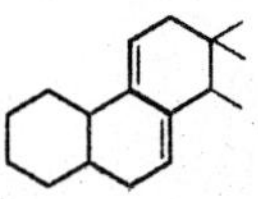

Base value (Heteroannular diene) = 217 nm
Ring residues 4 × 5 = 20 nm
Exocyclic double bonds 2 × 5 = 10 nm

Calcl. λ_{max} = 247 nm

(viii) 

Base value (Heteroannular diene) = 217 nm
Ring residues 2 × 5 = 10 nm
Exocylic double bonds 1 × 5 = 5 nm

Calc. λ_{max} = 232 nm

(ix) 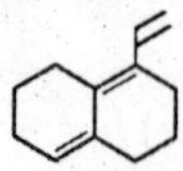

Base value (Heteroannular diene)	=	217 nm
Double bond extending the conjugation 1 × 30	=	30 nm
Ring residues 4 × 5	=	20 nm
Exocyclic double bond 2 × 5	=	10 nm
Calc. λ_{max}	=	277 nm

(x) 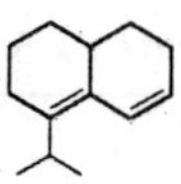

Base value (Heteroannular diene)	=	217 nm
Ring residues 3 × 5	=	15 nm
Alkyl substituent 1 × 5	=	5 nm
Exocyclic double bond 1 × 5	=	5 nm
Calc. λ_{max}	=	242 nm

(xi) 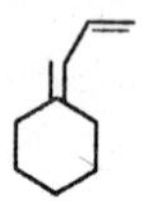

Base value (Acyclic diene)	=	217 nm
Ring residues 2 × 5	=	10 nm
Exocycluc double bond 1 × 5	=	5 nm
Calc. λ_{max}	=	232 nm

(xii) 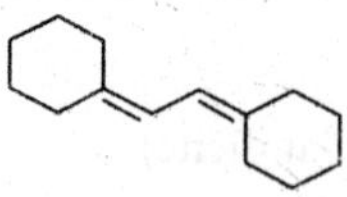

Base value (Acyclic diene)	=	217 nm
Ring residues 4 × 5	=	20 nm
Exocyclic double bond 2 × 5	=	10 nm
Calc. λ_{max}	=	247 nm

(xiii) 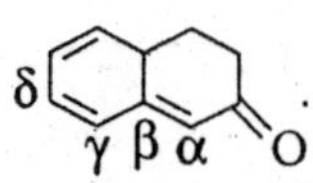

Base value (Enone)	=	215 nm
β substituent 1 × 12	=	12 nm
ω-substituent 1 × 18	=	18 nm
Two double bonds Extending conjugation 2 × 30	=	60 nm
Homodiene component 1 × 39	=	39 nm
Exocyclic double bond 1 × 5	=	5 nm
Calc. λ_{max}	=	349 nm

(xiv) 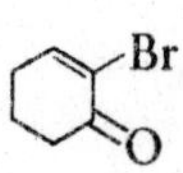

Base value	=	215 nm
(α, β -unsat ketone)	=	25 nm
α-bromo subst. 1 × 25	=	25 nm
β-subst. (ring residue) 1 × 12	=	12 nm
Calc. λ_{max}	=	252 nm

(xv) 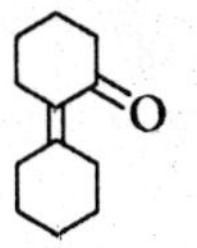

Base value (α, β-unsat. ketone)	=	215 nm
α-substituent 1 × 10	=	10 nm
β-substituents 2 ×12	=	24 nm
Exocylic double bond 2 × 5 (with respect to both the rings)	=	10 nm
Calc. λ_{max}	=	259 nm

(xvi) 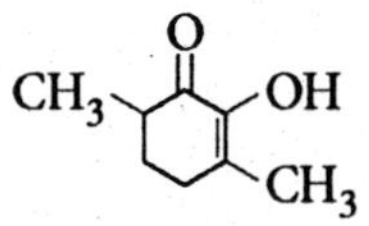

Base value	=	215 nm
OH at α	=	35 nm
β-substituents 2 × 12 (one alkyl, one ring residue)	=	24 nm
Calc. λ_{max}	=	274 nm

(xvii) 

Base value	=	215 nm
One β-substituent 1 × 12	=	12 nm
β substituents 1 × 18	=	18 nm
ω substituents 2× 18	=	36 nm
Exocyclic double bond 2 × 5	=	10 nm
Two double bonds 2 × 30 (Extending conjugation)	=	60 nm
Calc. λ_{max}	=	351 nm

(xviii) 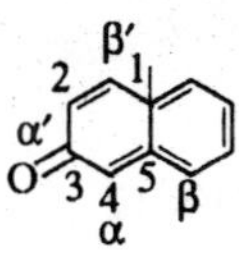

It is a cross-conjugated system

(a)	Base value ($\Delta^{4,5}$ enone)		215 nm
	Two β-substituents 2 × 12	=	24 nm
	One exocyclic double bonds 1 ×5	=	5 nm
	Calc. λ_{max}	=	244 nm
(b)	Base value ($\Delta^{1,2}$ enone)	=	215 nm
	One β′-substituent 1 × 12	=	12 nm
	Calc. λ_{max}	=	227 nm
	Expected λ_{max} (Longest wavelength)	=	244 nm

(xix) 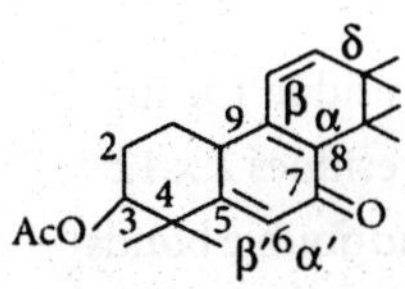

(a)	Base value ($\Delta^{8,9}$ enone)	=	215 nm
	One α-ring residue	=	10 nm
	One β-ring residue	=	12 nm
	One double bond extending conjugation 1× 30	=	30 nm

One δ-substituent 1 × 18	=	18 nm
One homoannular diene 1 × 39 (C-ring)	=	39 nm
Calc. λ_{max}	=	324 nm

(b)

Base ($\Delta^{5,6}$ enone)	=	215 nm
Two β′-substituent 2 × 12	=	24 nm
One exocyclic Double bond 1 × 5	=	5 nm
Calc. λ_{max}	=	244 nm
Expected λ_{max}	=	324 nm

(xx) 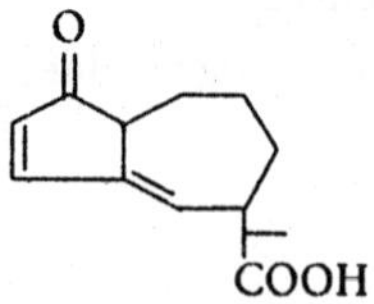

Base value	=	215 nm
One β-substituent 1 × 12	=	12 nm
One γ-substituent 1 × 18	=	18 nm
One δ-substituent 1 × 18	=	18 nm
One exocyclic double bond 1 × 5	=	5 nm
Calc. λ_{max}	=	268 nm

(xxi) $(CH_3)_2C{=}CH{-}C({=}O){-}CH_3$

Base value	=	215 nm
Two β-substituents 2 × 12	=	24 nm
Calc. λ_{max}	=	239 nm

(xxii) 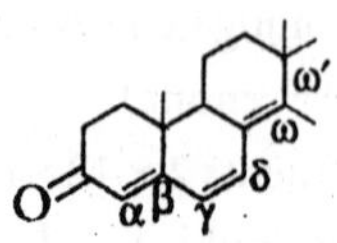

Base value	=	215 nm
One β-substituents 1 × 12	=	12 nm
One ω-substituents 1 × 18	=	18 nm
Two ω′-substituents 2 × 18	=	36 nm
Two double bonds extending conjugation 2 × 30	=	60 nm
Two exocyclic double bonds 2 × 5	=	10 nm
Calc. λ_{max}	=	351 nm

(xxiii) 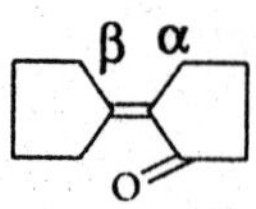

Base value	=	215 nm
one α-ring residue 1 × 10	=	10 nm
two β-ring residues 2 × 12	=	24 nm
two exocyclic double bonds 2 × 5	=	10 nm
Calc. λ_{max}	=	259 nm

(xxiv) 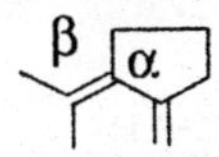

Base value	=	215 nm
One α-substituent (ring residue)		
1 × 10	=	10 nm
Two β-substituents 2 × 12	=	24 nm
One exocyclic double bonds 1 × 5	=	5 nm
Calc. λ_{max}	=	254 nm

(xxv) 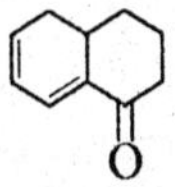

Base value	=	215 nm
One α-substituent 1 × 10	=	10 nm
One δ-substituent 1 × 18	=	18 nm
One double bonds extending		
conjugation 1 × 30	=	30 nm
One exocyclic double bond 1 × 5	=	5 nm
Calc. λ_{max}	=	278 nm

(xxvi) 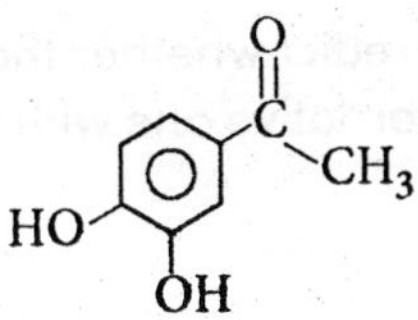

Base value for acyl benzene	=	246 nm
OH at *meta* position	=	7 nm
OH at *para* position	=	25 nm
Calc. λ_{max}	=	278 nm

(xxvii) 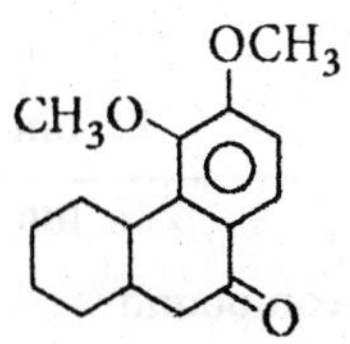

Base value	=	246 nm
(for acyl benzene)		
OCH_3 at *meta*	=	7 nm
OCH_3 at *para*	=	25 nm
alkyl group at *ortho* (ring)	=	3 nm
Calc. λ_{max}	=	281 nm

(xxviii) 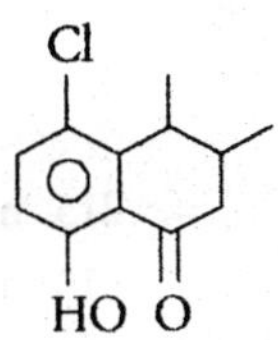

Base value	=	246 nm
OH at *ortho*	=	7 nm
Cl at *meta*	=	0 nm
o-ring residues	=	3 nm
Calc. λ_{max}	=	256 nm

(xxix) 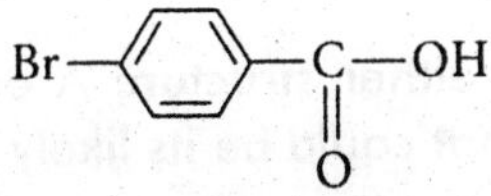

Base value for Ar—C(=O)—OH	=	230 nm
Br at *para* position	=	15 nm
Calc. λ_{max}	=	245 nm

(xxx) 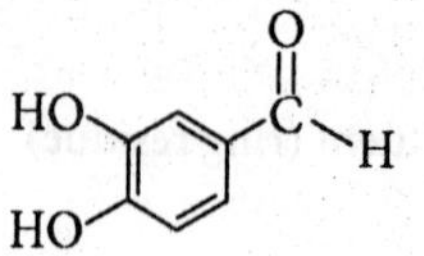

Base value for Ar—C(=O)—H	=	250 nm
OH at *meta* position	=	7 nm
OH at *para* position	=	25 nm
Calc. λ_{max}	=	282 nm

(xxxi) 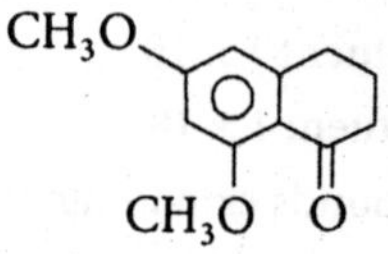

Base value	=	246 nm
o-ring residue	=	3 nm
o-OCH_3	=	7 nm
p-OCH_3	=	25 nm
Calc. λ_{max}	=	281 nm

55. A neutral compound showed λ_{max} rat 251 nm. Predict whether the compound has the following structure. If not, propose an alternative one with similar basic skeleton.

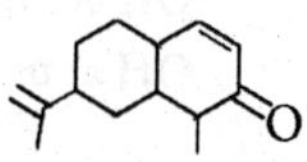

➜ The proposed structure will have λ_{max} as below:

Base value	=	215 nm
One β-substituent 1 × 12	=	12 nm
Calc. λ_{max}	=	227 nm

So the proposed structure is not the actual structure of the compound. An alternative structure may be as below:

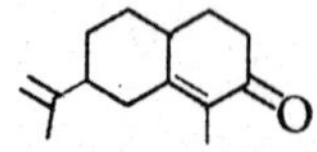

This alternative structure will have λ_{max} as calculated below:

Base value	=	215 nm
One α-substituent 1 × 10	=	10 nm
Two β-substituents 2 × 12	=	24 nm
One exocyclic double bonds 1 × 5	=	5 nm
Calc. λ_{max}	=	254 nm

This value is within 5 nm of the observed value.

56. An unknown compound is beleived to have either structure A or B. Its UV-spectrum shows λ_{max} at 320 nm (Ethanol). What could be its likely structure?

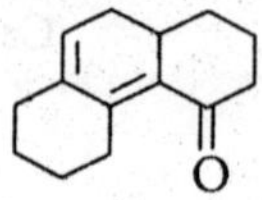

(A)

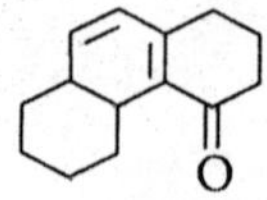

(B)

➡ Calculated λ_{max} for (A)

Base value	=	215 nm
One α-substituents 1 × 10	=	10 nm
One β-substituents 1 × 12	=	12 nm
2 δ-substituents 2 × 18	=	36 nm
One double bond extending conjugation 1 × 30	=	30 nm
One homodiene 1 × 39	=	39 nm
3 exocyclic double bonds 3 × 5	=	15 nm
λ_{max}	=	357 nm

Calculated λ_{max} for (B)

Base value	=	215 nm
One α-substituent 1 × 10	=	10 nm
One β-subsituents 1 × 12	=	12 nm
One δ-substituents 1 × 18	=	18 nm
one double bond extending conjugation 1 × 30	=	30 nm
One homodiene 1 × 39	=	39 nm
λ_{max}	=	324 nm

λ_{max} for (B) is very close to that of the observed value. Hence, (B) is the likely structure of the unknown compound.

CHAPTER 4

NUCLEAR MAGNETIC RESONANCE SPECTROSCOPY

1. (a) What is nuclear magnetic resonance (NMR) spectroscopy?

(b) Describe briefly the theory of NMR spectroscopy.

→ (a) The phenomenon of nuclear magnetic resonance was first observed in 1946 and since 1960 it has been routinely applied in organic chemistry. It has now grown enormously in power and versatility and has almost become an outstanding instrumental technique at the hands of organic chemists for the structure deduction of highly complex organic molecules. This technique is concerned with the absorption of electromagnetic radiation in the radio wave region by some spinning nuclei in a molecule when placed in a magnetic field.

(b) The basic principle of this technique is based on the determination of magnetic moment of nuclear magnet. Some nuclei whose spin quantum number is greater than zero are magnetic in character. The magnetic property of such nuclei arises from their spin. Due to the spin, such nuclei are rotating. Again, it is known that nuclei is surrounded by electrons which are electrical in nature. Thus, according to the principle of solenoid, nuclei become magnetic in character. Thus, those nuclei whose spin quantum number is greater than zero, may be regarded as tiny magnets. So they will have magnetic moment (μ) and nuclear magnets will be oriented in $(2I + 1)$ possible directions, where I is the spin quantum number of a nucleus. The method of determination of magnetic moment of a particular nucleus is based on the promotion of that nucleus from lower spin energy state to higher spin energy state and the amount of energy required for the promotion is equivalent to the magnetic moment of that nucleus. But under ordinary conditions all the $(2I + 1)$ orientations (spin energy states) are degenrate.

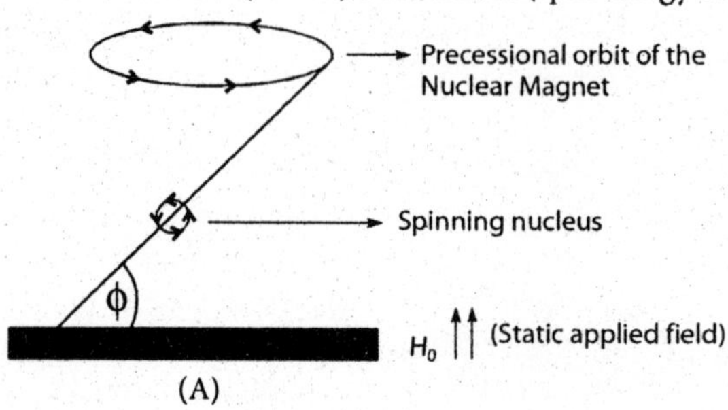

(A)

So under ordinary conditions promotion of nucleus to higher energy spin state is not possible. For this reason spinning nucleus is placed in a static applied magnetic field when the degeneracy of different spin states is lost (Zeeman effect) and different spin states are quantised. Since the nucleus is spinning (rotating) in nature, in such condition (in presence of magnetic field) it will undergo a circular motion around the axis of the applied static magnetic field. This circular motion is known as *precessional motion* (also known as *Larmor precession*) as shown in the figure (A). This precessional motion has a definite frequency termed as *Precessional frequency* or *Larmor frequency.*

The *Precessional frequency* is defined as the number of revolutions per second made by the nuclear bar magnet around the applied static magnetic field. Alternatively precessional frequency of the spinning bar magnet may be defined as equal to the frequency of the electromagnetic radiation in mega-cycles per second necessary to induce a transition from one spin state to another.

It may be mentioned that in the static applied magnetic field, the nucleus can precess in different orientations. As for example, in the case of proton, it can precess either aligned with the applied field (low energy) or opposed to the applied field (high energy) as shown in the figure below:

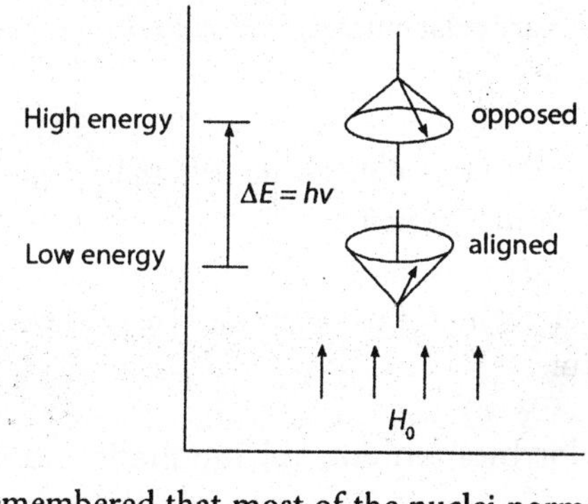

It may be remembered that most of the nuclei normally exists in the lower spin energy state making an angle ϕ (Figure A) with the axis of the applied field. Now if a magnetic field (second magnetic field) having frequency exactly identical to that of precessional nucleus is applied, energy may be absorbed and the precessional nucleus may be promoted to the higher energy spin state. But in order to promote the nucleus to higher energy spin state, the angle ϕ must be changed. This is possible if the second applied magnetic field is perpendicular to the static applied magnetic field and is rotating in the same phase as the precessional nucleus. When the frequency of this second applied magnetic field becomes exactly identical to that of the precessional nucleus, then it is said to be in the state of resonance. Only under this condition, energy is absorbed by the precessional nucleus and the transition of the precessional nucleus from lower to higher energy spin state takes place. This transition from one state to another is known as *flipping of the nucleus*.

As soon as the resonance is attained, a signal is obtained; for this reason, this spectral technique is termed as *nuclear magnetic resonance.* The amount of energy required for promotion to higher energy spin state is the measure of the magnetic moment of the tiny nuclear bar magnet. Thus, when $I = ½$ as in the case of proton, there will be only two orientations, viz., lower one termed as α-state (aligned to static magnetic field) and higher one designated as β-state (opposed to the applied static field) and energy required for promotion may be calculated as follows:

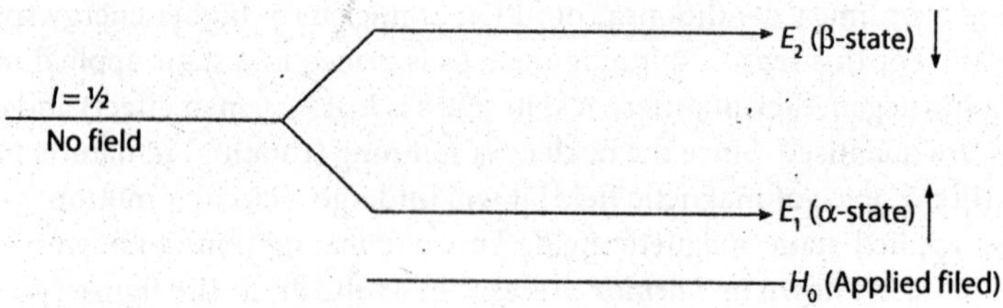

$$E_2 - E_1 = 4E = h\nu = \frac{\mu.B_N.H_0}{I}$$

[where B_N = Bohr's magneton

μ = magnetic moment of the nuclear bar magnet

I = spin quantum number]

$$\text{or}, \qquad \nu = \frac{\mu.B_N.H_0}{hI} = \frac{\mu.B_N.2\pi}{hI} \times \frac{H_0}{2\pi}$$

$$= \frac{\gamma.H_0}{2\pi}\left(\gamma = \frac{\mu.B_N.2\pi}{hI}\right).$$

γ is a constant for a particular nucleus and is known as *gyromagnetic ratio* and is different for different nucleus.

This frequency, ν, is in the order of 3×10^8 Hz = 300 MHz and falls in the range of radio frequency. Thus the second magnetic field to be applied is the electromagnetic radiation of radio frequency region.

It may be mentioned in this context that this small amount of energy cannot change any other (vibrational, rotational or electronic) energy level—only transition of spin energy state can occur.

2. Define (a) spin angular momentum, (b) magnetic moment and (c) Zeeman effect.

(a) Certain nuclei can spin like electrons, around their axes and hence possess spin angular moment. Its magnitude (p) is given by

$$p = \sqrt{I(I+1)} \cdot \frac{h}{2\pi},$$

where I is a number called spin quantum number and h is the Planck's constant. The value of I may be $0, \frac{1}{2}, 1, \frac{3}{2}$, etc. depending upon the nucleus. For proton nucleus, $I = \frac{1}{2}$, $^{12}C_6$ has $I = 0$, etc.

(b) The spinning nucleus behaves like a tiny bar magnet. The strength of any magnet is determined in terms of magnetic moment and it is indicated by μ. It is given by

$$\mu = g.\beta_N.\sqrt{I(I+1)},$$

where g is a number characteristic of the nucleus called *nuclear factor* whose value can only be determined experimentally. For proton, $g = 5.58$; β_N is another constant known as *Bohr's nuclear magneton.* For proton its value is 5.05×10^{-27} JT^{-1} (J = joule and

T = tesla, the unit of magnetic field strength). The magnetic moment is directly proportional to the angular momentum, p, i.e.,

$$m = \gamma p,$$

where, γ is a constant known as *gyromagnetic ratio*; γ for proton is 2.675×10^4 gauss^{-1}.

(c) *Zeeman effect*: It is a process of removal of the degeneracy of different spin energy states by the application of magnetic field and each energy state corresponding to different orientations of the nucleus is called *Zeeman energy level.*

3. What happens when a spinning nucleus is placed in a magnetic field? Or what is zeeman effect?

➜ See answers against Question No. 2(c).

4. (a) What does quantisation of spin state mean?
(b) What do you mean by (i) saturation and (ii) relaxation?

➜ (a). It must be remembered that although we are considering a single nucleus but according to quantum mechanical approach we are actually handling with an assembly of nuclei, when such assembly of nuclei is placed in a static magnetic field, there is distribution of nuclei in all the possible orientations $(2I + 1)$. Again, it has been stated that the amount of radio frequency required to promote nucleus from lower to higher spin energy level is the measure of magnetic moment of the tiny nuclear magnet. Further, promotion of nuclei from lower to higher spin energy level requires the absorption of energy by the nuclei. As there is absorption of energy, it is obvious that there must always be greater number of nuclei in the lower energy state or in other words, the assembly of nuclei in a static magnetic field is not equally distributed between the different possible spin states but rather they take up a Boltzmann distribution with a small but finite excess in favour of the lower level indicated by the Boltzmann factor, 0.001%. It is this small but finite excess of nuclei present in the lower energy state that gives rise to net absorption of energy for *nuclear magnetic resonance phenomenon.*

This unequal distribution of nuclei in different spin energy states is termed as quantisation of spin states of nuclei. This quantisation of the spin states of nuclei is highly significant, otherwise there should not have any absorption of energy and no nuclear magnetic resonance.

(b) (i) When to an assembly of nuclei in static magnetic field (quantised) energy from radio frequency region is supplied, it has been observed that there is continuous absorption of energy. So it is obvious that the excess of nuclei (by factor 0.001%) originally present in the lower state may gradually diminish and ultimately a situation will arise when the number of nuclei in the lower and higher energy states will be equal and there will be no further absorption of energy. This phenomenon is known as *Saturation.*

(ii) As we know absorption of energy is a continuous process, there must always be excess number of nuclei in the lower energy state than that of in the higher energy state even after the attainment of the state of saturation. Therefore, there must be some mechanism by which the initial excess of nuclei in the lower energy state is maintained. Such mechanism is known as *Relaxation Process.* Thus, relaxation process is defined as the radiationless process by which a nucleus from higher spin energy state returns to lower energy spin state. There are two types of relaxation process as discussed below:

(a) *Spin-spin relaxation*: It is caused by the mutual exchange of spins by two precessing nuclei in close proximity to each other. There is a magnetic field component associated with each of the precessional nucleus rotating in a plane perpendicular to the main field. This small rotating magnetic field can induce a transition in the neighbouring nucleus. This relaxation process (transverse relaxation) although shortens the lifetime of nucleus in higher energy spin state, yet is not so much effective to maintain the small but finite number of excess of nuclei in lower energy spin state.

(b) *Spin-lattice relaxation*: The term 'lattice' means the frame work of the molecules (sample and solvent) containing precessional nucleus. All the molecules in the lattice are undergoing translational, rotational and vibrational motions and have magnetic properties. So a variety of small magnetic fields are present in the lattice. Out of all these different magnetic fields if a particular small magnetic field is properly oriented in the lattice it can induce a transition in a particular precessing nuclear magnet from higher to lower energy spin state. The energy released due to this transition is transferred to the components of the lattice as additional translational, rotational and vibrational energy and ultimately dissipated as thermal energy. Thus a nucleus returns to lower energy spin state from upper energy spin state. This process is able to maintain an excess of nuclei in the lower state which is the necessary condition for nuclear resonance (NMR) phenomenon.

4. What property of certain atomic nuclei is involved in NMR phenomenon.

➻ Spin active property of certain atomic nuclei is involved in this phenomenon. It has been observed that a nucleus having odd atomic number or odd mass number has a nuclear spin which can be observed by NMR spectrometer. So such atomic nuclei are spin active. A large number of spin active nuclei such as ^{1}H, ^{13}C, ^{15}N, ^{19}F and ^{31}P, have permanent magnetic moments and can be studied by NMR spectroscopy.

5. Why in NMR technique other changes like electronic, vibrational and rotational do not occur?

➻ NMR spectroscopy utilises electromagnetic radiation of radiowave region. The energy content of this region is very small (6.6×10^{-25} erg). With this small amount of energy, only spin energy states are changed but other transitions which require more energy do not occur.

6. How will you determine the relative number of nuclei in the higher and lower energy states?

➻ The relative number of nuclei in the higher and lower energy states can be ascertained at thermal equilibrium from the Boltzmann distribution law as follows:

$$\frac{N_{upper}}{N_{lower}} = \exp\left(\frac{-\Delta E}{kT}\right)$$

where $\Delta E = E_{upper} - E_{lower}$ (E = energy), T is the temperature in kelvin (k) is an universal constant and N is the number of nuclei.

7. What is the signifance of nuclear resonance phenomenon in chemistry?

➻ The nuclear resonance phenomenon originates from the magnetic moment of spinning nucleus. But it must be remembered that there is no existence of 'bare nucleus'. Nucleus is always present in an atom, i.e., this nucleus is always surrounded by electrons. Again, electrons are magnetic in character. Hence, when we are measuring the magnetic moment of nuclear magnet, we are actually measuring the combined magnetic moment of nuclear magnet and surrounding electron magnet. Greater the electron density surrounding the nucleus, higher will be this magnetic moment. Electrons, particularly extra nuclear electrons, are connected to the chemical bond. Thus from NMR study the nature of the chemical bond may be ascertained and in fact, NMR spectroscopy is used in this respect by the chemists.

8. Explain why in nuclear magnetic resonance spectroscopy, proton magnetic resonance has been extensively studied?

➻ Chemical compounds usually consists of a large number of different kinds of atoms with different nuclear spin and hence with different magnetic moments. So to determine the magnetic moment of each atomic nucleus is practically impossible. But if the magnetic moment of only one very common nucleus is determined, then the difference in magnetic moment values of that particular nucleus in the different positions of the same compound as well as in different compounds must be due to the difference in the electron environment (electron density) at different situations. So in that circumstances, from magnetic moment value, a fruitful information regarding the structure of the compound may be obtained. For this reason, the magnetic moment of an extremely common nucleus is generally determined. A scrutiny of chemical compounds particularly organic compounds reveals that H atom is invariably present in most of the organic compounds. Other atoms that may be present are C, O, N, S, halogens, etc. These atoms form different types of bonds among themselves as well as with H-atoms. So the electron environment (electron density) surrounding H-atom at different positions in a particular compound will be different. Thus an organic compound may be regarded as an assembly of large number of proton magnets with different electron environment and hence with different magnetic moments. So difference in magnetic moment between any two different proton magnets gives a measure of the difference in electron environment between these two H-atoms and from this difference in electron environment of H-atoms, an information regarding the structure of the molecule may easily be obtained. For this reason, the general principle of nuclear magnetic resonance has been reduced to proton magnetic resonance (PMR) and is now indicated as ^{1}HNMR.

9. (a) How would you measure the ^{1}HNMR spectrum?
(b) How many methods are used for recording NMR spectra?

➻ (a) The ^{1}NHMR spectrum is usually taken in a solution. Normally 50 mg or more of the sample is dissolved in an appropriate solvent. The solution is taken in a precession-ground tube. The most commonly used tube is of 5 mm diameter with a depth of 2-3 cm. The solution should be free from any paramagnetic and insoluble impurities and it should not be viscous. The tube containing the solution of the sample is kept into a probe placed between the poles of the magnet in the strongest and most homogeneous part of the magnetic field. The probe has transmitter coil to supply radio frequency radiation of appropriate frequency, say, 100 or 200 MHz. The probe has also another coil, the receiver coil, wrapped at right angles to the transmitter coil. This is the optimum angle for detection of resonance signals. The magnet that is used is connected with a sweep generator to change the magnetic field precisely and continuously over a small range. Now the switch of the instrument is made on when different protons, i.e., protons with different electron environments will absorb energy at different magnetic fields for their resonances and resonance signals will be obtained at different magnetic fields. These signals are automatically recorded and expressed as a graph which is printed automatically using a pen and a chart paper. This graph is a plot of intensity of signals against radio frequency absorbed and is known as ^{1}HNMR or PMR spectrum. Different signals occupy different positions in the spectrum. It may be mentioned that *the equivalent protons* (protons with identical electron environment) will give a single signal—only *non-equivalent* protons (protons with different electron environment) will give different signals. The position of signals for different non-equivalent protons in the spectrum is indicated with respect to the position of the proton-signal of a reference compound. The commonly used reference compound is tetramethyl silane, $(CH_3)_4Si$ (TMS) and it is used internally. The proton-signal of TMS is placed at the extreme right-hand side of the spectrum while the proton signals of different non-equivalent protons of the organic compound under study occupy different positions in the left-hand side of proton signal of TMS as shown below.

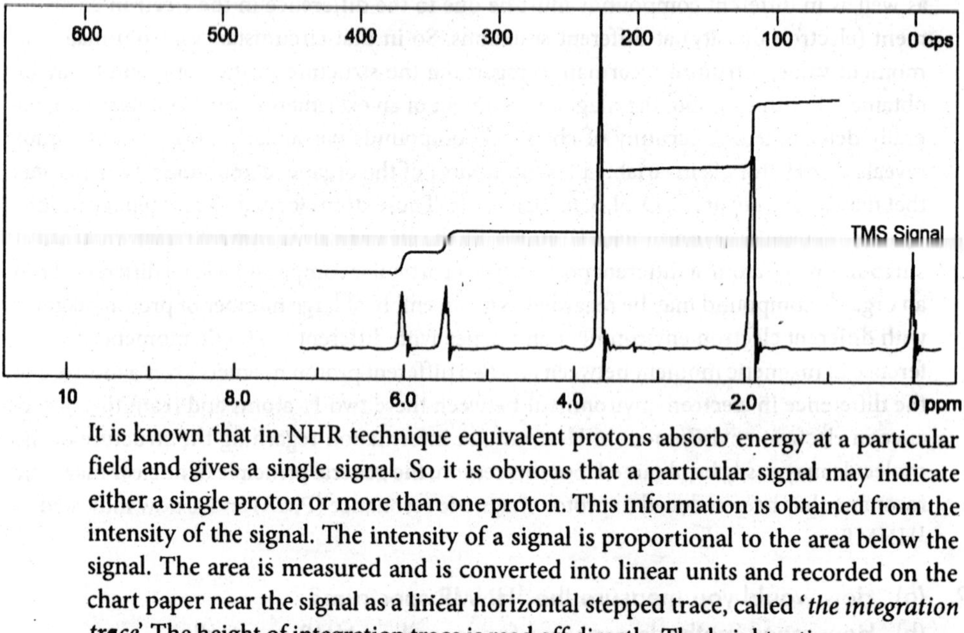

It is known that in NHR technique equivalent protons absorb energy at a particular field and gives a single signal. So it is obvious that a particular signal may indicate either a single proton or more than one proton. This information is obtained from the intensity of the signal. The intensity of a signal is proportional to the area below the signal. The area is measured and is converted into linear units and recorded on the chart paper near the signal as a linear horizontal stepped trace, called '*the integration trace*'. The height of integration trace is read off directly. The height ratios are proportional to the intensities of the signals. Since the intensity of a signal is directly proportional to the number of protons giving that signal, the integration data give the relative number of protons giving different signals.

The instrument, the Nuclear Magnetic Resonance (NMR) spectrometer used in this technique is an automatic recording spectrometer and consists essentially of the following four parts.

(i) a magnet capable of producing very strong homogenious field.
(ii) a means of continuously varying the magnetic field over a small range.
(iii) a radio frequency source.
(iv) a radio frequency detector to indicate that energy is being transferred from radio frequency beam to the nucleus.

A schematic diagram of the instrument is given below:

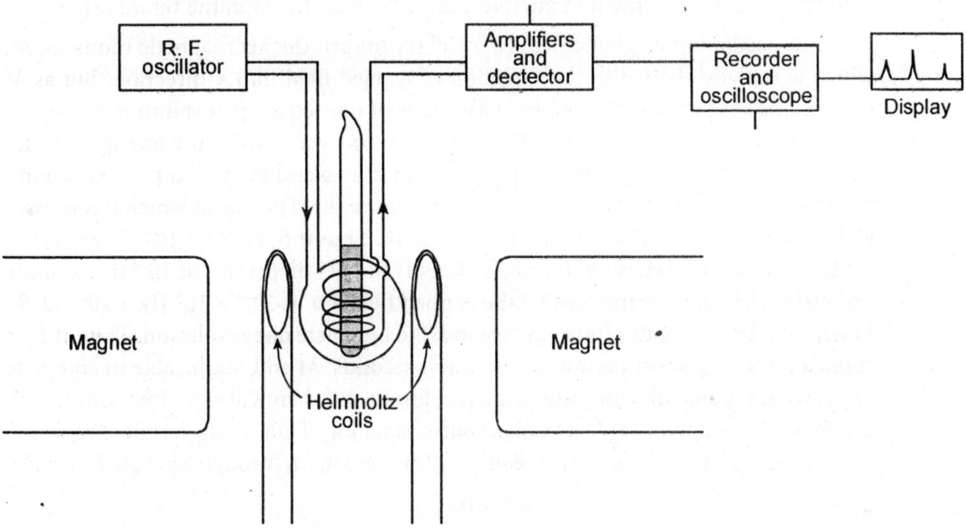

(b) There are two methods for recording NMR spectra, viz.,
(i) Traditional continuous wave (C.W.) method and (ii) Pulse technique.

(i) *Continuous wave (C.W.) method:* In general, NMR spectra showing chemical shifts of nuclei having appreciable abundance like proton are recorded by continuous wave method. However, because of difference in chemical shifts for different nuclei (protons), not all nuclei (protons) come to resonance exactly at a fixed magnetic field. So there must have provision in the instrument for detecting these different frequencies. For this purpose, C.W. method is operated in two modes—either by varying magnetic field keeping the frequency of the transmitter (radio frequency) constant or by varying the radio frequency keeping the magnetic field constant. The former mode is known as **Field sweep** while latter is termed as *frequency sweep*. It may be mentioned that the field sweep method is by far the simplest technically because it is easy to change a magnetic field smoothly and is fairly easy to maintain a precise frequency by using a carefully thermostatted crystal oscillator.

In the field sweep continuous wave mode, magnetic field is set to particular field strength and it is slowly raised by a current produced in the sweep generator. When a chemical compound is placed, since the magnetic environment of different nuclei are different, varying amounts of energies are required for their resonances. So starting, with small proportion, the applied field is slowly increased. Each kind of assembly of equivalent nuclei absorbs energy in promoting some of the nuclei in the lower energy spin state to the higher energy state and resonance occurs. Then relaxation takes place and the emitted energy is collected by the detector coil wound around the sample tube, amplified and passed to the recorder. The frequency range being studied is scanned steadily from one extreme to the other. The spectrum is plotted directly as it is being taken as absorption against frequency as in UV-IR spectroscopy.

(ii) *Pulse technique:* In pulse technique, the sample is subjected to a broad band radio frequency signal as a single powerful pulse which effectively covers the whole range of frequencies and lasts for a very short time (t_p) typically for a few microseconds. This pulse generates an oscillating magnetic field (B_1) along x-axis, conventionally through which radio frequency transmitter is aligned. This field is at right angle to the applied field (H_0) which is assumed to operate along z-axis as shown in the figure (a).

This oscillating magnetic field (B_1) will try to turn the net magnetic moment, M, which is initially aligned in the direction of applied field into x-direction. But as M being composed of a large number of assembly of non-equivalent spinning nuclei behaves like a gyroscope and cannot be turned in this way. But as the energy of radio frequency is absorbed, M gets displaced from z-axis and begins to precess around both the axes in the XY plane as shown in the figure (b). The rate at which it precesses will be directly proportional to the strength of B_1. Thus if B_1 be 2.5×10^{-4} T (where 2.5 T (T, Tesla) is the strength of the applied field) the M will precess at 10^{-4} times of the precessional frequency induced by the applied field. Or at $10^{-4} \times 10^8$ Hz = 10^4 Hz. So M will take 10^{-4} seconds (100 microseconds) to complete one revolution. Thus, if B_1 is applied for a very short period, say 50 microseconds, M will not be able to complete one revolution but will complete half a revolution and then will stop precessing. Such a pulse is referred to as 180° (π-radian) pulse radiation. Thus it may be concluded that the net effect of the pulse is to tip the magnetic moment, M, through an angle ϕ given by

$$\phi = \gamma . B_1 t_p$$

Normally, time (t_p) is so chosen that the angle ϕ becomes 90° and such pulses are termed as $\pi/2$ pulses.

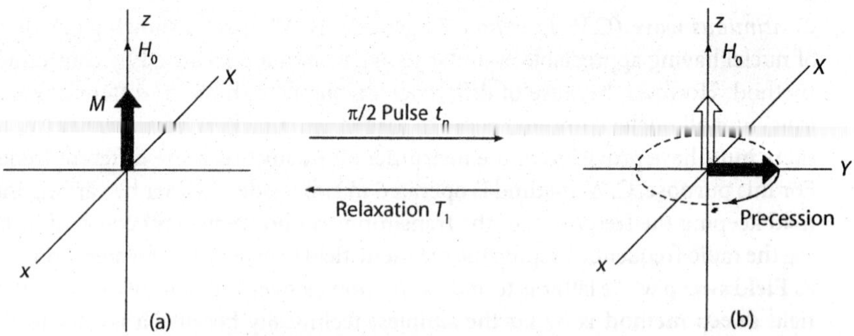

Passing for the desired time (t_p), the pulse is stopped for a short period during which magnetic moment is measured. The nuclei relax to their original equilibrium states and emit radio frequencies at the same frequency and intensity as the absorbed radio frequency and these intensities called *free induction decay* (FID) are picked up by the receiver coil, conventionally aligned along y-axis and fed into a computer. The pulse is applied for a short time again, stopped and the process is repeated many times. The data are presented as total intensity vs. time graph which is designated as the *time domain spectrum* as shown in the figure below:

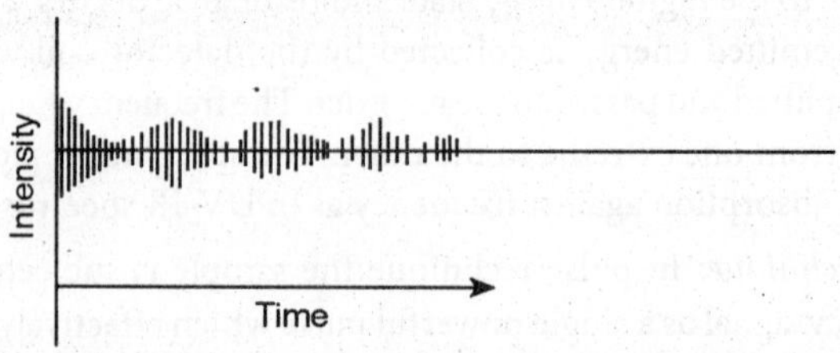

The Fourier transformation of the time domain spectrum converts it into a spectrum which is known as frequency domain spectrum. The spectrum is then plotted exactly in the same form as a continuous wave (C.W.) spectrum. It may be mentioned that pulse technique is more advantageous than C.W. method due to the following reasons:

(i) Normally energetically the difference between the lower and higher spin states is very small particularly for protons in lower filed as applied in C.W. method. But in higher field strength as usually applied in the case of pulse technique such difference is greater which means the pulse technique is more sensitive.

(ii) Pulse technique has the further advantages that all the nuclei of a given type can be excited simultaneously and the large number of pulses can be applied and FID signals are easily added together in a computer. As a result, absorption is intensified and noise which is random is largely cancelled out. Further accumulation of a large number of spectra improves signal intensity.

(iii) In the case of nuclei like 13C where the isotope abundance is very poor and the nuclei are inherently much less sensitive, pulse technique is much more needed.

10. (a) Why is an internal standard used in obtaining NMR spectrum?

(b) What is the most commonly used internal standard and why?

➠ (a) The standard reference to be used, in principle, may be external or internal. But the use of external reference suffers from certain drawbacks. Thus in the case of external reference it is necessary to measure the signal positions relative to the reference from the spectra of the compounds in solution of varying concentration and to extrapolate to infinite dilution. Such procedure requires a number of measurements at different concentrations for each compound with the same solvent throughout. This is obviously a disadvantage. Further, during the use of external reference, care should be taken with respect to the selection of the capillary in which the reference substance is taken. The capillary should be accurately cylindrical and should cause the sample to spin unevenly. It is natural that such requirement for capillary tube may not always be maintained. Once again, the radio frequency field experienced may not be the same for the sample solution and the reference. The radio frequency directely falls on the solution of the sample while the radio frequency must cross through the wall of the capillary before falling on the reference. (In practice, capillary containing reference is dipped into the solution of the sample).

All the above difficulties are best avoided using an internal reference. An internal reference is actually dissolved in the sample. The reference substance, therefore, experiences exactly the same field as the compound under investigation and the question of other difficulties does not arise. For this reason, internal standard is used in obtaining the NMR spectra.

(b) Since the reference is used internally, it must fulfil the following criteria.

(i) It should be chemically inert.

(ii) It should be miscible with most of the organic solvents and other organic liquids.

(iii) It should be highly volatile and

(iv) It must give a single, sharp and readily recognisable absorption signal.

The most commonly used reference for proton spectroscopy is tetramethyl silane $(CH_3)_4Si$ (TMS). It is chemically inert, volatile (bp 27 °C), miscible with organic solvents and gives a sharp single absorption signal. That is why TMS is commonly used as internal reference.

11. Mention some of the commonly used NMR solvents.

➜ The most useful and commonly used NMR solvent is deutero chlroform ($CDCl_3$). In this solvent most of the organic compounds—polar and non-polar—are soluble. It is volatile and can be easily removed if recovery of the sample is necessary. A small residual $CHCl_3$ signal seldom causes any problem.

The other solvents that may be used are:

(i) Carbon tetrachloride (CCl_4): It is useful for certain acid-sensitive compounds.

(ii) Deutero dimethyl sulphoxide (DMSO-d_6): It is relatively non-volatile and so may be suitable for working at high temperature up to 140 °C.

(iii) Deutero-water (D_2O): It is highly polar and mainly used for salts. It can readily exchange all acidic protons.

(iv) Deuterobenzene (C_6D_6): It is not normally used in routine work. It is often added to $CDCl_3$ solution.

(v) Trifluoroactic acid (F_3CCOOH): It is specially useful for amine and heterocyclic compounds.

12. What do you mean by chemically equivalent protons?

➜ Protons which are in identical chemical environments are called chemically equivalent or simply equivalent protons. Equivalent protons exhibit same signal amd appear at same position in the spectrum. Thus, all the protons in TMS, [$(CH_3)_4Si$], benzene, acetone, cyclopentane give rise to a single absorption signal in their ^{1}HNMR spectra. All these are said to be chemically equivalent.

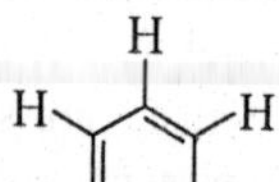
Benzene

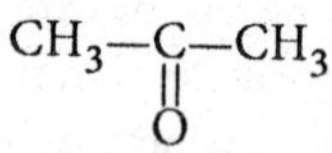
Acetone

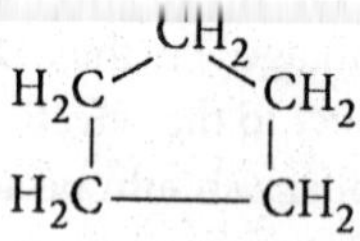
Cyclopentane

$(CH_3)_4Si$
Tetramethyl Silane

On the other hand, molecules which have sets of equivalent protons, chemically distinct from each other, may give rise to an absorption peak for each set. Thus,

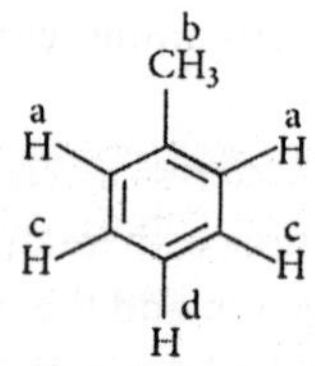
Four sets of equivalent protons (a, b, c & d) (Four NMR signals)

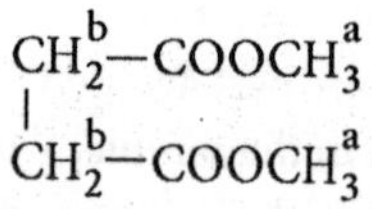
Two sets (a, b) of equivalent proton (two NMR peaks)

$\overset{c}{CH_3}-\overset{b}{CH_2}-O-\overset{a}{H}$
Three sets of equivalent protons (a, b, c) (Three NMR signals)

It may be mentioned that chemically equivalent protons may or may not be magnetically equivalent.

13. How many kinds of protons are there in the following compounds and predict the number of PMR signals to be expected.

Compound	Number of kinds of protons and signals
(i) $\overset{d}{CH_3}-\overset{c}{CH_2}-\overset{b}{CH_2}-O-\overset{a}{H}$	Four kinds of protons (a, b, c, d) and four signals.

Compound	Number of kinds of protons and signals
(ii) $(\overset{c}{CH_3})_2-\overset{b}{CH}-O-\overset{a}{H}$	Three types of equivalent protons (a, b, c) and three signals.
(iii) $(\overset{a}{CH_3})_2O$	Only one type of equivalent proton, '*a*' and single peak
(iv) 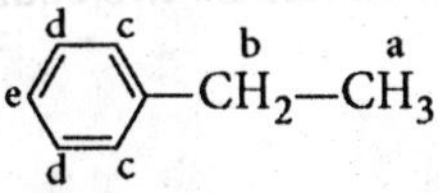	Five kinds of equivalent protons (a, b, c, d, e) and five signals.
(v) $\overset{c}{CH_3}-\overset{b}{CH_2}-O-\overset{a}{H}$	Three kinds of equivalent protons (a, b, c) and three peaks.
(vi) $\overset{a}{CH_3}-\overset{b}{CH_2}-\overset{a}{CH_3}$	Two types of equivalent protons (a, b) and two signals.
(vii) $\overset{b}{CH_3}-\overset{a}{CH_2}-Cl$	Two signals as there are two kinds (a, b) of equivalent protons.
(viii) $\overset{a}{CH_3}-\overset{a}{CH_3}$	Only one signal because all the protons are equivalent.
(ix) $Cl-\overset{a}{CH_2}-\overset{a}{CH_2}-Cl$	Only one signal—all are equivalent.
(x) $\overset{a}{CH_2}=\overset{a}{CH_2}$	Only one kind of equivalent protons and so only one signal will be obtained.
(xi) $\overset{c}{CH_3}-\overset{b}{CH}=\overset{a}{CH_2}$	Three kinds (a, b, c) of protons, so three signals.
(xii) 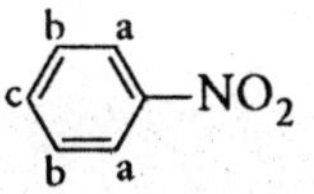	Three kinds (a, b, c) of protons and so three signals.

14. Write the structural formulae for the compounds with the following molecular formulae that show only one signal in their PMR spectra.

(i) C_5H_{12}, (ii) C_3H_6, (iii) C_2H_6O, (iv) C_3H_4, (v) $C_2H_4Br_2$, (vi) C_4H_6, (vii) C_8H_{18}.

➜ Since only one signal appears in the spectra of these compounds all the protons in each molecule are equivalent.

(i) All the twelve protons must be, in four equivalent methyl groups; thus its structure is

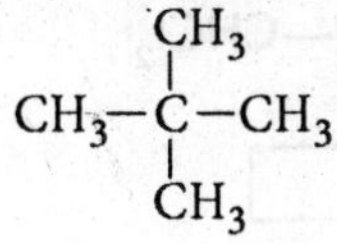

Tetramethyl methane

(ii) The molecular formula, C_3H_6, suggests that there must be three equivalent methylene ($—CH_2—$) groups in the molecule. This is possible if the molecule is cyclopropane.

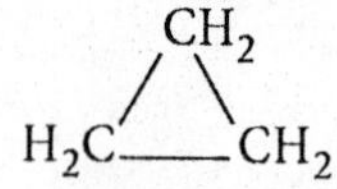

Cyclopropane

(iii) The molecular formula is C_2H_6O. This suggests that there should be two equivalent methyl groups. So the molecule is dimethyl ether.

$$CH_3—O—CH_3$$

Dimethyl ether

(iv) The molecular formula, C_3H_4 of the compound reveals that there should be two equivalent methylene groups attached to the third carbon. So probably the compound is allene.

$$CH_2=C=CH_2$$

Allene

(v) The molecular formula, $C_2H_4Br_2$ of the compound is a clear indication that it contains two equivalent methylene groups and as the presence of bromine atoms does not hamper the equivalency of the two methylenes, probably the compound is 1,2-dibromo ethane.

$$Br—CH_2—CH_2—Br$$

1,2-Dibromo ethane

(vi) The compound has molecular formula, C_4H_6. Since all the hydrogen atoms are equivalent, the probable structural formula of the compound may be as below.

$$CH_3—C\equiv C—CH_3$$

Dimethyl acetylene

(vii) The compound has molecular formula, C_8H_{18}. In order to have all eighteen hydrogens equivalent, the compound must have six methyl groups attached to two carbon atoms as shown below:

$$(CH_3)_3C—C(CH_3)_3$$

Hexamethyl ethane

15. Which member of each of the following structural isomers would be expected to display a single peak in PMR spectrum?

(i) 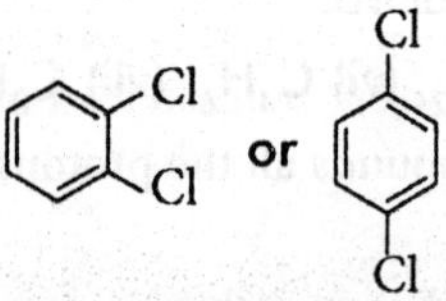

(ii) $CH_3—CCl_3$ or $Cl—CH_2—CHCl_2$

(iii) $\begin{matrix}CH_3\\CH_3\end{matrix}\!\!>C=CH_2$ or 

(iv) $Cl—CH_2—CH_2—Cl$ or $CH_3—CHCl_2$

(v) $CH_2=C=CH_2$ or $CH_3-C\equiv CH$

➜ Since the particular member will have to display single peak in PMR spectrum, all the protons must be equivalent. Scrutiny of the structures of each member of each pair reveals that the following members will exhibit single peak in the PMR spectrum because all the protons in these compounds are equivalent.

(i) 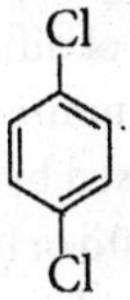 (ii) CH_3CCl_3 (iii)  (iv) $Cl—CH_2—CH_2—Cl$ (v) $CH_2=C=CH_2$

16. Which of the following compounds will show a single peak in PMR spectrum?
(i) Acetone (ii) 1,2-Dibromoethane (iii) Methyl acetate (iv) Acetaldehyde.

➜ The structural formulae of these compounds may be written as below:

(i) $CH_3-\underset{\underset{O}{\|}}{C}-CH_3$ (ii) $Br-CH_2-CH_2-Br$ (iii) $CH_3-\underset{\underset{O}{\|}}{C}-O-CH_3$

(iv) $CH_3-\underset{\underset{O}{\|}}{C}-H$

From the structures it is seen that in acetone and 1,2-dibromo ethane all the protons present are equivalent. So these compounds will show single peak in their PMR spectra.

17. For nuclear magnetic resonance, there are two options: sweeping magnetic field at constant radio frequency or sweeping radio frequency at constant magnetic field. Show that two are equivalent. Which one is commonly employed?

➜ We know that the frequency of nuclear magnet in resonance is given by the equation

$$\nu = \frac{\gamma . H_0}{2\pi}.$$

This indicates that to cause the NMR transition we have two choices, viz.,

(i) Radio frequency is kept constant and the magnetic field is varied until resonance occurs. This procedure is known as *field sweep*.

(ii) The magnetic field is kept constant and radio frequency is varied till resonance takes place. This procedure is known as *frequency sweep*. Both these choices are equivalent as the magnetic field can be expressed in terms of radio frequency or vice versa through the above equation.

However, the former procedure (field sweep) is experimentally more satisfactory and is commonly adopted. This is because the radio frequency is a very small quantity and so its variation is very difficult while it is easier to vary magnetic field which is relatively a bigger quantity. In fact, a small variable magnetic field is superimposed on the main field. This is done by fixing a pair of Helmholtz coils in the pole faces of the main field. The magnetic field induced by these coils can be varied by varying the current flow through them and this field acts in the same direction as the main field. The sample is exposed to both the fields and they appear as one field to the molecule.

18. (a) What is meant by the term 'chemical shift'?
(b) Why is its unit dimensionless?

↔ (a) The distance between the position of the proton signal of the reference (normally TMS) and that of the unknown proton(s) is the difference in the radio frequency required for the resonance of these two protons. This difference gives a measure of difference in electron environment (electron density) between these two types of non-equivalent protons. This distance is usually termed as 'chemical shift'. It is expressed by δ (delta) or τ (tau) unit. Normally PMR spectrum has ten divisions, i.e., either 10δ or 10τ. In the case of δ unit, the position of the proton signal of the reference is considered as zero, while in τ unit it is 10. So the relation between δ and τ may be indicated as

$$\delta = 10 - \tau.$$

It may be mentioned that since the signal of the protons of the reference is placed at the extreme right-hand side of the spectrum and the proton(s) signals of the unknown compound appear at the left-hand side of the spectrum, this scale (δ or τ) is a negative scale. Thus in δ unit, the positions of the different signals will appear at 0(TMS), −1, −2, −3, ··· , −10δ, etc. But for convenience negative sign is omitted and is written as 1δ, 2δ, 3δ etc. In τ unit, the positions are indicated as 10τ (TMS), 9τ, 8τ, 7τ ... 0τ. Thus, higher value of δ or smaller value of τ indicates less electron density.

(b) The unit of chemical shift (δ or τ) is dimensionless. It is purely a number. This is because of the following reasons.

We know that the precessional frequency of the nucleus (proton) is directly proportional to the applied static magnetic field strength—greater the strength of the applied magentic field, higher will be the precessional frequency of the nucleus. Again, the determination of magnetic moment of the precessional nuclear magnet requires the creation of resonance with applied radio frequency. So we can say that chemical shift (H_S—H_R) should also be proportional to the strength of the applied static magnetic field. Therefore, instruments with staitc magnets having different field strengths will give different values of chemical shifts for the same proton(s). As a result, confusion will be created. So it is desirable to express chemical shifts (δ or τ) in such a form which is independent of the field strength of the applied magnet. For this reason this unit is expressed as ratio as indicated below:

$$\delta\,(\text{or }\tau) = \frac{(H_S - H_R)\,(\text{in Hz})}{H_0\,(\text{MHz})}$$

where H_R = field strength at which resonance of protons of reference occurs.
H_S = field strength at which resonance of protons of unknown sample takes place
H_0 = field strength of the applied static magnet
Hz → hertz, MHz → megahertz

The above relation may be illustrated with an example. If the observed difference (H_S—H_R) from reference (TMS) is 200 Hz, and the operating frequency of the applied field is 100 MHz, then chemical shift (δ) will be

$$\delta = \frac{200 \text{ Hz}}{100 \times 10^6 \text{ Hz}} = 2 \times 10^{-6}, \text{ i.e., 2 parts per million or 2 ppm.}$$

Thus, if the ratio (2×10^{-6}) is multiplied by 10^6 an easily handled number ($2 \times 10^{-6} \times 10^6 = 2$ ppm) may be obtained. Thus the expression of chemical shift may be written as

$$\delta = \frac{(H_S - H_R) \text{ in Hz}}{\text{operating frequency of the instruments (MHz)}} \times 10^6 \text{ ppm},$$

It is evident from the ratio that it is *dimensionless* and *is a pure number* as the quantities in both numerator and denominator are in units of Hz (or MHz). In order to express chemical shift in Hz, δ value in ppm is to be multiplied by the the number of megahertz of the operating frequency. Thus, if the operating frequency is 60 megahertz, then the ppm value is to be multiplied by 60. Similarly, for 40 MHz, the ppm value is to be multiplied by 40. Therefore, in general,

δ in Hz = observed ppm value multiplied by x where operating frequency is x MHz (x is the number of MHz).

19. Using a 60 MHz NMR machine, the difference in frequency between absorption by a proton in a compound and that by protons of TMS was found to be 430 Hz. What is the chemical shift in ppm?

➜ We know that

$$\delta = \frac{\text{observed difference in Hz}}{\text{operating frequency in MHz}}$$

$$= \frac{430 \text{ Hz}}{60 \text{ MHz}} = \frac{430 \text{ Hz}}{60 \times 10^6 \text{ Hz}} = \frac{430}{60} \times 10^{-6} = 7.17 \times 10^{-6}, \text{ i.e., 7.17 ppm.}$$

20. The observed chemical shift of a proton is 300 Hz from TMS and the operating frequency of the machine is 100 MHz. Calculate the chemical shift in ppm.

➜ Chemical shift (δ) $= \frac{300 \text{ Hz}}{100 \text{ MHz}} = \frac{300 \text{ Hz}}{100 \times 10^6 \text{ Hz}} = 3 \times 10^{-6} = 3.0 \text{ ppm}.$

21. Using a 60 MHz instrument the chemical shift of a proton was found to be 180 Hz. What would be the chemical shift if a 40 MHz instrument is used?

➜ δ in ppm when 60 MHz instrument is used $= \frac{180 \text{ Hz}}{60 \times 10^6 \text{ Hz}} = 3 \times 10^{-6} = 3.0 \text{ ppm}.$

Now, if a 40 MHz instrument is used,

Chemical shift = δ × operating frequency

= 3 ppm × 40 MHz

= $3 \times 10^{-6} \times 40 \times 10^6$ Hz = 120 Hz.

22. How many hertz does 1 ppm correspond to for a spectrometer operating at a radio frequency of (a) 60 MHz and (b) 100 MHz?

➜ We know that,

$$\text{chemical shift in ppm value} = \frac{\text{chemical shift in Hz}}{x\,\text{MHz}}$$

$$\text{So, } 1 \times 10^{-6} = \frac{\text{chemical shift in Hz}}{60\,\text{MHz}} \qquad [\because\ 1\text{ ppm} = 1 \times 10^{-6}]$$

or, chemical shift in Hz = $1 \times 10^{-6} \times 60 \times 10^{6} = 60$ Hz and

$$1 \times 10^{-6} = \frac{\text{chemical shift in Hz}}{100\text{ MHz}}$$

or, chemical shift in Hz = $1 \times 10^{-6} \times 100 \times 10^{6} = 100$ Hz.

23. Define chemical shifts as dimensionless number with examples.

➜ See answer against Question No. 18(b).

24. Furnish a chart of characteristic chemical shifts of some common protons with respect to TMS as standard?

➜ Characteristic chemical shifts of some common protons with respect to TMS as standard reference may be tabulated as under:

Type of proton		Chemical shift (δ), ppm
Cyclopropane	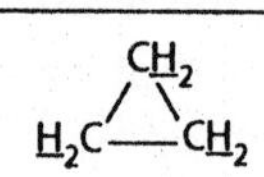	0.2
Primary	R—$\underline{CH_3}$	0.9
Secondary	R_2—$\underline{CH_2}$	1.3
Tertiary	R_3—C—$\underline{H}$	1.5
Vinylic	C=C—$\underline{H}$	4.6–5.9
Acetylenic	C≡C—$\underline{H}$	2.3
Aromatic	Ar—$\underline{H}$	6–8.5
Benzylic	Ar—C—$\underline{H}$	2.2–3.00
Allylic	C=C—C—$\underline{H}$	1.7
Alcohol	R—O—$\underline{H}$	3–5.5
Carbonyl compounds	$\underline{H}$—C—C=O	2–2.7
Aldehyde	$\underline{H}$—C(R)=O	9–10
Ether	$\underline{H}$—C—O—R	2.3–4.3
Acid (Carboxylic)	R—C(=O)—O—$\underline{H}$	10.5–12
Esters	(i) R—C(=O)—O—C—$\underline{H}$	3.7–4.0
	(ii) $\underline{H}$—C—COOR	2–2.2
Phenolic	Ar—O—$\underline{H}$	4–12
Ketomethyl	R—C(=O)—$\underline{CH_3}$	2–2.5

Type of proton		Chemical shift (δ), ppm
Vinylic methyl	$C{=}C{-}CH_3$	1.80–2.00
O-methyl	$O{-}CH_3$	3.5–4.50
N-methyl	$N{-}CH_3$	2–2.50
C-methyl	$C{-}CH_3$	0.9–1.20

25. Describe with examples the various factors which control the magnitude of chemical shift.

➜ Primarily, the chemical shift of a particular proton depends upon its surrounding electron density—greater the electron density surrounding the proton, more it will be protected (i.e., shielded) from the influence of the applied field under which it moves in a precessional motion and so higher field has to be applied for its resonance with the radio frequency, i.e., the proton will appear at higher field (less δ value) in the spectrum. Reversely, if the surrounding electron density is less, the proton will be less protected, i.e., deshielded from the influence of the applied field and so the proton will appear at lower field (higher δ values) in the spectrum. So it is logical to assume that the factors which are responsible for the variation of electron density should primarily be responsible for the variation of chemical shifts. Such factors are as mentioned below:

(i) *Inductive effect*: If a proton is attached to an electronegative atom or group the proton will be deshielded. Greater the electronegativity of the atom, more is the deshielding. So the δ value of the proton will be more. Thus,

Compound	δ value
$CH_3{-}F$	4.25
$CH_3{-}Cl$	3.10
$CH_3{-}Br$	2.64

Again, greater the distance of the electronegative atom or group from the proton, less will be deshielding effect on it.

Further when electron releasing groups are attached to a proton, it is shielded. Thus, in TMS since silicon is electropositive, it pushes electrons into the methyl groups by + 1 inductive effect and due to this powerful shielding effect TMS protons (all are equivalent) resonate at a very high field defined as zero.

(ii) *Type of atomic orbital by which the proton is connected to carbon*: The electron density also depends upon the nature of hybridisation of the atomic orbital of the carbon to which proton is attached. Thus in alkane the σ-bond of C–H is formed by sp^3 atomic orbital of carbon while in alkene (=C—H) the atomic orbital is sp^2. Thus the % of *p*-orbital in alkane σ-bond is more (or *s*-character is less) but in alkene the % of *p* is slightly less (i.e., *s* character is high). Due to this higher character of *s*, alkene proton is deshielded and absorbs at downfield than alkane proton. The observed chemical shifts in ethylene and ethane are 4–5 δ and 1–2 δ respectively.

(iii) *Resonance*: Resonance can also affect the chemical shift. Thus,

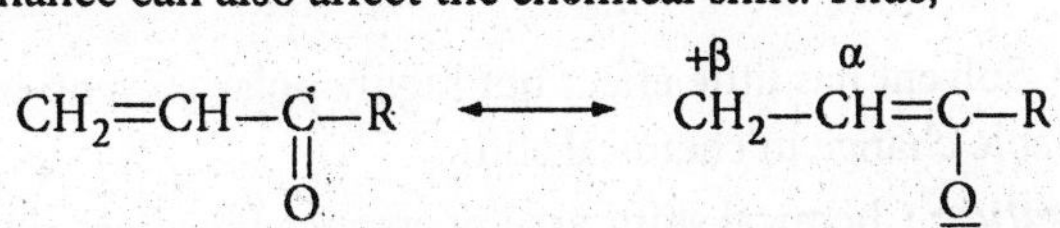

Due to positive charge β protons appear at lower field than expected.

(iv) *Nature of substituent in the carbon atom(s) adjacent to proton*: Chemical shift of a proton also depends upon the nature of substituent present in the adjacent carbon. This effect has been studied by Shoolery who has predicted an empirical rule for predicting the chemical shift in X–CH_2–Y type of structural moiety. Thus,

CH_3– Cl	CH_2Cl_2	$CHCl_3$
3.05 δ	5.29 δ	7.82 δ

Polar groups attached directly to a benzene ring cause upfield and downfield shifts. Thus,

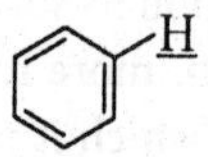
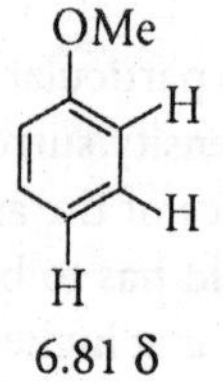
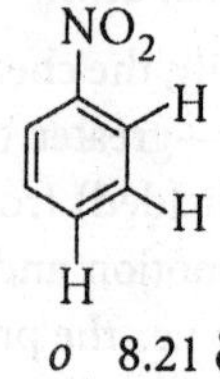

7.27 δ

o 6.81 δ
m 7.17 δ
p 6.86 δ

o 8.21 δ
m 7.45 δ
p 7.66 δ

(v) *H-band*: A hydrogen atom involved in hydrogen bonding shares its electrons with two electronegative atoms as shown below:

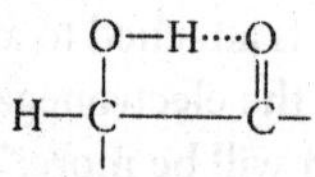

As a result, its electron density is decreased. So it is deshielded and resonates at low field. Thus in water, in very dilute solution in $CDCl_3$, hydrogen bonding is practically nil, protons of water resonate at 2.00 δ. But if droplets of water is suspended in $CDCl_3$, molecules form intermolecular hydrogen bond and the protons of water resonate at ~4.50 δ. Much stronger intermolecular hydrogen bonding in carboxylic acid dimer leads to very low-field absorption in 9–15 δ. In the case of intra-molecular hydrogen bonding as in the case of enols, protons resonate at a very low field ~ 15.4 δ.

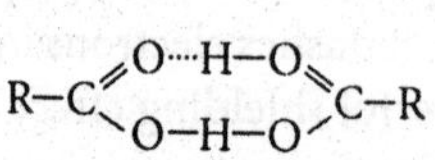

Carboxylic acid dimer

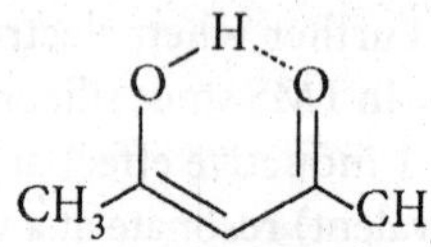

Enolised β-diketone

(vi) *Structural feature*: Structure of the molecule plays an important role in ascertaining chemical shift. The structure may be such that it can involve steric hindrance or van der Waal's force. Thus,

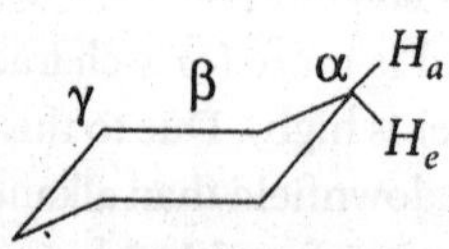

In this structure β γ bond may shield to some extent the equatorial proton than axial proton.

(vii) *Solvent*: Solvent has little effect but highly polar solvent; in some cases, may cause appreciable change in chemical shift.

(viii) *Temperature*: Chemical shifts are not appreciably affected by temperature. But running NMR spectrum at higher temperature inter-molecular H-bonding by polar solvents may be avoided. Thus protons of OH, NH and SH resonate at higher field at higher temperature due to the reduction of degree of hydrogen bonding.

26. Discuss in detail the shielding and deshielding effects involved in NMR spectroscopy.

or

(a) What is the cause of shielding and deshielding effects in NMR spectroscopy?

(b) Discuss the different types of mechanisms of shielding and deshielding.

(a) It is known that chemical shift of a proton primarily depends upon the electron density surrounding it. But if electron density becomes the only factor for chemical shift, then acetylenic proton should resonate at lower field (higher δ) than the ethylenic protons and aromatic protons should have chemical shift value equal to that of ethylenic protons. This deviation arises because of the fact that in this technique electron density is not so important—the most important factor is the strength of the secondary magnetic field generated due to circulation of surrounding electrons and the direction of this secondary magnetic field generated, towards the applied field under which the proton along with its surrounding electrons has been placed. Sometimes the direction of the secondary magnetic field generated be such that it opposes the applied field as a result, the proton experiences less field than actually applied, i.e., proton is shielded. In other cases direction of the secondary magnetic field may be such that it always align (added) with the applied field. So the field experienced by the proton becomes more than the field actually applied, i.e., proton is deshielded. The extent of shielding and deshielding is expressed in terms of a parameter, α, known as *shielding parameter.* Thus the situations may be expressed as under:

$$H' = H_0 - \alpha \cdot H_0 = H_0(1-\alpha) \quad \text{(when shielding)} \qquad \text{(i)}$$

$$H'' = H_0 + a \cdot H_0 = H_0(1+\alpha) \quad \text{(when deshielding)} \qquad \text{(ii)}$$

where H' and H'' are the actual fields experienced by the proton and H_0 is applied field.

In the field sweep procedure the frequency of the radio wave is kept constant which is directly proportional to H_0 i.e.

$$\nu = \frac{\gamma \cdot H_0}{2\pi}$$

So in order to resonate at frequency ν, applied magnetic field must be changed in both the cases to overcome the shielding and deshielding effects occurred. Thus it may be concluded that the shielding and deshielding effects originate from the induced magnetic field generated due to circulation of surrounding electrons in the applied field.

(b) *Mechanistically shielding and deshielding effects have been considered under three headings*: (i) Diamagnetic effects, (ii) Paramagnetic effects, (iii) Anisotropic effect.

(i) *Diamagnetic effects*: Diamagnetism is an atomic property and is observed when all the electrons in the system are paired. Diamagnetic effect arises from the interaction of the applied magnetic field with the field induced in completely filled orbitals. It has been observed that this effect is normally important when the electron distribution surrounding the nucleus of the atom is spherically symmetrical. When such spherically symmetrical electron-distribution is placed in an uniform magnetic field, electrons will circulate in a direction perpendicular to the applied magnetic field around the nucleus and will create a secondary magnetic field surrounding the nucleus. This secondary magnetic field always opposes the applied field. As a result nucleus will experience less applied magnetic field, i.e., due to this diamagnetic shielding, the apparent magnetic field experienced by the nucleus is reduced. So to cause resonance of the nucleus with the radio frequency, higher field is to be applied or in other words such nucleus gives signal at higher field. Since this shielding effect arises from the diamagnetism of the atom, it is known as *diamagnetic shielding effect* and is dependent upon electron density around the nucleus. More the electron density, higher is the shielding and the nucleus (proton) will appear at higher field (less δ value). Pictorially this effect may be shown as below:

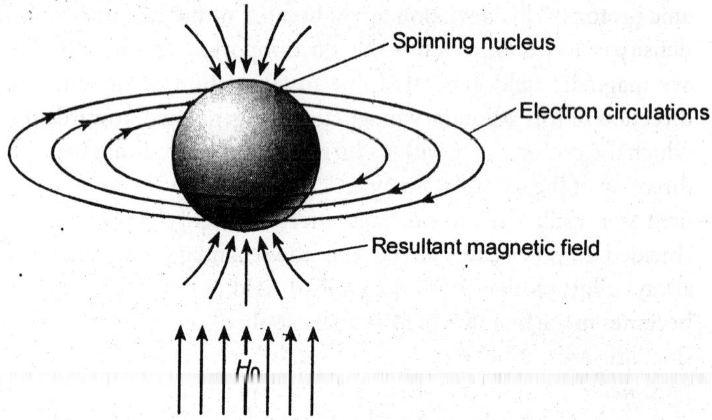

It may be mentioned that this effect primarily finds its origin from σ-electrons because such electrons are in close proximity with the nucleus and hence exerts maximum shielding effect. Inductive effect is the most important factor in controlling this effect.

(ii) *Paramagnetic effect*: This effect arises from the circulation of electrons within the molecule not in close proximity but in the immediate vicinity of the nucleus when they are specifically oriented with respect to the magnetic field. Circulation of such electrons about the nucleus generates a secondary magnetic field. This field is parallel to the applied field at the nucleus. As a result shielding of the nucleus is decreased. Thus such field do not contribute to the shielding of the protons, but such field generated about other atoms such as carbon, nitrogen and oxygen may contribute to the shielding of an adjacent or neighbouring proton. This situation develops in the case of acetylene molecule when it is perpendicular to the applied magnetic field. In this situation the induced fields generated are paramagnetic at carbon but diamagnetic at proton and so positive shielding occurs. This effect is generally observed in the case of linear molecules like acetylene, hydrogen cyanide, hydrogen halide, etc. when they are oriented perpendicular to the applied magnetic field. The shielding effect of acetylenic proton may be shown as under:

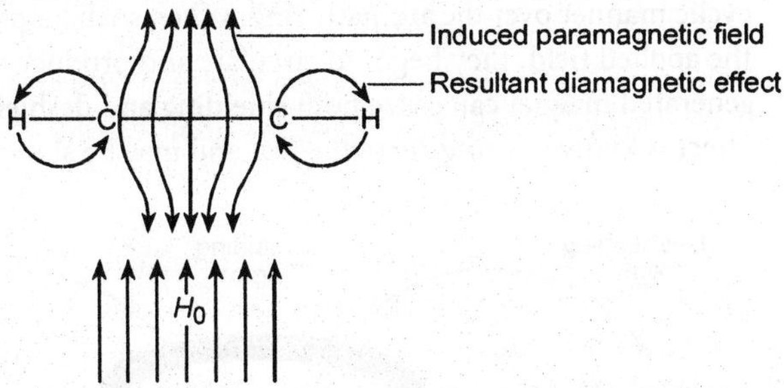

(iii) *Anisotropic effects*: Anisotropic effect originates from the circulation of electrons in other parts of the molecule. Normally, such effect is appreciable in relatively rigid molecule. The field generated by the circulation of electrons in other parts of the molecule is directional in nature and depends upon the orientation of the nucleus in question. In general, electrons of π-bonds are specially important as has been observed in alkenes, alkynes, carbonyl aromatic etc. Thus, a neighbouring nucleus which may not be directly chemically bonded to the π-system may be shielded or deshielded depending on its orientation in space with respect to π-system. Thus when an akene group is so oriented that the plane of the double bond is 90° to the direction of the applied field as shown in the figure (a), the induced circulation of π-electrons generates a secondary magnetic field. This field is diamagnetic around carbon atoms and paramagnetic in the region of protons. So the alkene protons feel greater field strength and consequently resonate at lower field (higher δ value).

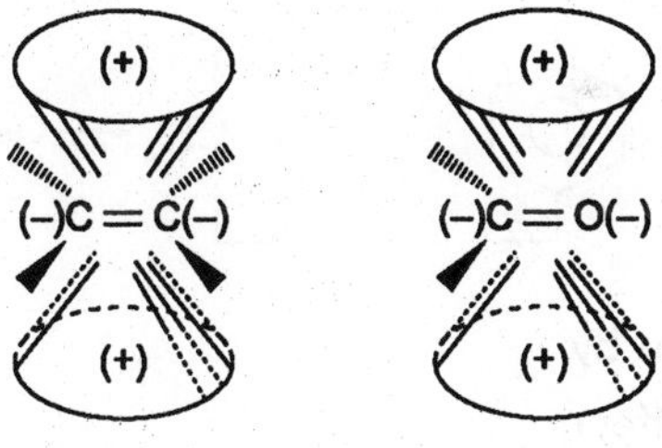

(a) (b)

Again, when the carbonyl group of an aldehyde is oriented in such a way that the plane of trigonal carbon atom is perpendicular to the field, the ciruclation of π-electrons in carbonyl group produces an anisotropic effect at the proton that causes the deshielding of aldehydic proton as shown in figure (b). Further, when linear acetylene molecule is oriented parallel to the applied field, the electronic circulation within the *cylindrical π-electron* cloud induces a diamagnetic shielding at the acetylenic proton as shown in the figure. As a result, acetylenic protons appears at higher field than expected.

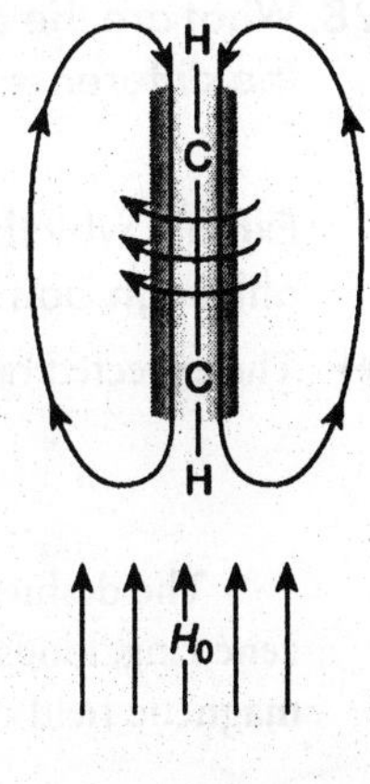

In the molecule of benzene and other aromatic compounds in general π-electrons are delocalised and move in cyclic manner over the aromatic ring. When such loops of electrons are placed in the applied field, they begin to circulate and produce a secondary magnet. This generated magnet can cause both shielding and deshielding of the nucleus. This effect is known as *ring current effect* and may be shown as below:

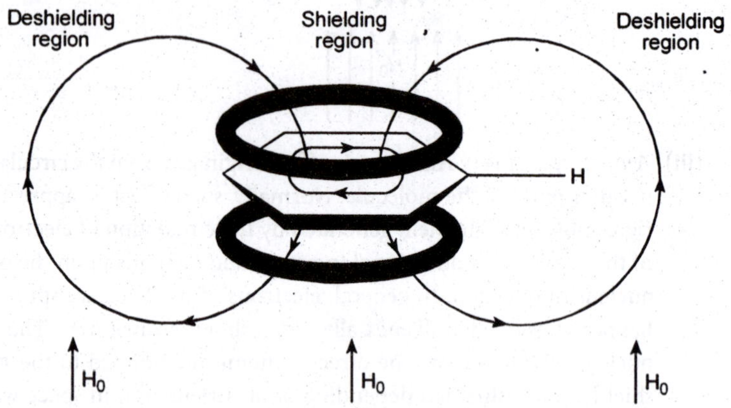

From the figure it is evident that this effect causes deshielding to aromatic protons (shifted to higher δ value) and protons of any group which is in the plane of aromatic ring experience slightly reduced effect. Further, from the figure it is seen that the protons lying in the central position of the aromatic ring are shielded, as has been observed in the case of methylene protons of polymethylene benzene which appear at higher field than expected.

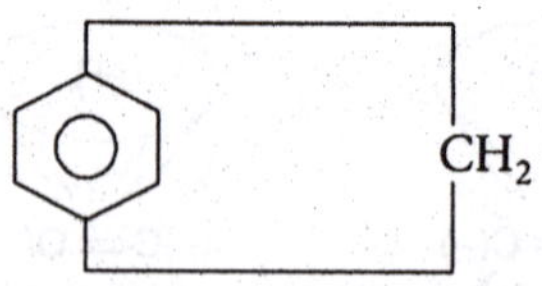

27. What do you mean by

(a) Diamagnetic effect

(b) Paramagnetic effect and (c) Anisotropic effect in NMR spectroscopy?

➜ See answer against Question No. 26(b).

28. What are the expected ranges of absorptions of ethylene and benzene? Explain the difference in their positions of absorption bands in NMR spectrum.

or

Explain why the aromatic protons are more deshielded than ethylenic protons although both the types of protons are attached to *sp*² hybridised carbons?

➜ The expected ranges of absorption of these two types of protons are:

Ethylene	δ 4.60–5.90
Aromatic	δ 6.0–8.50

The deshielding of aromatic protons originates from the ring current effect. In benzene ring, loops of π-electrons are delocalised over the aromatic ring and in presence of magnetic field circulate to produce ring current. This induced ring current is paramagnetic outside the ring. As a result, aromatic protons, around the periphery of the ring, experience a much greater magnetic field than applied and so they (aromatic protons) resonate at lower values of applied field than ethylenic protons where no other factors excepting the anisotropic effect offered by π-electrons are operating.

29. Justify the statement: "Acetylenic protons appear at much higher field than ethylenic protons although former are attached to more electronegative *sp*-carbon."

➜ An ehthylene molecule when placed in an applied magnetic field, the plane of the π-bond lies at right angle to the direction of the applied field. The induced field generated is paramagnetic in the region of the protons and so ethylenic protons appear at lower field.

But acetylenic protons, on the otherhand, when placed in the applied magnetic field, becomes oriented in such a way that the plane of triple bond lies parallel to the direction of the applied field and the induced magnetic field generated due to the circulation of cylindrical electrons of triple bond opposes the applied field. Even when acetylene molecule is perpendicular to the applied field, the induced generated field is diamagnetic towards acetylenic protons. Thus acetylenic protons experience smaller field strength (shielding) and so resonate at higher field (lower δ value).

30. (a) Explain with examples what do you mean by "splitting of signals". Why is it called spin-spin coupling?

(b) Discuss how the nature of multiplicity and intensity of a signal are related to the number of adjacent non-equivalent protons.

➜ (a) We know that under normal condition equivalent protons absorb energy at a particular field for their resonance and give a single signal (peak) in PMR spectrum. It is actually found to be so under low resolution. Thus, in ethanol, CH_3—CH_2—OH, there are three sets of equivalent protons (—C$\underline{H}_3$, —C$\underline{H}_2$ and O—$\underline{H}$) and under low resolution three signals appear in the spectrum as shown below:

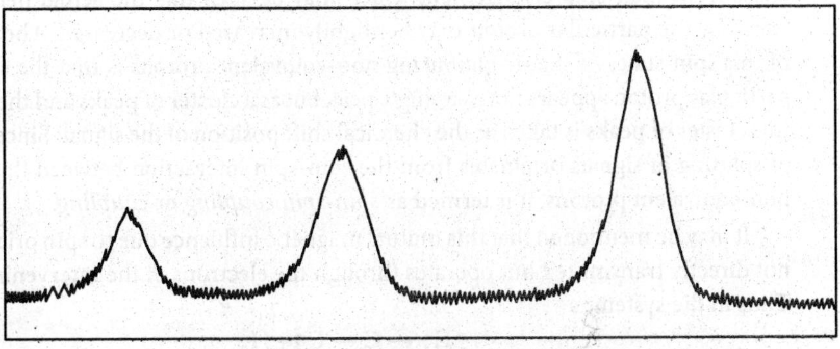

The low resolution spectrum of ethanol (40 MHz)

But when the spectrum is taken under high resolution, each signal appears as a bunch or multiple of peaks, i.e., the signals are splitted into several small peaks within the peak areas as seen in the figure below.

The middle position of the cluster is taken as the actual position of the signal.

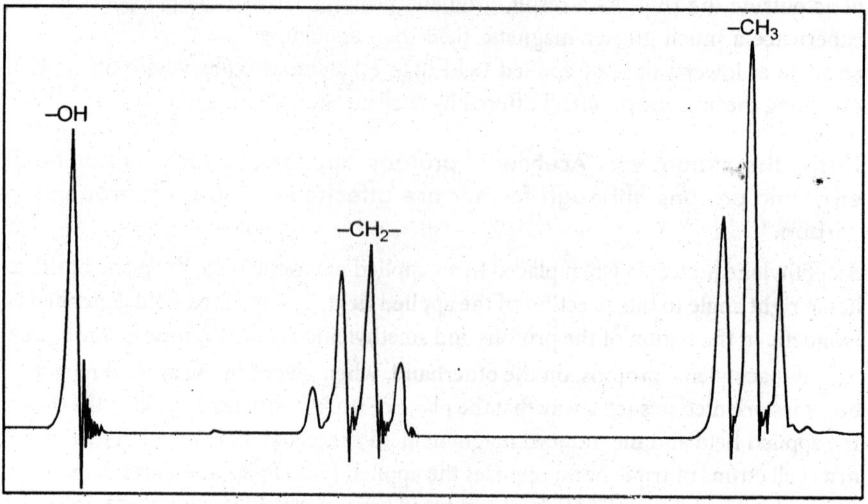

The high resolution spectrum of ethanol obtained at 60 MHz

The cause of such splitting of signals may be rationalised by the proper interpretation of magnetic environment. It is known that the chemical shift, i.e., the magnetic field at which resonance of a particular proton (nucleus) occurs depends upon the total magnetic environment generated in the adjacent position of that proton. So it is obvious that in this total magnetic environment, magnetic environment of the adjacent *non-equivalent* proton(s) must be involved. The direction of the magnetic environment of the adjacent non-equivalent proton(s) may be such that it may align (reinforce) or may oppose the applied field. As result, the actual field experienced by a particular proton may be slightly increased or decreased. So the signal may be slightly shifted to higher position or to the lower position. It has been observed that this directional effect of the magnetic environment of the adjacent non-equivalent proton(s) arises from their spin arrangement which may be such that the actual field experienced by the particular proton may be slightly increased or decreased. The net effect of this spin states of the neighbouring non-equivalent protons is that the signal of a particular proton appears not as a single peak but as a cluster of peaks and the centre of this cluster of peaks is taken as the chemical shift position of the signal. Since this type of splitting of signals originates from the spin-spin interaction between the adjacent non-equivalent protons, it is termed as *spin-spin coupling* or *coupling*.

It may be mentioned that this mutual magnetic influence due to spin orientation is not directly transmitted but operates through the electrons of the intervening bonds. Thus in the system,

$$H_A \leftarrow C \leftarrow C \leftarrow H_B$$

Nuclear spin of H_A couples with the spin of electrons of C—H_A bond which in turn couples with the spin of C—C bonding electrons and next with the of C—H_B bonding electrons. It has also been observed that this spin-spin interaction is maximum up to three bonds and then gradually decreases. However, in a few exceptional cases up to four or more bonds this effect is found to be operative but with very low strength.

(b) It has been observed that the nature of multiplicity of a signal and its intensity are directly related to the number of adjacent non-equivalent protons. This may be elaborated by considering the following examples.

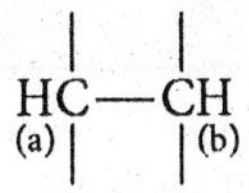

(I)

In this case, both $H_{(a)}$ and $H_{(b)}$ are non-equivalent and the field experienced by the proton $H_{(a)}$ is influenced by the spin states of the adjacent proton $H_{(b)}$. Proton $H_{(b)}$ may have two types of spin as shown below:

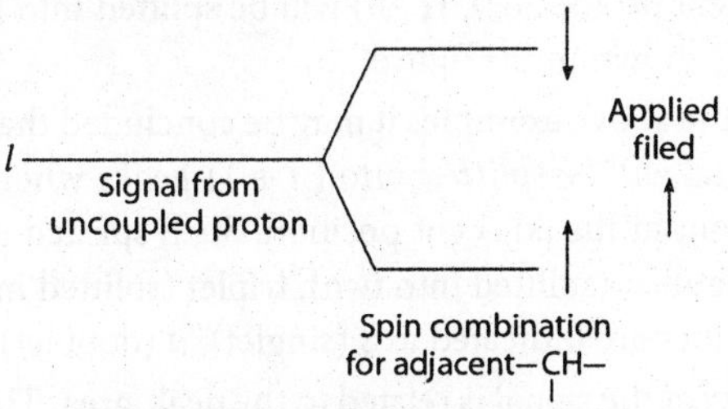

Spin-spin coupling; coupling with one proton gives a 1:1 doublet

So the signal of $H_{(a)}$ will be splitted into two small signals. Similarly, the signal of $H_{(b)}$ will be splitted into two peaks.

Again, in the example below (II), the signal of $H_{(a)}$ proton will be splitted into three small peaks as the two adjacent $H_{(b)}$ (present as CH_2) protons may have three types of spin arrangements as shown below:

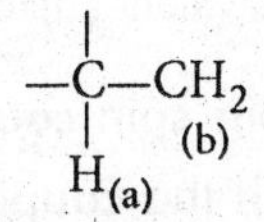

(II)

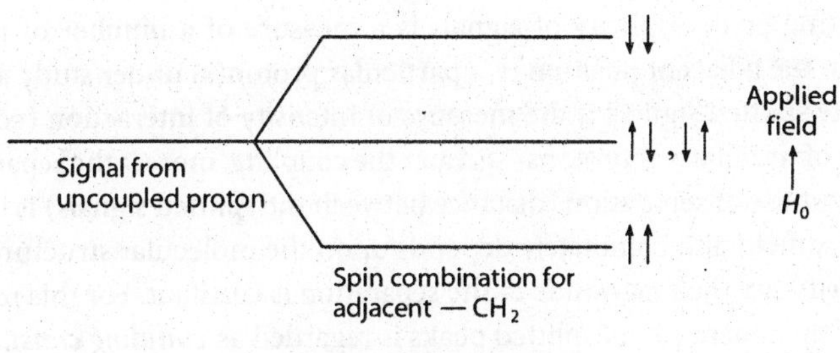

Spin-spin coupling; coupling with two protons gives 1 : 2 : 1 triplet

On the other hand, two $H_{(b)}$ protons of CH_2 will be splitted into two peaks due to only two spin arrangements of only one $H_{(a)}$ proton. Let us consider example (III)

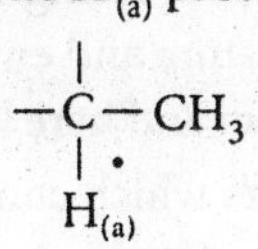

(III)

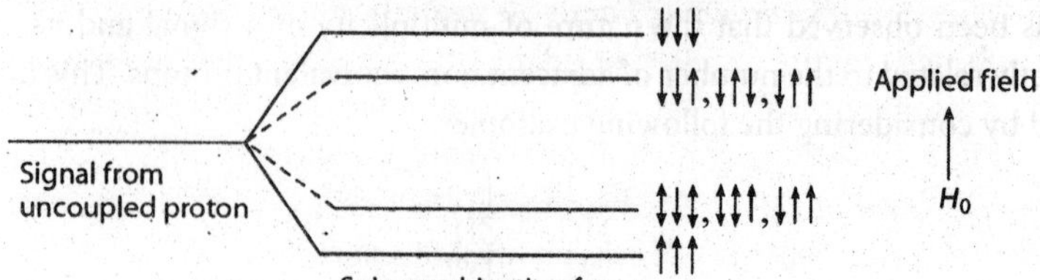

Spin combination for adjacent —CH_3

Spin-Spin coupling; coupling with three protons gives 1 : 3 : 3 : 1 quartet

In this case, the signal of $H_{(a)}$ proton will be splitted into four peaks but the signal of three equivalent protons of CH_3(b) will be splitted into two signals due to the spin arrangements as shown in the figure.

Thus from the above examples it may be concluded that a *particular non-equivalent* proton signal *will be splitted into* $(n + 1)$ *peaks* where n is the number of non-equivalent protons at the adjacent position. Such splitted peaks are known as singlet (no splitting), doublet (splitted into two), triplet (splitted into three peaks) and so on. These multiplicities are indicated as *s* (singlet), *d* (doublet), *t* (triplet), etc.

The intensity of the signal is related to the peak-area. The peak-area depends upon the number of splitted signals. Thus the relative ratios of the peak-areas are doublet (1 : 1), triplet (1 : 2 : 1), quartet (1 : 3 : 3 : 1). In general, the peak-area is given by the expression $(1 + x)^n$ where 'n' is the number of adjacent protons. The co-efficient of indices of x are the peak-area ratio. Thus for doublet $(n = 1)$, $(1 + x)^1 = 1 + x$, i.e., 1 : 1; for triplet $(n = 2)$, $(1 + x)^2 = 1 + 2x + x^2$, i.e., 1 : 2 : 1. Similarly, for quartet, $(1 + x)^3 = 1 + 3x^2 + 3x + 1x^0$, i.e., 1 : 3 : 3 : 1.

Thus, in this way, both the nature of multiplicity and intensity of signal are directly related to the number of non-equivalent adjacent protons.

31. (a) What is the significance of spin-spin coupling and how is it expressed ?

(b) What are the factors on which the coupling depends?

(c) Discuss different types of coupling.

➧ (a) The splitting or multiplicity of signals is a measure of a number of non-equivalent protons at the adjacent position of a particular proton(s) under study and the separation of the splitted signals is the measure of intensity of interaction (coupling) by the spin state of the adjacent protons—greater the coupling, more is the separation of peaks. The magnitude of separation (distance between the splitted signals) is independent of the applied field and completely depends upon the molecular structure. So for a particular structure the magnitude of the separation is constant. For this reason the magnitude of the separation of splitted peaks is regarded as *coupling constant* and is indicated by J. Since this constant is independent of the strength of applied magnetic field it can be safely expressed in units of hertz rather than ppm. Thus, coupling constant (J) serves as a indicator of coupling between adjacent non-equivalent protons and its value reveals the magnitude (intensity) of coupling. Hence, from the value of coupling constant information regarding the nature and environment of the proton, i.e., the structure of the molecule may be known and here lies its significance.

(b) There are a large number of factors which control the coupling between non-equivalent protons. These factors may be stated as below:

(i) *Number of intervening bonds between the interacting protons*: It has been observed that the coupling or interaction between the interacting non-equivalent protons is very strong when they are adjacent to each other and it gradually decreases as the number of intervening bonds increases.

(ii) *Dihedral angle*: The dihedral angle between the interacting protons plays an important role in determining the magnitude of coupling.

(iii) *Number and nature of substituents*

(iv) *Structure of the molecule* on C-atoms *bearing the coupling protons.*

(v) *As stated, coupling is independent of applied magnetic field.*

(c) It is customary and convenient to classify the couplings on the basis of the number of intervening bonds between the interacting protons. Each class of the coupling is indicated either by numerals or by appropriate terms. Different classes of coupling may be discussed in sequel.

(i) *Geminal Coupling* (^2J): When coupling protons are separated by two bonds, i.e. when the interacting non-equivalent protons are attached to the same carbon atom, the coupling involved is known as *geminal coupling* and is indicated by ^2J.

Thus, 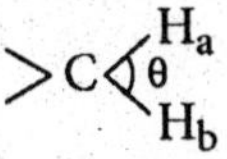

This type of coupling is very weak and can have positive or negative value of coupling constant. The value of coupling constnat generally ranges from –20 to + 44 Hz. The magnitude of coupling constant depends upon bond angle (θ) and nature of substituents. It has been observed that when the angle is 105°, ^2J is –25 Hz while its value becomes –12 Hz when the angle is 125°. Again, when the angle is more than 125°, its value becomes positive. The relationship between the value of coupling constant (J) and the bond angle (θ) may be shown by a plot as below:

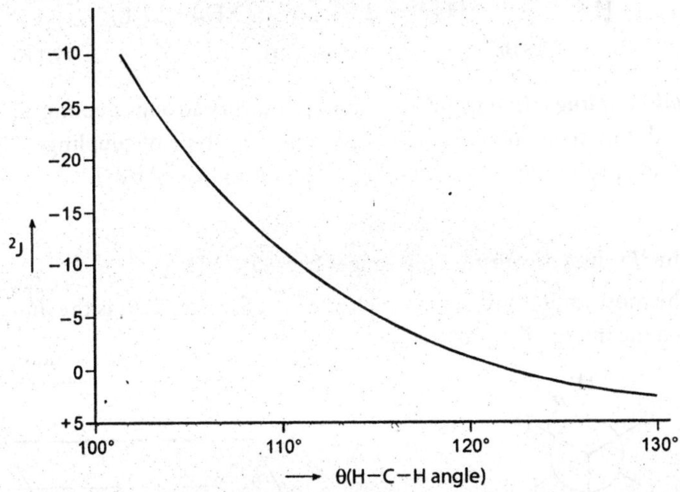

From the curve it is evident that with the increase in the H–C–H angle coupling increases. The cause of this change is due to the change in the mode of hybridisation. An increase in H—C—H angle increases the *s* character of the orbital and so the coupling becomes more positive. Thus, in methane (sp^3-hybridisation), ^2J is –12.4 Hz while it is +2.5 Hz in ethylene (sp^2 – hybridisation).

Geminal coupling is highly dependant upon the nature of substituents present on carbon atom.

(a) When electronegative atom is present at α-position ^2J value is increased as seen from the examples.

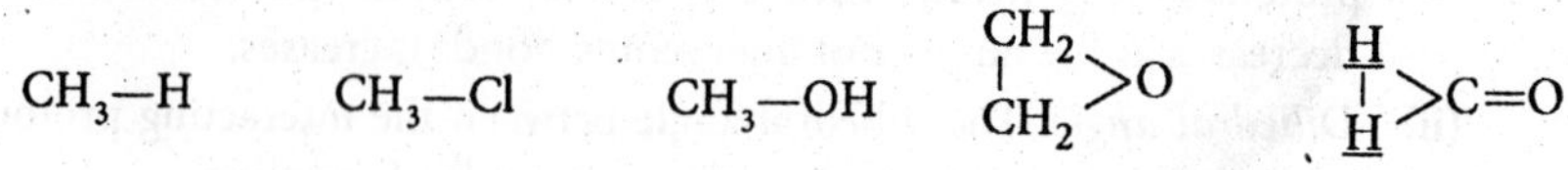

CH_3—H: −12.4 Hz; CH_3—Cl: −10.4 Hz; CH_3—OH: −10.4 Hz; ethylene oxide: +5.5 Hz; formaldehyde: +42.2 Hz

The influence of oxygen in ethylene oxide results in a positive coupling constant. The specially large positive coupling constant in formaldehyde is due to the fact that the non-bonded electron pair on oxygen atom has the correct symmetry for hyper conjugative donation. As a result, reinforcement of σ and π effects brings about the very large and positive coupling constant in formaldehyde.

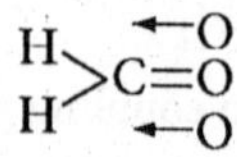

But if there is hyperconjugative withdrawal of electrons at the α-position, ^2J value is decreased as observed in toluene, and allene derivatives.

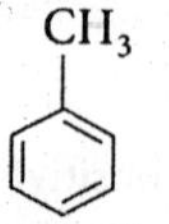

−14.4 Hz

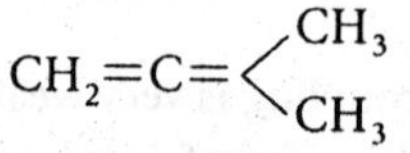

-9 Hz

(b) When electronegative substituent is present at β-position ^2J value decreases while the presence of electropositive substituent increases it. This effect is, however, mainly observed in π-bonded compounds. Thus,

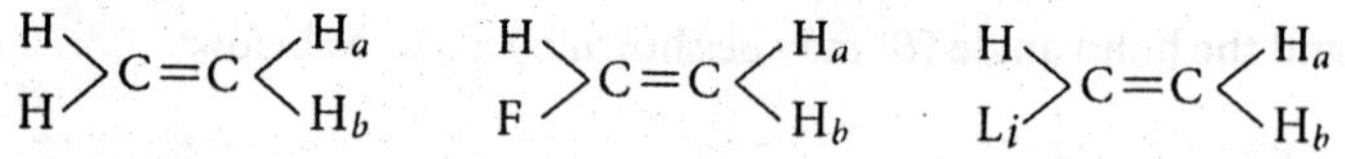

^2J: + 2.3 Hz; −3.8 Hz; +7.1 Hz

(ii) *Vicinal Coupling* (^3J): It is the most important, very strong coupling and is frequently observed if there are non-equivalent protons. This type of coupling occurs when the interacting protons are three bonds apart and is indicated by ^3J.

$$H_a \overset{1}{-} C \overset{2}{-} C \overset{3}{-} H_b$$

The ^3J values are postive and ranges from 0–20 Hz.

The most important factor which controls this coupling is the dihedral angle (θ) between the interacting protons.

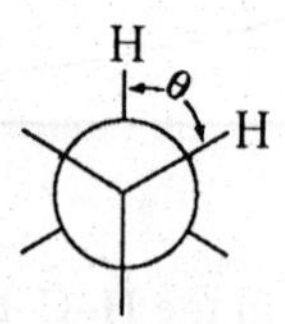

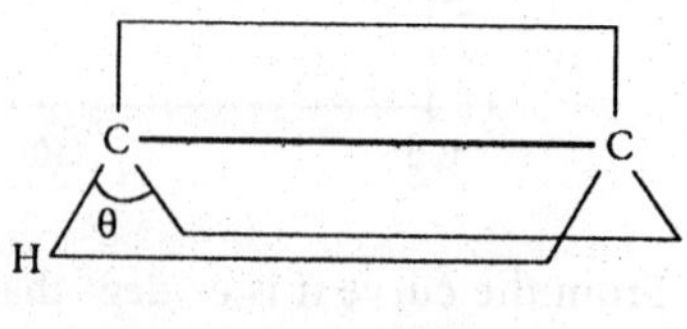

The relationship between ^3J and dihedral angle has been studied by Martin Karplus. He is the first worker to develop an equation known as Karplus equation which may be expressed as

$$^3J = A + B\cos\theta + C\cos\theta$$

where A = 7, B = −1, C = 5

The variation of vicinal coupling constant with dihedral angle is represented graphically as under. This graph is known as *karplus curve*.

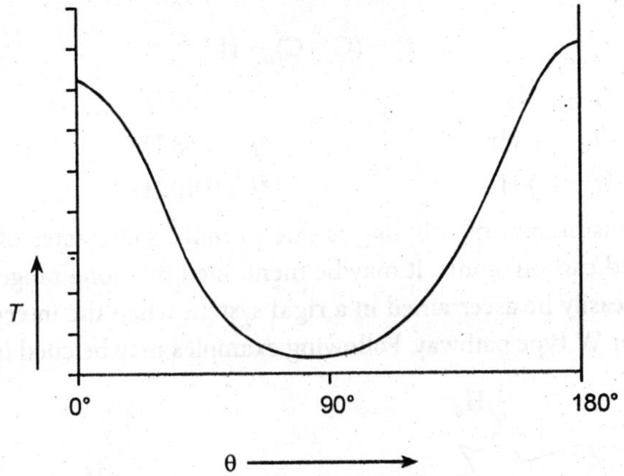

It has been observed that ^3J value is maximum when dihedral angle is either zero or 180° and becomes slightly negative when the dihedral angle is 90°. This is because at zero degree side-side overlap of two C—H bond orbitals is at maximum and minimum at 90°. Consequently karplus equation may be modified as

$^3J = 8.5 \cos\theta - 0.28$ when $0 \le \theta \le 90^\circ$

$^3J = 9 \cos\theta - 0.28$ when $90^\circ \le \theta \le 180^\circ$

Other factors that may have some role in determining the value of ^3J are

(a) Bond length of central C–C bond.

(b) Nature of the substituents

(c) Structure of the molecule.

It has been observed that ^3J value is linearly related with C--C bond length and it decreases when electron withdrawing substituents are present at the central carbon atoms. It has also been observed that (a) diaxial protons display ^3J 8–14 Hz and axial-equatorial or diequatorial protons exhibit ^3J 1–5 Hz (b) in olefinic system $^3J_{trans}$ is more than $^3J_{cis}$ (c) in the β-form of glucose, the anomeric proton possesses, in addition to higher shielding a large ^3J value than in the α-form.

(iii) *Long range coupling*: Protons separated by four or more bonds normally do not show any coupling. However, if there is multiple bond in between the interacting protons long range coupling generally occurs. The most important long range coupling is coupling between allylic protons, i.e.,

$$H_a \overset{4}{—} C \overset{3}{—} C \overset{2}{=} C \overset{1}{—} H_b$$

In this type of allylic coupling ^4J value is very low, normally ranges from 0.5–2.5 Hz. But when the allylic C—H bond is parallel to p_z atomic orbitals which form the π-bond, ^4J may be high as is evident from the following examples.

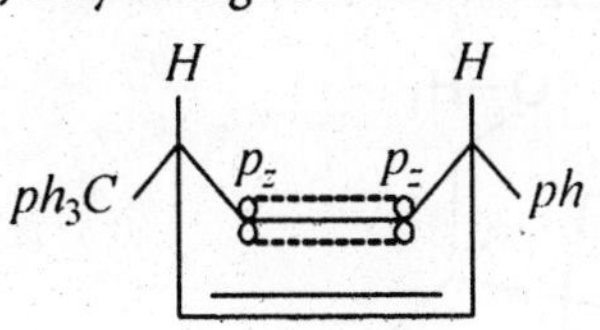

$^5J = 11.0$ Hz

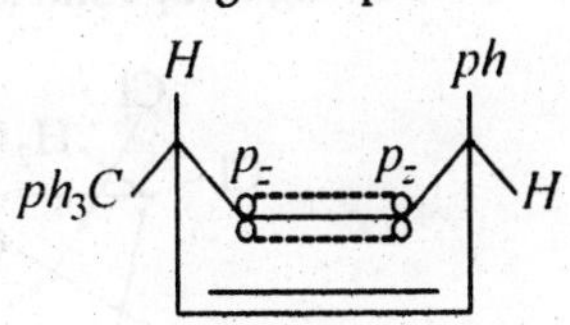

$^5J = 7.5$ Hz

It may be mentioned that if triple bond(s) are present in the intervening bonds, coupling may be transmitted over more number of bonds and even ^9J has been observed, Thus,

$$H-(C=C)_n-H$$

$^4J = 3$ Hz	$^7J = 1.3$ Hz
$^5J = 2.2$ Hz	$^8J = 0.68$ Hz
$^6J = 1.3$ Hz	$^9J = 0.40$ Hz

This transmission of coupling is due to more s-character of σ-bonds between *sp*-hybridised carbon atoms. It may be mentioned that long range coupling, particularly ^4J, can easily be ascertained in a rigid system when the intervening bonds form zig-zag, M or W type pathway. Following examples may be cited in this context.

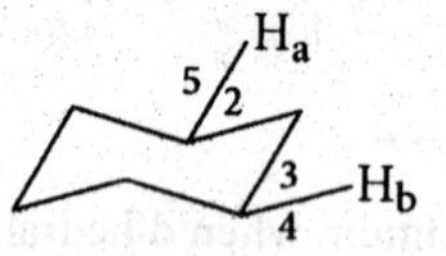

^4J= 1–2 Hz

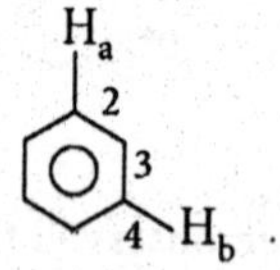

^4J= 2–3 Hz

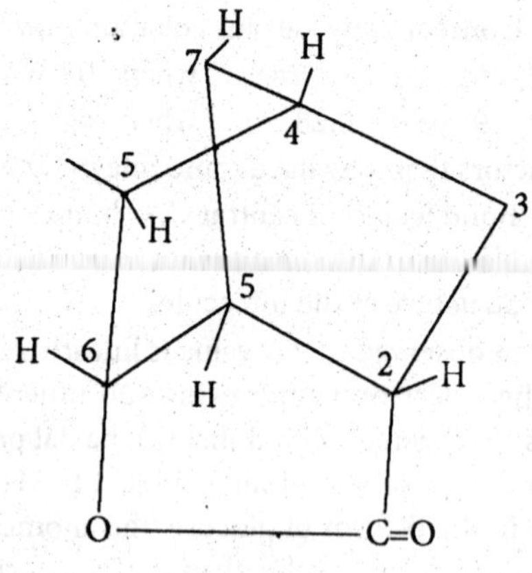

Hz	Path
$^4J_{1\text{-}4} = 1.6$	$H-C_1-C_7-C_4-H$
$^4J_{5\text{-}7} = 2.7$	$H-C_5-C_4-C_7-H$
$^4J_{2\text{-}6} = 1.00$	$H-C_2-C_1-C_6-H$
$^4J_{4\text{-}6} = 1.00$	$H-C_4-C_5-C_6-H$

In some cases spin-spin coupling appears to involve an unpaired electron pair that transmits the spin interaction through space to a rigidly held neighbouring nucleus without the influence of intervening bonds. For example, the coupling between H_1 and H_2 in the following compound is 1.1 Hz.

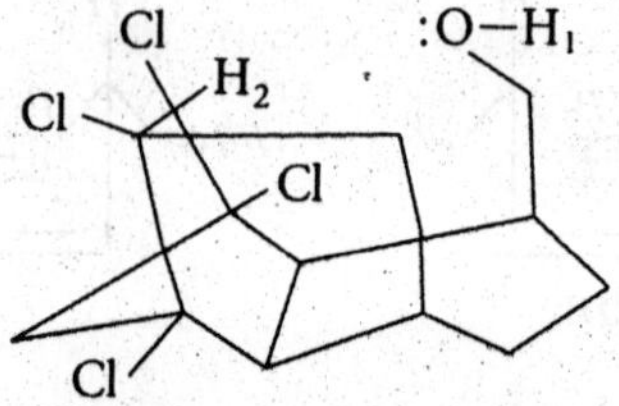

(iv) *Coupling in fused ring systems*: The coupling in fused ring system needs special mention. It is known that cyclohexane ring normally exists in chair conformation. In such conformation there are two bonds at each carbon atom—one is equatorial (e) and other one is axial (a). As conformationally these two bonds are different, the corresponding protons are non-equivalent. Hence it is expected that these two protons will have different chemical shifts and will couple to each other. The chemical shift difference between these two non-equivalent protons is about 0.5δ and the corresponding geminal coupling constant ^2J is about 4-5 Hz. The signals appear as broad band. Again, vicinal coupling may also occur. The values of such coupling constants are $^3J(a, a)$ = 10-14Hz. $^3J(e, e)$ = 4-5Hz and $^3J(a, e)$ = 4-4.5 Hz. However, such observation in ^{1}HNMR spectrum is possible for cyclohexane derivatives which are conformationally rigid. But in the case of unsubstituted cyclohexane such situation is not observed, only a single signal around 1.40δ appears. This is due to the interconversion of axial and equational hydrogen atoms by flippling of the ring as shown below:

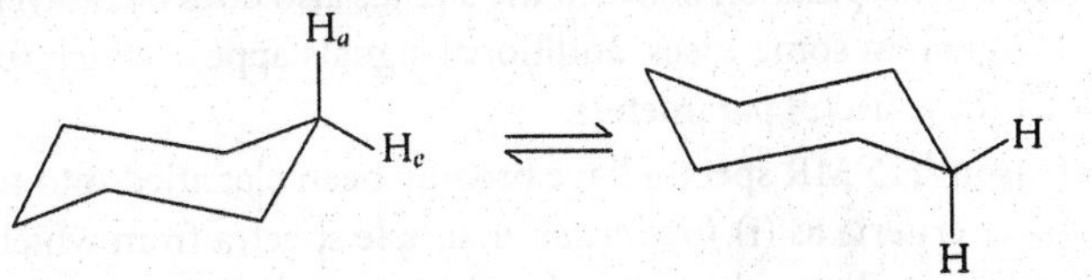

At room temperature the rate of flipping is very high as a result these two kinds of protons become indistinguishable.

On the otherhand, the fused ring system like decalin may be regarded as *a* 1, 2-disubstituted cyclohexane. So it is expected that in such case separate signals as well as J values may be obtained. The fusion of the two rings can occur in two ways. viz in *cis* or in *trans* manner as depicted below:

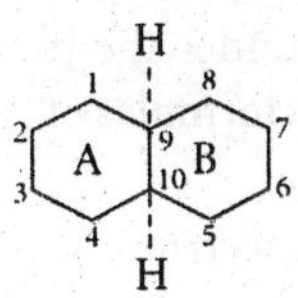

cis

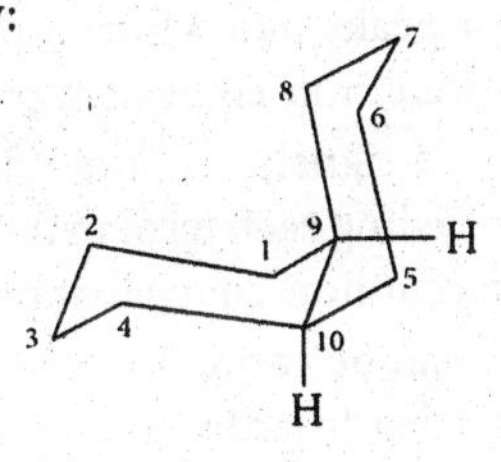

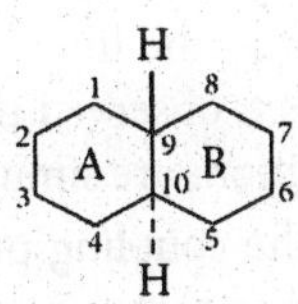

trans

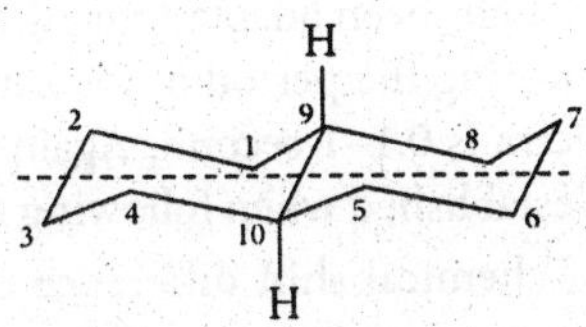

In *cis*-decalin two rings are fused through *a, e* (axial-equatorial) bonds and can easily undergo inversion by inter changing *a, e* bonds in the ring juncture i.e.

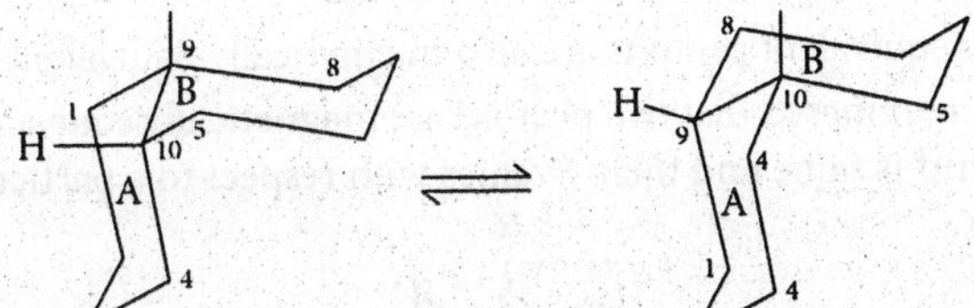

So in *cis*-decalin equational and axial protons can not be differentiated. But in trans-decalin two rings are fused through equational-equational (*e, e*) bonds and hence it has rigid structure. It can not undergo ring inversion because in such situation highly strained (*a*, *a*) ring fusion will occur.

32. What is meant by the term 'order of NMR spectrum'?

➜ If we look at any NMR spectrum three most fundamental, important and characteristic features of spectral pattern will reveal. These are chemical shifts (δ or τ), peak intensities and spin couplings (J). All these features are measurable quantities and are considered as the *first criteria* or *first order* of the pattern of the spectrum. Since these parameters are measurable quantities, it may appear that these parameters can be deduced in a simple and straight forward manner from NMR spectra. But a vast number of organic compounds give spectra from which these three useful parameters cannot be easily deduced. The reason for this difficulty is the complex or strong coupling between the interacting protons. In these cases, the centre (mid-point) of the cluster of signals does not correspond to the chemical shift and the separation between the signals also does not correspond to the coupling constant. Again, in some cases, additional signals appear which further complicate the deduction of these useful parameters.

For this reason ^{1}HNMR spectra have broadly been classified into two groups on the basis of first order criteria as (i) *first order* or simple spectra from which the three useful criteria (first order) can be easily deduced and (ii) *second order* or complex spectra from which the above useful parameters cannot be easily ascertained.

33. What are the essential criteria for first order spectrum?

➜ The essential criteria which determine the first order NMR spectrum may be formulated as below:

(i) The number of peaks into which a band split is equal to $(n + 1)$, where n is the number of adjacent non-equivalent protons causing the split.

(ii) The intensities of signals, i.e., the relative area under the peaks, are given by the coefficients preceding each term in the expansion formulas of $(x + 1)^n$ where n is the number of adjacent non-equivalent protons.

(iii) Chemical shift can be easily deduced from the spectrum. It may be mentioned in this connection that in order to ascertain these three parameters, the essential assumption that has been adapted that the coupling protons do not change their spin orientation during the period of recording spectra—the time usually taken for recording spectra is 0.1–1 second. Again, it has been observed that this assumption can only be established when following two conditions are simultaneously satisfied.

(a) The ratio of chemical shift difference between the coupling protons expressed in hertz and coupling constant (usually expressed in hertz) should be equal to or greater than 10, i.e.,

$$\frac{\Delta\delta \text{ in Hz}}{J \text{ in Hz}} \geq 10.$$

(b) Chemically equivalent protons are also magnetically equivalent.

It may be remembered that two protons are magnetically equivalent only when their chemical shift is same and their J values with respect to a particular proton are also same. Thus,

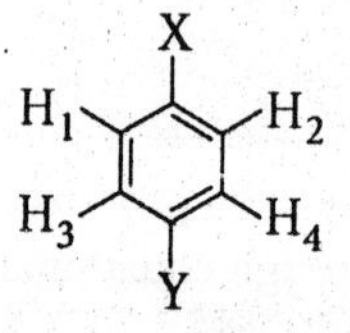

H_1 & H_2 and H_3 & H_4 are chemically equivalent. Let us consider the coupling of H_1 with H_3 and H_4. The two coupling constants, JH_1H_3 (ortho) and JH_1H_4 (para) are different. So H_1 and H_2 as well as H_3 and H_4 are magnetically non-equivalent.

34a. Applying first order criteria of NMR, determine for (a) CH_3—CH_2—Cl and (b) CH_3—CH_2—O—H (i) number of NMR peaks for CH_3 protons and ratio of their intensities. (ii) number of NMR peaks for CH_2 protons and ratio of their intensities. Sketch the complete ^{1}HNMR spectrum showing chemical shift and spin-spin splitting.

➞ (a) CH_3—CH_2—Cl

(i) ^{1}HNMR band of CH_3 in CH_3—CH_2—Cl

Number of adjacent non-equivalent protons (—CH_2—) which couple = 2. So the number of splitted peaks to be observed is 2 +1 = 3.

The ratio of peak areas will be $(1 + x)^2 = 1 + 2x + x^2$, i.e., 1 : 2 : 1.

(ii) ^{1}HNMR band of CH_2 in CH_3—CH_2—Cl

Number of adjacent non-equivalent protons =3. So, the number of splitted signals will be 3 + 1 = 4. The ratio of peak areas will be $(1 + x)^3 = 1 + 3x + 3x^2 + x^2$, i.e., 1 : 3 : 3 : 1. Hence the first order ^{1}HNMR spectrum of CH_3—CH_2—Cl may be drawn as below:

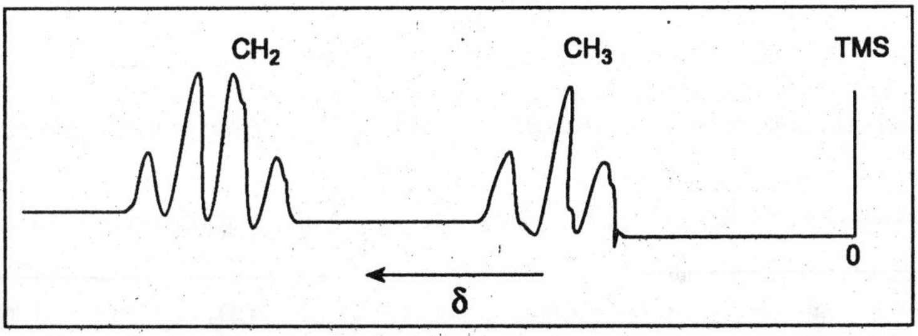

(b) CH_3— CH_2— O — H (Ethanol):

For CH_3, the number of splitted peaks will 2 + 1 = 3 and for CH_2, the number of splitted signals will be 3 + 1 = 4. The corresponding peak area ratio will be $(1 + x)^2 = 1 + 2x + x^2$, i.e. 1 : 2 : 1 for CH_3 and $(1 + x)^3 = 1 + 3x + 3x^2 + x^3$ i.e., 1 : 3 : 3 : 1 for CH_2 protons.

For OH, theoretically there should be splitting but in fact, the single peak of OH proton remains unsplit due to the presence of traces of water in ethanol the proton of which rapidly exchanges with protons of OH. Thus the nature of the spectum may be as shown below:

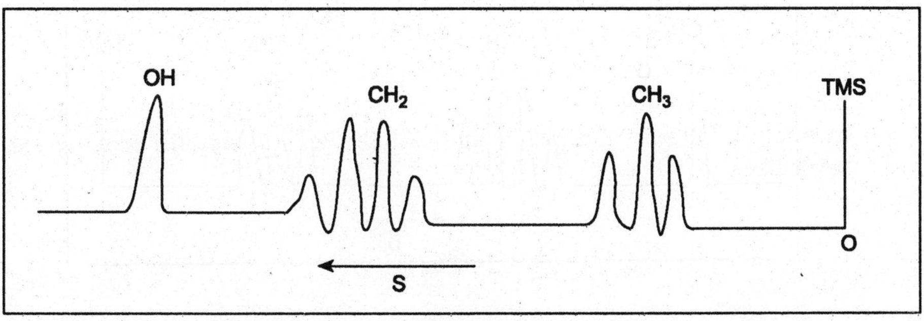

34b. Sketch the first order spectra of methanol, ether, ethylacetate, toluene and ethanol.

→ CH_3–O–H (Methanol)

CH_3 signal:

No. of non-equivalent adjacent proton = I

So the number of splitted peaks to be observed = 1+1 = 2

The peak areas will be $(1 + x)^1 = 1 + x$, i.e, 1 : 1

O—H signal:

No. of non-equivalent protons = 3

So number of splitted peaks to be observed = 3 +1 = 4

The ratio peak areas will be $(1 + x)^3 = 1 + 3x + 3x^2 + x^3$, i.e., 1 : 3 : 3 : 1

Thus the sketch of the spectrum wil be as under:

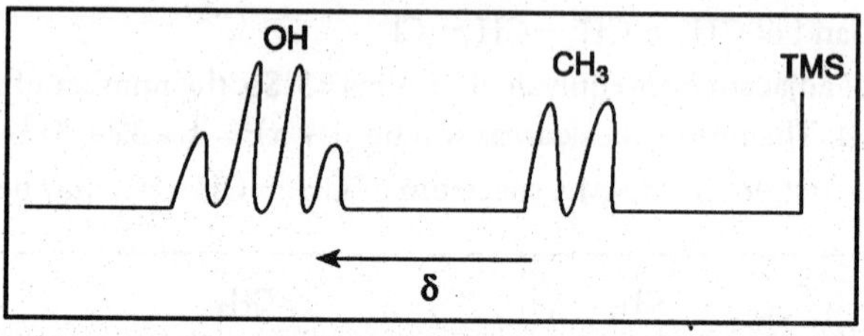

Similarly the nature of ^{1}HNMR spectra of ether, ethylacetate, toluene and ethanol may be drawn as below:

For ether, $CH_3—CH_2—O—CH_2—CH_3$, the nature of the spectrum will be as under:

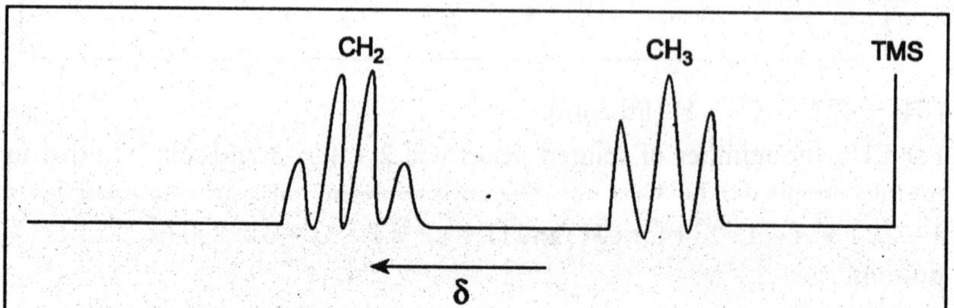

For ethyl acetate, $CH_3—\underset{\underset{O}{\|}}{C}—O—CH_2—CH_3$, the spectrum may be sketched as below:

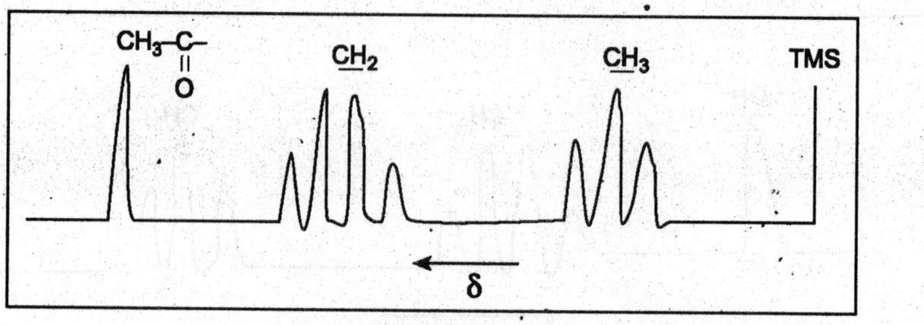

^{1}HNMR spectrum of $C_6H_5—CH_3$ (Toluene) may be drawn as below:

In toluene, there are two types of protons — aromatic and methyl. The nature of the ^{1}HNMR spectrum is as below:

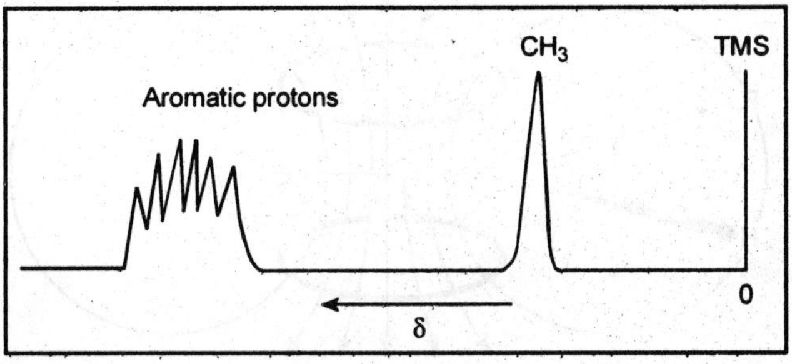

For ethanol, $CH_3—CH_2—OH$ spectrum may be as below:

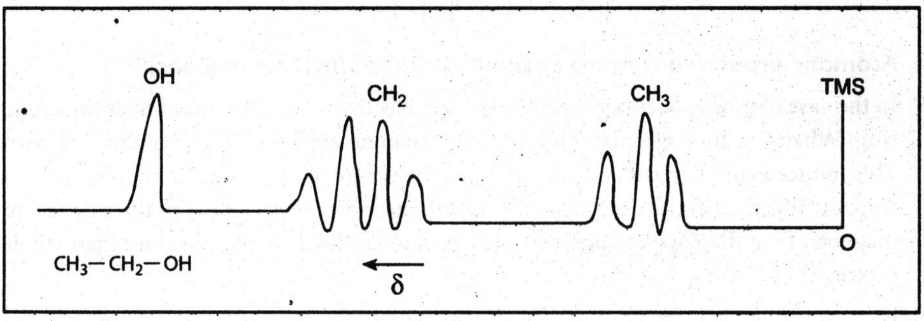

35. Explain:

(a) Acetylenic protons appear at higher field than ethylenic protons in PMR spectrum.

➜ In external magnetic field ethylene molecule is so oriented that the plane of its double bond is at right angle to the applied field and so the induced magnetic field developed due to the circulation of π-electrons is paramagnetic in the region of protons.

But acetylene molecule when placed in an external magnetic field its plane of triple bond lies parallel to the direction of the applied field i.e., the molecule assumes a vertical orientation and the induced field originated due to circulation of the π-electrons of the tripple bond opposes the applied magnetic field in the region of protons. As a result protons experience less field strength than applied and so resonance occurs at higher field (lower δ value).

(b) Aldehydic proton appears at much lower field in PMR spectrum.

➜ Aldehydic proton is highly deshielded because it lives in the deshielding zone of carbonyl group as shown below:

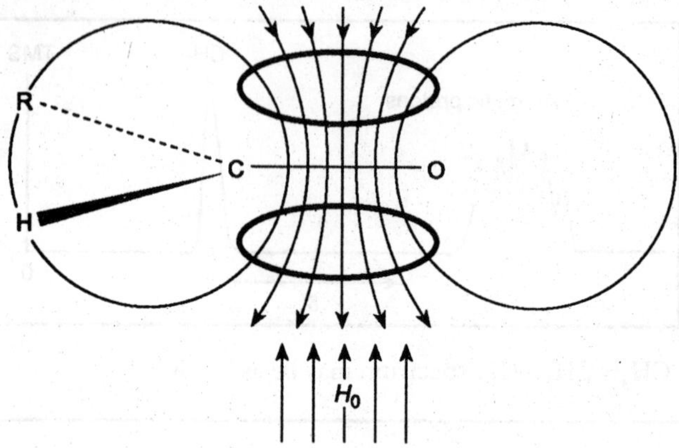

(c) Aromatic protons are more deshielded than ethylenic protons.

➜ In the case of aromatic compounds loops of π-electrons are delocalised over the aromatic ring. When this loop circulates in the presence of an applied field produces ring current. This induced ring current is diamagnetic at the centre of the aromatic ring but paramagnetic outside the ring. So, the aromatic protons around the periphery of the ring feel more magnetic field than applied and so resonate at lower field (higher δ value) than ethylenic protons.

(d) Broad signals are often observed in PMR spectra associated with OH and NH resonances.

➜ Relaxation times of these types of protons are frequently shortened by chemical exchange processes.

(e) ^{1}HNMR spectrum of dimethyl formamide shows two signals at δ 2.84 and 3.0 for methyl protons at room temperature but a single sharp line appears at high temperature.

➜ At room temperature, the conjugation between the lone pair of electrons on nitrogen atom and amide carbonyl group increases sufficiently the double bond character of C–N bond to restrict rotation and two different methyl (*cis* and *trans*) signals appear. But at high temperature, the rotation around C — N bond is so rapid that each methyl group experiences the same time-averaged environment.

(f) *cis*-Decalin shows a single resonance line but *trans*-decalin gives a broad band in their ^{1}HNMR spectrum.

➜ The *trans* has a relatively rigid framework and so all the hydrogen atoms are either axial or equatorial to the rings. The broad band in *trans* isomer is due to the presence of these two types of protons. But *cis*-decalin is a more flexible system, one set of chair form can easily be converted into alternate set of chair form by ring transformation. So each axial hydrogen becomes equatorial and vice versa. When this 'flipping-over' is sufficiently rapid any particular proton is at any moment neither axial nor equatorial and hence a single resonance line results.

(g) In 2-Methyl propene (I) splitting is observed but not in neopentyl chloride (II).

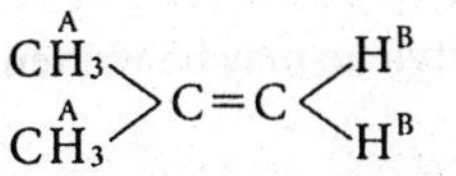

I

$(\overset{A}{CH_3})_3-C-\overset{B}{CH_2}-Cl$

II

➜ In (I) protons H_A and H_B are close enough (allylic) to couple each other and hence splitting occurs. But in (II) H_A and H_B are not adjacent and far away to interact. So no splitting is observed.

(h) The monoterpene, α-pinene has three methyl groups which exhibit signal at different positions in ¹HNMR spectrum.

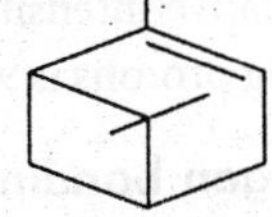

α-Pinene

➜ Three methyl groups in α-pinene may be marked as 'a', 'b', and 'c'.

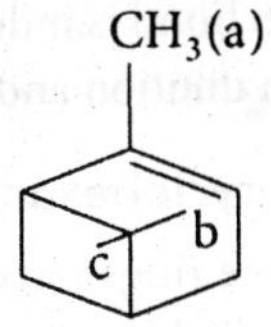

The CH_3(a) is attached to a π-bond and is deshielded and appears at a low field around 1.80 δ. Out of geminal methyl groups, CH_3 group marked 'b' lies in the shielding zone of double bond and appears at higher field (~0.90 δ).

(i) Methyl protons of I-methyl cyclohexene appear at higher field (δ1.70) than methyl protons of toluene (δ 2.30).

➜ The structural formulae of I-methyl cyclohexene may be written as shown below:

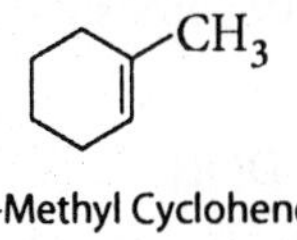

I-Methyl Cyclohene

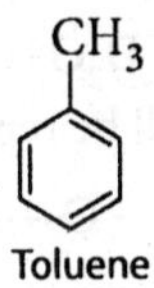

Toluene

The deshielding effect induced by ring current in aromatic ring extends beyond hydrogen attached directly to the aromatic ring and so methyl protons of toluene appear at lower field than that of I-methyl cyclohexene.

(j) The unit of chemical shift is dimensionless but that of coupling has dimension.

➜ Coupling constant is a molecular property and depends only on the structure of the molecule. It is independent of the strength of the applied field. So it can be safely expressed in the units of hertz. But chemical shift depends upon the strength of the applied magnetic field. So it cannot be expressed by units with dimension.

36. How will you distinguish the following from ¹HNMR spectra?

(a) Primary, Secondary and Tertiary alcohols

➜ The three types of alcohols can be differentiated by taking the ¹HNMR spectra in DMSO-d_6 when proton of OH group appears as a triplet, doublet and singlet respectively in primary, secondary and tertiary alcohols.

(b) p-Xylene and mesitylene

➜ The structural formualae of *p*-Xylene and Mesitylene may be written as shown below:

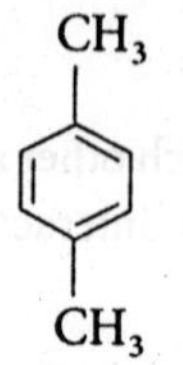

p-Xylene

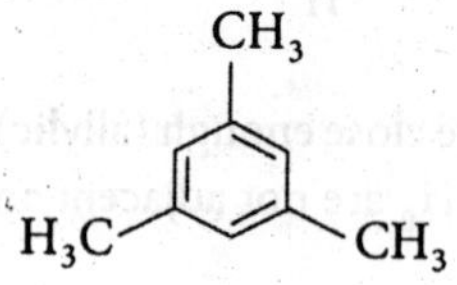

Mesitylene

Although both the compounds will display two singlet one for aromatic protons and other for methyl protons but the relative intensities of peaks will be different. In mesitylene, ratio of methyl protons and ring protons is 9:3 or 3:1 while in *p*-Xylene it is 6:4 = 3:2.

(c) Inter-and intramolecular hydrogen bonding.

➜ Intermolecular hydrogen bonding is temprature and concentration dependent. So this type of hydrogen bonding will change on changing temperature and concentration. On the other hand, intramolecular hydrogen bond is independent of concentration and temperature. So it will remain unchanged on dilution and changing temperature.

(d) Equatorial and axial protons of cyclohexane ring.

➜ The equatorial protons in cyclohexane rings resonate at about 0.5 δ higher than the axial protons. This is due to anisotropic deshielding by σ-electrons in βγ-bonds as shown below:

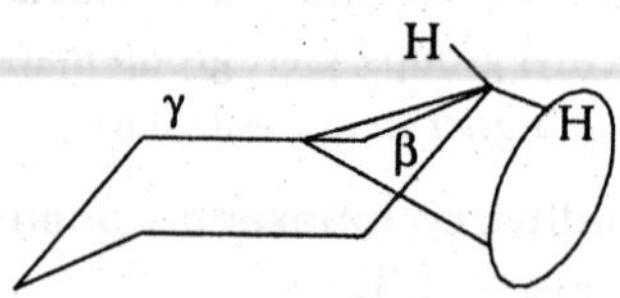

(e) Proton of OH and other proton in any compound.

➜ Excess amount of D_2O is added to the solution, shaken vigorously and spectrum is retaken. Proton of OH will be replaced by deuterium.

$$R—O—H + D_2O \rightleftharpoons R—O—D + DOH$$

So the original proton peak of OH will disappear and a new signal for DOH appears.

(f) Cyclohexanol and chlorocyclohexane.

➜ Structural formulae of cyclohexanol and chlorocyclohexane may be written as

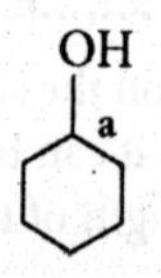

Cyclohexanol

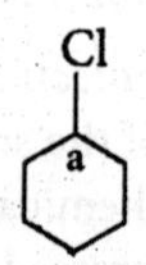

Chlorocyclohexane

Chemical shifts of 'a' protons in both is practically same. Again, D_2O exchange will not be reliable in this case. But can be differentiated by acetylation with $(CF_3CO)_2O$ when only cyclohexanol will undergo acetylation.

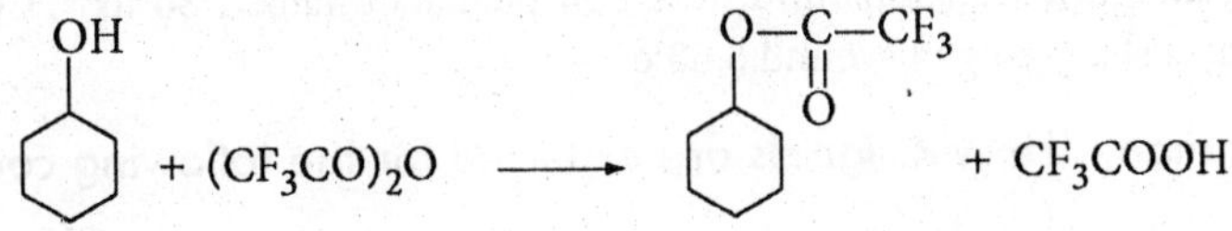

Now the spectrum will be different and hence may be differentiated

(g) *trans*-Stilbene and *cis*-Stilbene.

➜ Let us look at the structures of this two isomers

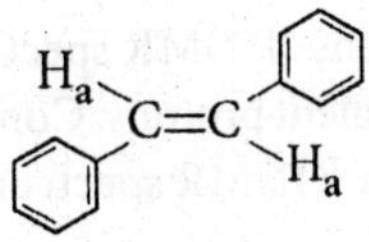

trans-Stilbene

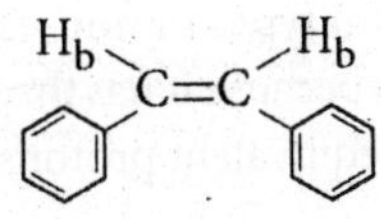

cis-Stilbene

In *trans* isomers two protons (H_a) are deshielded by both the aromatic rings but in *cis* isomers each hydrogen (H_b) is deshielded by only one aromatic ring. So H_a of *trans* will appear at lower field (~ 7.00 δ) than H_b of cis (~ 6.50 δ).

(h) *p*-dichlorobenze and *o*-dichlorobenzene.

➜

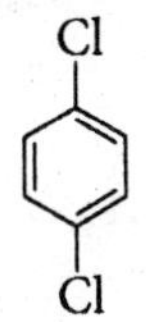

p-Dichlorobenzene

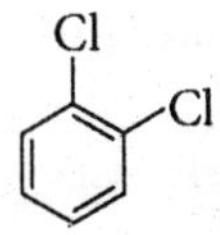

o-Dichlorobenzene

In *p*-dichlorobenzene all the four hydrogen atoms are equivalent. So only one peak will appear in ¹HNMR spectrum whereas in *o*-dichlorobenzene, there are two types of equivalent protons. Hence, in its spectrum two signals will be observed.

(i)

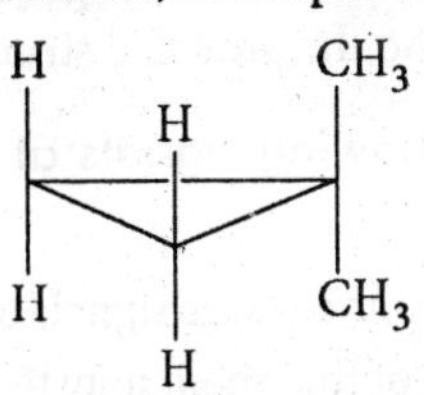

(A)

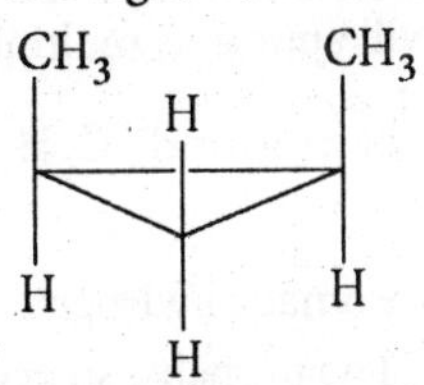

(B)

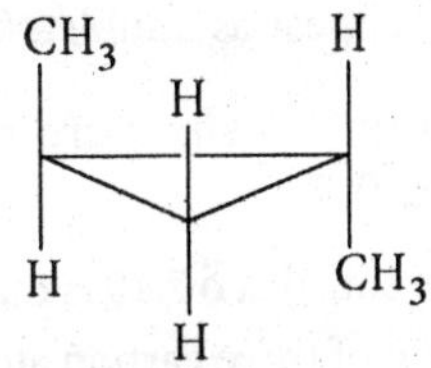

(C)

In (A) six protons of two methyl groups are equivalent and give rise to quartet. But in (B) and (C) two methyl groups are attached to two different carbon atoms. Hence, the methyl protons of each methyl appear as two separate doublets. Again, in (B) two methyl groups are *cis* while in (C) they are *trans* to each other. Hence, *J* value in (C) will be higher than that of (B).

(i) Benzene and Cyclo-octatetraene.

➜

Benzene

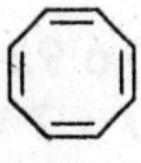

Cyclo-octatetraene

Benzene is an aromatic compound and all its six hydrogen atoms are equivalent. So it will show a singlet in the aromatic region ~ 8.00 δ. On the other hand, cyclo-octatetraene is not an aromatic compound. It is non-planar in nature. So its ^{1}HNMR spectrum will display a sharp singlet around 5.60 δ.

37. How many ^{1}HNMR signals are expected for the following compounds.

$CH_3{-}CH_3$ (I) $\quad CH_3{-}C({=}O){-}H$ (II) $\quad CH_3{-}CH_2{-}CH_2{-}Br$ (III) $\quad CH_3{-}CH(Cl){-}C({=}O){-}CH_3$ (IV)

➜ (I) will show only one signal as all the hydrogen atoms are equivalent while (II) will exhibit two since it has two types of non-equivalent protons. The ^{1}HNMR spectrum of (III) will show three signals because it has three sets of non-equivalent protons. Compound (IV) has three sets of non-equivalent protons. So three signals in ^{1}HNMR spectrum are expected.

38. (i) An organic compound having molecular formula C_3H_7Cl exhibits following signals in ^{1}HNMR spectrum.

(i) a triplet at 0.9 δ

(ii) another triplet at downfield than previous one

(iii) a multiplet.

Suggest the probable structure.

➜ The possible structures are

$\overset{C}{CH_3}{-}\overset{B}{CH_2}{-}\overset{A}{CH_2}{-}Cl$ (I) $\qquad \overset{A}{CH_3}{-}\overset{B}{CH}(Cl){-}\overset{A}{CH_3}$ (II)

In (I) CH_3 (C) and CH_2(A) protons are split into individual triplet by two protons of CH_2(B). But (A) proton is more downfield. CH_2 protons marked (B) are split both by (A) and (C) protons to give multiplet. But in (II) protons of two methyl groups are equivalent and appear as doublet. (B) will appear as multiplet. So the probable structure is (I).

(ii) Suggest a structure of a compound, C_9H_{12} showing signals at δ 7.1, 2.2, 1.5 and 0.9.

➜ The singal at δ 7.10 is due to aromatic hydrogen. The other three signals are due to an alkyl chain of three carbon atoms. The probable structure of this chain may be as shown below: $CH_3{-}CH_2{-}CH_2{-}$

Thus, the structure of the compound is 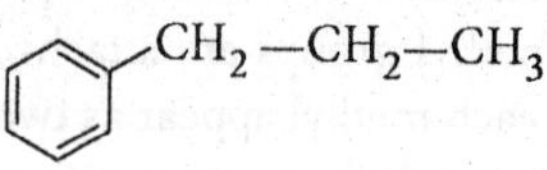

(iii) ^{1}HNMR spectrum of ethyl crotonate, $CH_3{-}CH{=}CH{-}C({=}O){-}CH_2CH_3$ exhibits the following signals, (i) 6.95 δ (*dq; J* = 16, 6.8 Hz, 1H); (ii) 5.81 δ (*dq, J* = 16, 1.7 Hz, 1H); 4.13 δ (*q; J* = 7 Hz, 2H); 1.88 δ (*dd, J* = 6.8, 1.7 Hz, 3H) and 1.24 (*t; J* = 7 Hz, 3H). Assign these signals to various hydrogen and tell if the double bond is *cis* and *trans* substituted.

➜ 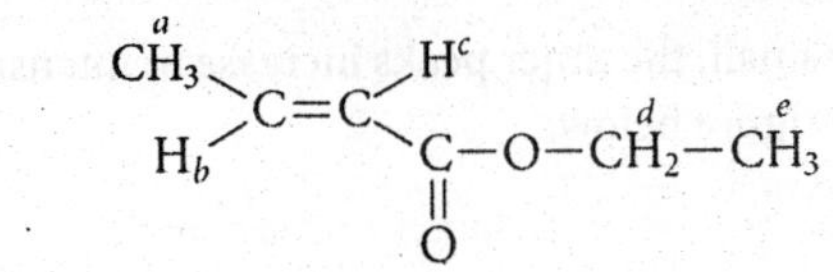

$a = 1.88\ \delta$, $b = 6.95\ \delta$, $c = 5.81\ \delta$, $d = 4.13\ \delta$ and $e = 1.24\ \delta$.

The double bond has *trans* geometry as is evident from $J = 16.0$ Hz.

39. What is meant by second order or non-first order NMR spectra?

➜ We know that the NMR spectra from which three essential characteristics, viz., chemical shifts, coupling and intensities of signals can be easily deduced are termed as *first order NMR spectra*. It has been observed that deduction of these characteristics is only possible when $\frac{\Delta v}{J} \geq 10$ (where Δv is chemical shift difference of interacting protons in Hz and J is coupling constant in Hz) and chemically equivalent protons are also magnetically equivalent. But there are a vast number of organic compounds whose spectra are highly complicated and the above essential features cannot be easily ascertained from the spectra. A careful scrutiny of these complicated spectra reveals that the distortion, i.e., complication primarily originates from the non-equivalency of chemically and magnetically equivalent protons. As a result strong coupling between the interacting protons occurs and the centre of the group of signals does not correspond to chemical shift nor the separation between the signals corresponds to coupling constant. Also in some cases, additional signals appear complicating further the deduction of essential features. These spectra are termed as second order or non-first order NMR spectra.

40. How would you analyse second order NMR spectra?

➜ In order to analyse the complex NMR spectra, coupled protons are classified into sets of spin-systems on the basis of chemical shift difference and coupling constant values and the sets are designated by alphabets. Thus sets of protons separated by a small chemical shift are assigned letters adjacent in the alphabet such as AB, ABC, etc. The sets of protons widely differing in chemical shift are designated by letters that are well separated in alphabet leading to the systems like AX, AMX and so on. But when the coupled protons are chemically equivalent but magnetically non-equivalent they are differentiated by primes (AA′, BB′, etc).When the chemical shift differences approach the magnitude of coupling constant, then the letters adjacent in the alphabet are used. Thus, AX system becomes AB and AMX system becomes ABC and so on. Among different types of spin-systems, ^{1}HNMR spectra involving AB and ABC spin-systems are complex and cannot be easily analysed using first order rules. The analysis of AB and ABC spin-systems are discussed below:

40.1 Two-proton spin-system (AB-system)

In this system, there are two mutually coupled protons A and B. So, if AB would have been AX system, then a pair of well separated doublets (i.e., four signals) could be expected as shown in the figure below:

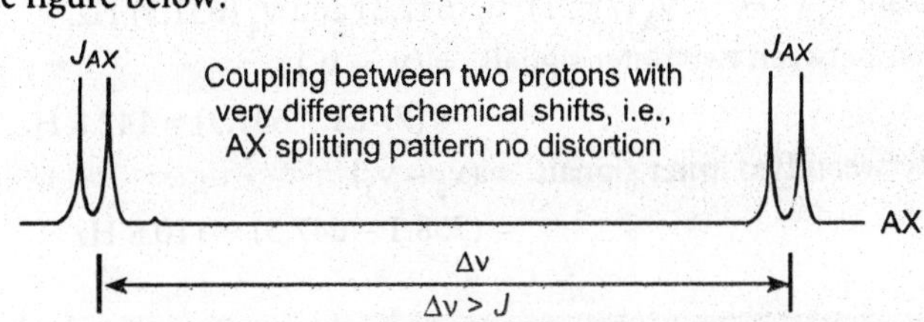

But as the chemical shift difference is very small, the inner peaks increase in intensity and outer ones diminish as is evident from the figure below:

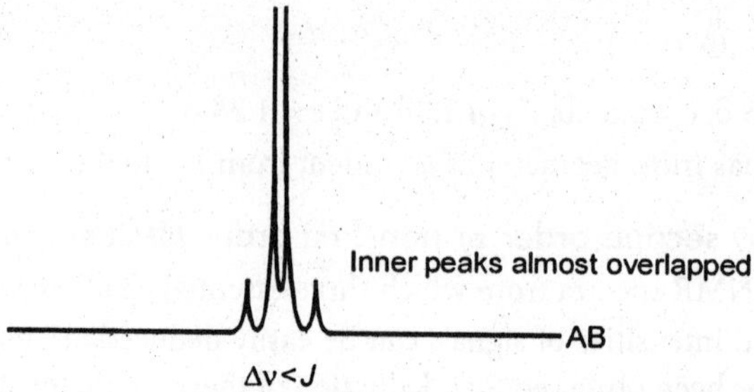

Splitting patterns

So the chemical shift positions are no longer found at middle points of the doublet but lie around the centre of gravity of the doublets. Therefore, it is apparent that deduction of chemical shifts and coupling constant is extremenly difficult. However, those parameters have been deduced from mathematical calculations. For such calculation four signals are marked as 1, 2, 3, 4. From these four signals chemical shifts of A (δ_A), B (δ_B) and coupling constant (J_{AB}) may be ascertained.

$$\delta_A = x + \frac{1}{2}\sqrt{L \times S}$$

$$\delta_B = x - \frac{1}{2}\sqrt{L \times S}$$

$$J_{AB} = (\nu_1 - \nu_2) = (\nu_2 - \nu_4)$$

where x is the middle point of these four signals which is calculated as

$$\frac{1}{2}(\nu_1 + \nu_4) = \frac{1}{2}(\nu_2 + \nu_3)$$

$$L = (\nu_1 - \nu_4) \text{ and } S = (\nu_2 - \nu_3)$$

ν_1, ν_2, ν_3 and ν_3 are the frequencies of four signals, L = separation between two outer signals in Hz and S = separation between two inner signals in Hz. Groups like $-CH_3-$,

$-CH{=}CH-$, 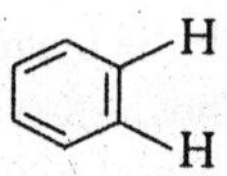 and 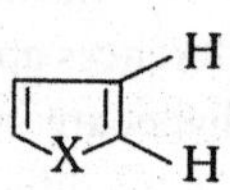 usually give AB spectrum.

40.1.1 Methyl *trans* cinnamate displays three groups of signals which can be assigned to CH_3, C_6H_5 and $CH{=}CH$ protons. The last two protons ($-CH{=}CH-$) show typical AB pattern and the positions of signals are at 774.1, 758.1, 647.3 and 631.3 Hz from TMS in 100 MHz spectrum. Calculate δ_A, δ_B and J_{AB}.

➜ The four signals are ν_1 (774.1), ν_2 (758.1), ν_3 (647.3) and ν_4 (631.3) Hz.

So, L = spearation between two outer signals $= (\nu_1 - \nu_4)$

$= (774.1 - 631.3) = 142.8$ Hz.

S = Separation between two inner signals $= (\nu_2 - \nu_3)$

$= (758.1 - 647.3) = 110.8$ Hz

$$\text{Mid-point } (x) = \frac{1}{2}(\nu_1 + \nu_4) = \frac{(774.1 + 631.3)}{2} = 702.7 \text{ Hz}$$

$$\text{or, } \frac{1}{2}(\nu_2 + \nu_3) = \frac{(758.1 + 747.3)}{2} = 702.7 \text{ Hz}$$

$$\text{Thus, } \delta_A = 702.7 + \frac{1}{2}\sqrt{142.8 \times 110.8} \text{ Hz}$$

$$= 702.7 + \frac{125.8}{2} = 765.68 \text{ Hz}$$

So, in 100 MHz, it is 7.656 ppm or δ.

$$\delta_B = 702.7 - \frac{1}{2}\sqrt{142.8 \times 110.8} \text{ Hz}$$

$$= 702.7 - \frac{125.8}{2} = 639.8 \text{ Hz} .$$

So, in 100 MHz it is 6.398 ppm or δ.

$J_{AB} = (\nu_1 - \nu_2) = (\nu_3 - \nu_4) = 774.1 - 758.1 = 16$ Hz.

or, (647.3 – 631.3) = 16 Hz

Thus, $\delta_A = 7.657\ \delta$

$\delta_B = 6.398\ \delta$

$J_{AB} = 16$ Hz.

40.1.2. Organic compound, ascaridole exhibits an *AB* quartet for vinyl protons at 395.5, 386.9, 385.5 an 376.9 Hz from TMS. Calculate δ_A, δ_B and J_{AB}.

→ $L = 395.5\ (\nu_1) - 376.9\ (\nu_4) = 18.6$ Hz.

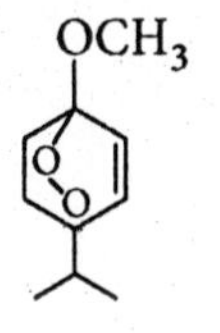

$S = 386.9\ (\nu_2) - 385.5\ (\nu_3) = 1.4$ Hz.

$x = \frac{1}{2}(395.5 + 376.9) = 386.2$ Hz.

or, $\frac{1}{2}(386.9 + 385.5) = 386.2$ Hz.

$\therefore\ \delta_A = 386.2 + \frac{1}{2}\sqrt{18.6 \times 1.4} = 386.2 + \frac{5.1}{2} = 388.75$ Hz.

$\delta_B = 386 - \frac{1}{2}(5.1) = 383.45$ Hz

$J_{AB} = 395.5 - 386.9 = 8.6$ Hz

or, 385.5 – 376.9 = 8.6 Hz.

Thus, $\delta_A = 388.75$ Hz

$\delta_B = 383.45$ Hz

$J_{AB} = 8.6$ Hz.

40.2 *Three-proton spin-systems*:

Complicated spectra of three-proton spin-systems may be of the types A_2B or AB_2, *ABX* and *ABC*. Out of these types two groups of equivalent protons (A_2B or AB_2) and *ABX* type of three non-equivalent protons are most important. Calculations of chemical shifts and coupling constants are discussed in sequel.

40.2.1 *A_2B or AB_2 system:*

If this spin-system be considered as first order, then five signals—two for two equivalent protons A_2 or B_2 and a triplet for other proton A or B will be observed as is evident from the figure below.

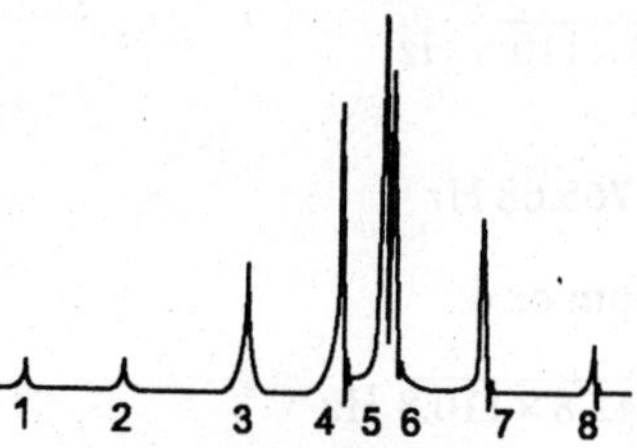

an AB_2 spectrum

But due to strong coupling, $\alpha\beta \neq \beta\alpha$, the triplet is split into four signals and the doublet is also split into four signals giving rise to a total of eight signals. Theoretically, an additional peak, the ninth is somethimes observed. The nature of the spectrum may be as seen in the figure shown below. (It may be mentioned that aromatic protons of some compounds like pyrogallol displays the pattern of an AB_2 spectrum.

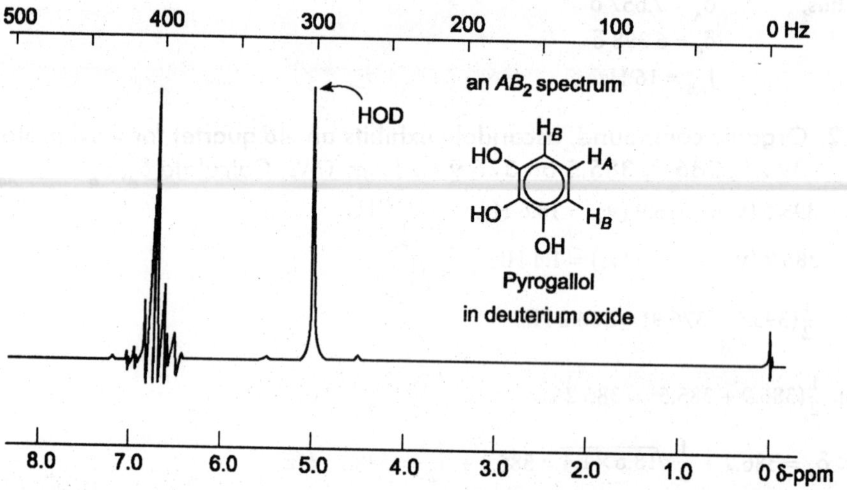

The eight signals are marked as 1, 2, 3, 4, 5, 6, 7 and 8 for the purpose of calculation. It has been observed that

(i) position of 3rd signal is the chemical shift of A, i.e., δ_A (which is present as single peak, i.e. system is AB_2)

(ii) Chemical shift of B proton, δ_B (system AB_2) is the centre of 5th and 7th signals

and (iii) $J_{AB} = \frac{1}{3}[(\nu_4 - \nu_1) + (\nu_8 - \nu_6)]$.

(Peak should be numbered in order of increasing frequency, where ν indicates frequency in hertz).

40.2.1.1 2, 6–dichlorotoluene gives eight signals at 725.20, 723.05, 716.7, 715.9, 703.75, 697.45, 694.5 and 688.1 Hz from TMS. The last four values correspond to one proton. Calculate chemical shift values and *J* values and assign them.

➜ The structural formula of 2,6–dichlorotoluene may be expressed as

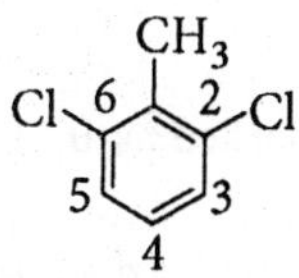

The above signals are for three ring protons (H_3, H_4 & H_5). As there are eight signals, it is most probably AB_2 system. The eight signals may be numbered as

Signal No.	8	7	6	5	4	3	2	1
Signal position ν in Hz	725.20	723.05	716.7	715.9	703.75	697.45	694.5	688.1

Now according to observed calculation, the position of the 3rd signal is the position of the single proton, i.e., for AB_2. Chemical shift of A, i.e., $\delta_A = 697.45$ Hz = 6.97 δ (in 100 MHz).

Now chemical shift of other proton B (AB_2) is the mid-point of 5th and 7th signal, i.e.,

$$\frac{1}{2}(715.9 + 723.05) = 719.48 \text{ Hz} = 7.19\ \delta \text{ (in 100 MHz)}$$

$$J_{AB} = \frac{1}{3}[(\nu_4 - \nu_1) + (\nu_8 - \nu_6)]$$

$$= \frac{1}{3}[(703.75 - 688.1) + (725.2 - 716.7)]$$

$$= \frac{1}{3}(15.65 + 8.5) = \frac{24.15}{3} = 8.05 \text{ Hz}$$

Thus the single proton is H_4, and H_3 and H_5 are two equivalent protons.

40.2.2 *ABX system*:

The important aspect of *ABX* system is the strong coupling between *A* and *B* and weak coupling between *X* and *A* and *B*. The chemical shifts of *A* and *B* are very close to each other. Otherwise the coupling between them would have been first order. The chemical shift of *X* is far away from those of *A* and *B* and may be on upfield or downfield of them. The signals appear in two separate groups, one for *AB* and other for *X*-protons. Let us consider these two parts separately.

40.2.2.1 *AB part of ABX spectrum*:

In this part eight signals are observed which appear as two sets of quartet. Out of eight signals, four are due to the coupling of two spin states of *A* with two spin states of *X* and other four signals originate from the coupling of two spin states of *B* with spin states of *X* proton. These eight signals appear as two sets. One set is due to one type spin state of *X* (say, α) with both the spin states (α and β) of *A* and *B* protons. Other set consists of signals due to another spin state (say, β) of *X* with both the spin states (α and β) of *A* and *B* protons; i.e. eight signals are as below:

One set (X in α orientation) (I)	Other set (X in β orientation) (II)
$A_\alpha X_\alpha = \theta_1$	$A_\alpha X_\beta = R_1$
$A_\beta X_\alpha = \theta_2$	$A_\beta X_\beta = R_2$
$B_\alpha X_\alpha = \theta_3$	$B_\alpha X_\beta = R_3$
$B_\beta X_\alpha = \theta_4$	$B_\beta X_\beta = R_4$

The four signals of set (I) are marked as θ_1, θ_2, θ_3 and θ_4 while those of set (II) are marked as R_1, R_2, R_3 and R_4.

Now $J_{AB} = (R_1 - R_2) = (R_3 - R_4) = (\theta_1 - \theta_2) = (\theta_3 - \theta_4)$

The positions of AX_β and BX_β signals are obtained as in the case of analysis of AB spin system. Thus,

$$AX_\beta = x + \frac{1}{2}\sqrt{L \times S} \qquad \text{(I)}$$

$$BX_\beta = x - \frac{1}{2}\sqrt{L \times S} \qquad \text{(III)}$$

where $x = \text{mid-point} = \frac{1}{2}(R_1 + R_4)$

$L = R_1 - R_4$ and $S = R_2 - R_3$

Simialrly, the positions of AX_α and BX_α signals are obtained as

$$AX_\alpha = x + \frac{1}{2}\sqrt{L \times S} \qquad \text{(II)}$$

$$BX_\alpha = x + \frac{1}{2}\sqrt{LXS} \qquad \text{(IV)}$$

where $x = \text{mid-point} = \frac{1}{2}(\theta_1 + \theta_4)$

$L = (\theta_1 - \theta_4)$ and $S = (\theta_2 - \theta_3)$

The signals (I), (II), (III) and (IV) are due to coupling of X with A and with B, i.e., A gives I and II and B gives III and IV.

$$\delta_A = \frac{1}{2}(\text{I} + \text{II}) \text{ and } \delta_B = \frac{1}{2}(\text{III} + \text{IV})$$

$J_{AX} = (\text{I} - \text{II})$ and $J_{BX} = (\text{III} - \text{IV})$.

So from AB part of ABX spectrum following five parameters can be obtained. δ_A, δ_B, J_{AB}, J_{AX} and J_{BX}.

40.2.2.2 *X part of the ABX spectrum*:

In this part generally four signals are observed. So as a whole in ABX spectrum a total of 12 signals are obtained. However, in some cases two more very weak signals may appear. The centre of all these twelve signals gives the chemical shift of X i.e. δ_X. The spectrum of styrene is a typical example of ABX type of spectrum.

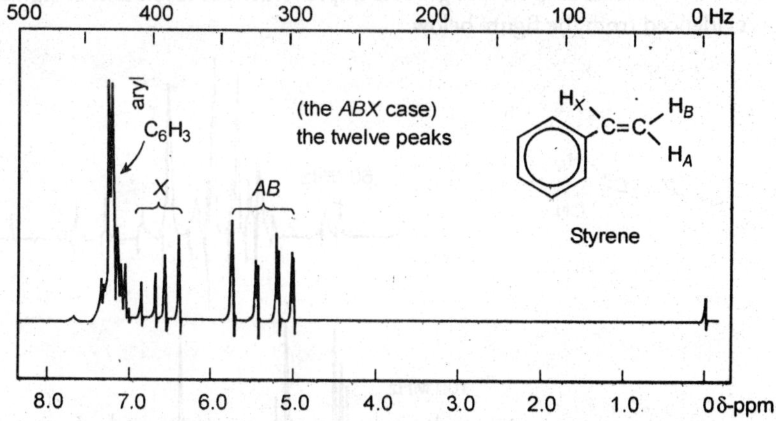

It must be mentioned in this connection that the success of the above procedure for the deduction of required parameters (chemical shifts, intensities and coupling constants) for the interpretation of non-first order spectra solely depends upon the identification of different signals. Unfortunately in many cases identification of all the signals practically becomes impossible due to highly complex nature of the spectra. Consequently, complete analysis of the spectra becomes extremely difficult. So for more accurate deduction of required parameters spectra should be simplified by the use of appropriate technique prior to going into the procedure for deduction of different parameters.

41. Discuss some experimental techniques for the simplification of non-first order NMR spectra.

➜ The complete analysis of a NMR spectrum becomes extremely difficult when the signals overlap each other and as a result useful information regarding the nature of the compound is not obtained. It has been observed that in many cases the over lapping of signals is due to closely related methylene groups in a molecule. In such situation an intense, broad and completely unresolved signal may appear in region 1–2 δ. Such signal is termed as '*methylene envelope*'. Complexity in spectra may originate when strong coupling occurs between interacting protons and the ratio of chemical shift difference and coupling constant is much less than 10, i.e., $\frac{\Delta\nu \text{ in Hz}}{J \text{ in Hz}} < 10$. A number of techniques have recently been developed which can be applied to simplify the complexity and enable more accurate analysis of chemical shifts and coupling constants. Some of the techniques are discussed below.

(a) *High field strength*: It has been observed that first-order spectrum is generally obtained when $\frac{\Delta\nu}{J} \geq 10$. It is also known that the chemical shifts in hertz are dependent upon the applied field but coupling constants are independent of field strength. So $\frac{\Delta\nu}{J}$ can be effectively raised for a given group of coupled protons by using higher magnetic fields and the complex coupling can be reduced to simple (or first-order) coupling. As result signals become easily recognisable. Thus an AB coupling pattern from a 60 MHz spectrum tends to become AX type by recording at 200 MHz or higher MHz. Thus, when spectrum of acrylonitrile is taken at different field strengths starting from 60 MHz to 220 MHz gradual improvement in resolution of signals occur as is evidenced from the figure below:

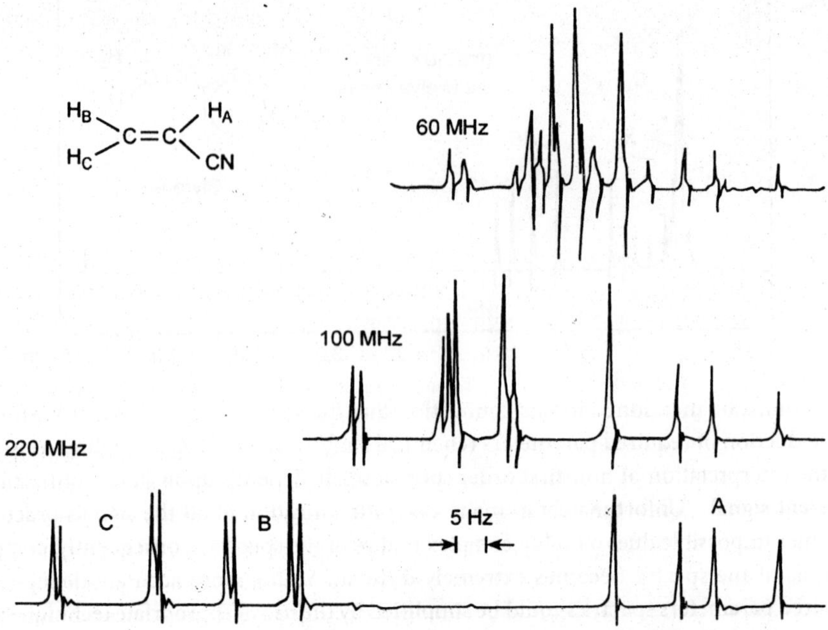

(b) *Lanthanide shift reagents*: It has been found that certain organometallic complex can shift the ^{1}HNMR signals without any significant broadening effect. These complex compounds contain rare earth elements (or lanthanides) as the metal components and the organic ligands are usually 1,3-dione, $-\underset{\underset{O}{\|}}{C}-CH_2-\underset{\underset{O}{\|}}{C}-$ (β-dicarbonyl). These complexes are known as paramagnetic shift reagents and have wide applications in NMR studies, particularly for complex molecules. Some important members of these complexes which are commonly used as shift reagents are as mentioned below.

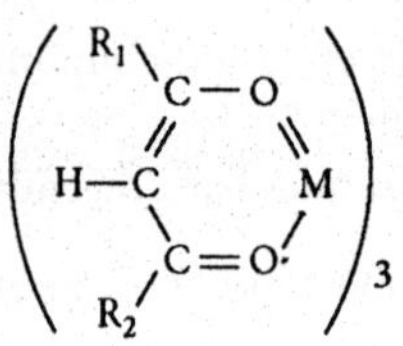

I $R_1 = R_2 = -C(CH_3)_3$ — tris-(dipivalomethanate) Complex, Eu (DPM)$_3$ when M = Europium

II $R_1 = R_2 = -C_2F_5$ — Decafluoroheptadione Complex Eu (FHD)$_3$ when M = Europium

III $R_1 = -C(CH_3)_3$, $R_2 = -C_3F$ — Heptafluoro dimethyl-octadione complex Eu (FOD)$_3$ when M = Europium

The presence of lanthanide ions makes these complexes mild Lewis acids and can co-ordinate with the functional groups having non-bonded electron pair such as $-OH$, $-NH_2$, $-CHO$, $>C=O$, $-COOR$, $-CN$, epoxides, etc of the compound under study forming stable complexes. The co-ordination takes place as under:

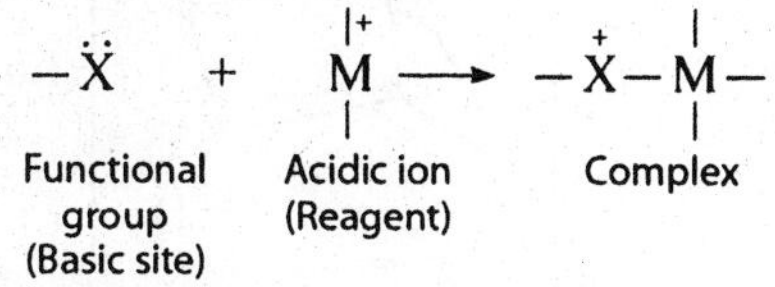

When such a reagent is added to the compound under study and the spectrum is taken, the signals of the protons are shifted and that is why these reagents are known as *shift reagent.* They are also paramagnetic and have the effect of changing substantially the magnetic field in their immediate environment. The result is a shift of signals coming from the protons near the basic site in the organic molecule. For this reason they are also known as *paramagnetic shift reagents*.

The shift of the signals may be in either direction (downfield or upfield) depending upon the nature of metal atom. Europium complexes shift the signals to the downfield while yttrium, holonium, erbium and praseodymium produce upfield shifts of signals. It has been observed that the magnitude of shift depends upon the distance between the group at which co-ordination takes place and the concerned proton—closer the distance wider is the shifting of the signal. The mechanism by which this shifting occurs is known as *pseudo-contact shift*, because this shifting is due to the generation of magnetic field by the spinning paramagnetic ions like europium and other rare earth metals around the protons under study.

An important example of the application of shift reagent may be cited in this context. 6-Methyl quinoline under normal condition gives the sepctrum-I, but after the addition of a soluble europium (III) complex to the solution the spectrum is spread over a wider range of frequencies and becomes almost first oder as seen in the spectrum-II.

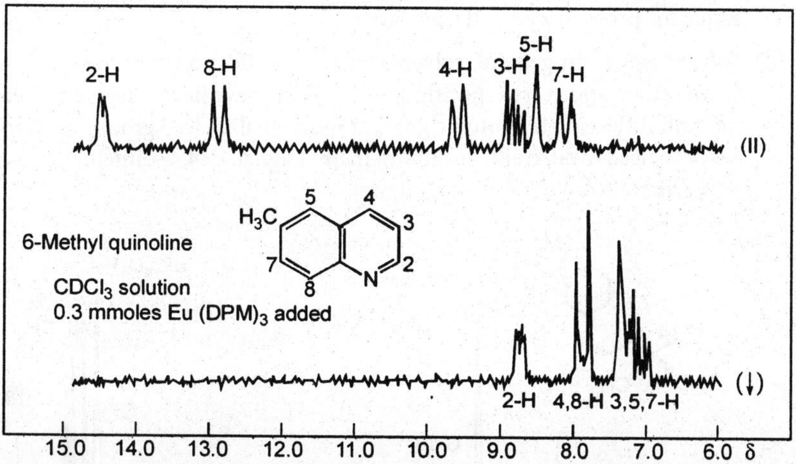

Another most useful application of shift reagents is in the determination of the proportion of R and S forms present in a mixture of enantiomers. If the chiral molecule has a basic site, the two enantiomers under study can have different bonding constants and can adopt different conformations. As a result, signals are shifted differently and can give proportions of enantiomers after integration. The most commonly used chiral lanthanide shift reagent is that of europium as shown below:

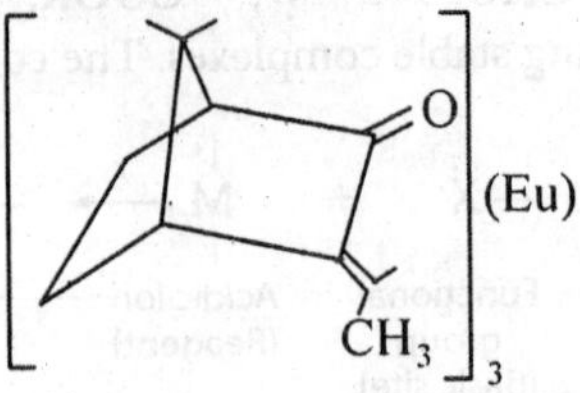

However, before using shift reagents following points must be remembered.

(i) The formation of stable complex between the reagent and the compound is necessary for effect to be observed

(ii) Reagent must be soluble in $CDCl_3$ and other common solvents.

(c) *Deuterium Exchange*: Deuterium D or ^2H is used extensively in ^{1}HNMR spectroscopy because it is easily introduced in the molecule and does not exhibit any signal in ^{1}HNMR spectrum. It has been observed that the hydrogen atoms attached to hetroatoms like OH, NH, SH, etc. resonance at variable fields and their signals are generally broad. These signals often may be buried in the multiplet and cause serious inconveniences in the interpretation of the spectra. Since they are exchangeable, the original signal is eliminated on shaking the sample with D_2O and so the spectrum is simplified. Thus,

$$ROH + D_2O \rightleftharpoons R{-}OD + H{-}OD$$

The signal for OH proton will be missing and instead a signal for proton in H–OD will appear. Similarly, if a little D_2O is added to RCOOH, then due to rapid exchange, it becomes RCOOD.

$$R{-}\underset{\underset{O}{\|}}{C}{-}OH + D_2O \rightleftharpoons R{-}\underset{\underset{O}{\|}}{C}{-}OD$$

Clearly, the signal for the proton in RCOOH will be missing or diminished and a signal for proton in H—OD appears.

(d) *Solvent effects*: In general, solvents have little effect on chemical shift and coupling. However, in some cases specific solvent effects have been observed, specially to separate and differentiate proton signals of individual methyl groups in steroids. Benzene has been used to differentiate differnt methyl signals of 4,4–dimethyl–5α–androstanone as is evident from the following figure.

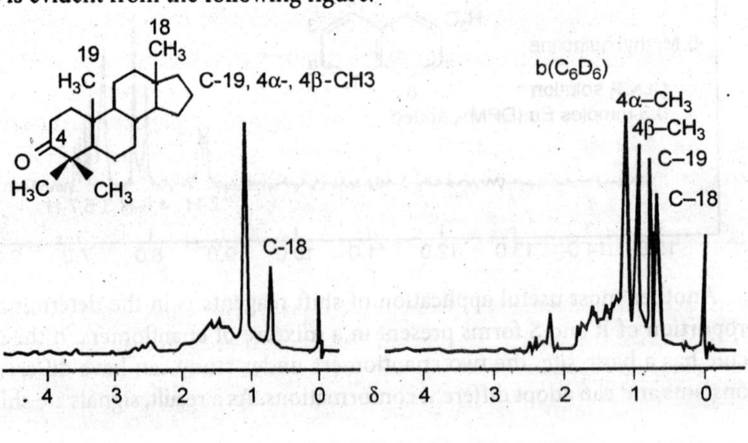

(e) *Trifluoroacetylation*: Trifluoroacetylation is relatively a quick and reliable technique for identification of —OH, —NH—, vNH$_2$ groups where deuteration does not function reliably. The identification of OH group in cyclohexanol has been achieved by this technique.

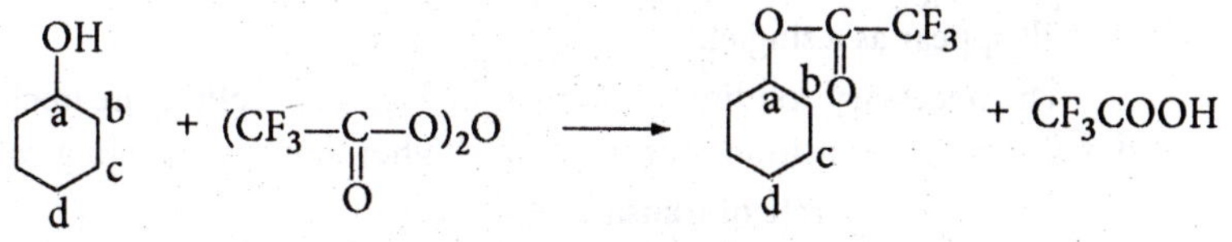

Cyclohexanol

The spectrum after trifluoroacetylation was found to be different from that taken in absence of trifluoroacetylation as seen from the figure below:

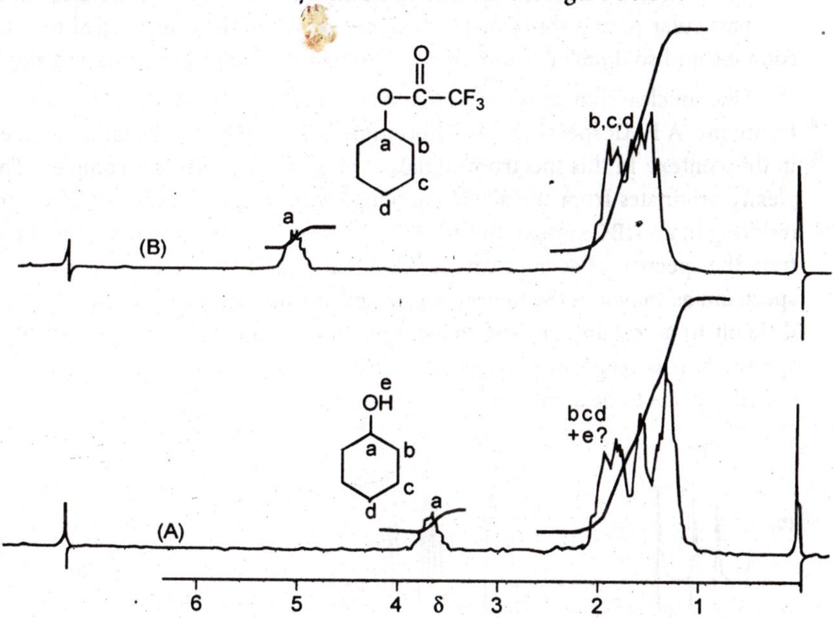

(A) Normal ¹HNMR spectrum of cyclohexanol run in CDCl$_3$;
(B) ¹HNMR spectrum of Trifluoroacetic ester of cyclohexanol.

(f) *Double Resonance* (*Spin Decoupling*): It is one of the most outstanding NMR techniques. It consists of simultaneous irradiation of a sample by two radiations. For this reason, this technique is known as *double resonance* or *double irradiation* and has wide range of applications.

We know that the multiplicity of NMR signals originates due to more than one spin states of the neighbouring non-equivalent nuclei. Thus, in the PMR spectra of

$$X - CH_A - CH_B - Y$$

both H_A and H_B appear as doublets separated by ^3J because H_A and H_B are non-equivalents. Now if a RF field having frequency equal to H_B is applied perpendicular to the static applied field and the PMR signal of H_A which is coupled to H_B, is observed it will be found that the signals are affected. Two signals of H_A merge into a singlet.

This happens because the rate of interconversion of α and β spin levels of H_B is increased and so, the lifetime of the nucleus (H_B) in any one spin state becomes too short to couple with H_A. As a result, the proton H_A finds only one time-averaged view of H_B and hence H_A will come to resonance only once but not twice. Thus the signal of H_A is simplified.

Similarly, if H_A is irradiated with correct RF in desired direction when the signal of H_B is being observed it will undergo rapid transition between two spin states and so H_B will appear as a singlet.

The necessary condition to be fulfilled for decoupling one nucleus from other is that the rate of transition must be much higher than the coupling constant, i.e.,

$$\text{rate of transition} >>> J$$

or, $$\frac{\gamma . H_{RF}}{2\pi} >>> J, \text{ where } \gamma \text{ is gyromagnetic ratio of irradiated nucleus.}$$

As a result of increase in the rate of transition between two spin states, the lifetime in a particular state is shortened. It becomes less than the time needed to resolve the coupling and so doublet cannot be resolved, consequently, it appears as a singlet.

Since nuclear spin during this process is reduced, it is also known as spin-decoupling technique. A PMR spectrum of allylbromide at 100 MHz may be taken as an example in this context. In this spectrum signals of vinyl protons are very complex. The complexity originates from the allylic couplings with protons of the $-CH_2-$ grouping resulting in an ABC system. But if allylic $-CH_2-$ protons are decoupled by irradiation, the spectrum becomes simple AMX pattern. From the normal and decoupled spectrum as shown in the figure we can easily visualise that while it would have been difficult to ascertain coupling interactions for various vinyl protons in the normal spectrum, it is relatively easy to determine the J-values in the decoupled spectrum which appears to be a much simpler splitting pattern.

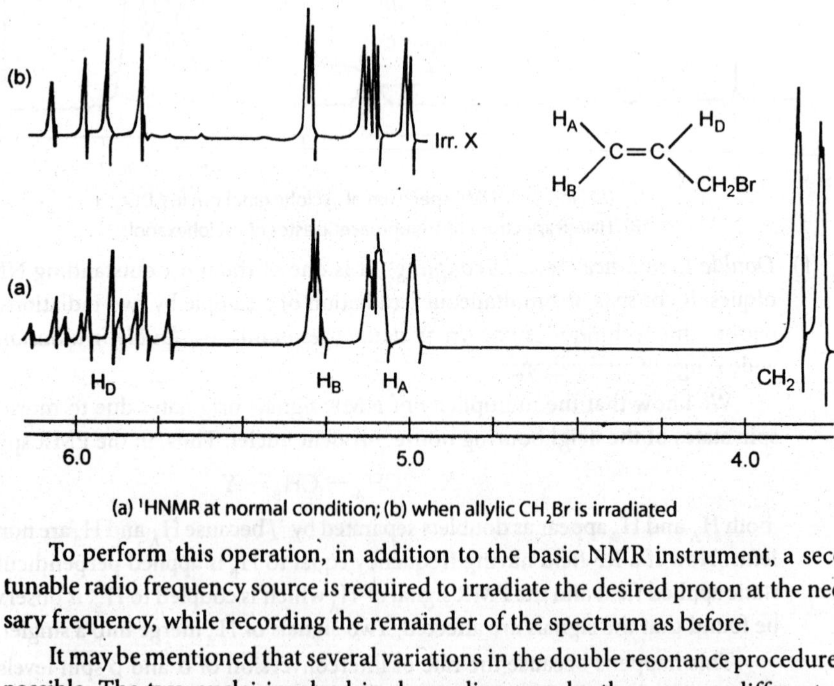

(a) ^{1}HNMR at normal condition; (b) when allylic CH_2Br is irradiated

To perform this operation, in addition to the basic NMR instrument, a second tunable radio frequency source is required to irradiate the desired proton at the necessary frequency, while recording the remainder of the spectrum as before.

It may be mentioned that several variations in the double resonance procedure are possible. The two nuclei involved in decoupling may be the same or different, i.e., double resonance may be *homonuclear* or *heteronuclear*. Of these two types hetronuclear is more advantageous because in this type; couplings are simple or first order, i.e., $\frac{\Delta\delta}{J}$ is high because different nuclei give signals at different frequencies which may not always be satisfied in homonuclear double resonance.

Another variation in the procedure is in the sweep. In any double resonance experiment, two of the three variables—recording frequency (ν_{obs}), irradiating frequency (ν_{RF}) and magnetic field (H_0) are kept constant and third one is varied. Creation of double resonance by varying any one of the variables is called either *frequency sweep* when ν_{obs} or ν_{RF} is varied or *field sweep* when magnetic field H_o is changed.

(g) *Spin Tickling*: Spin tickling technique is, in fact, a modified double resonance technique. In this technique a low intensity irradiation having power one-tenth of that applied for full decoupling is used to irradiate one line in a multiplet instead of irradiating the whole multiplet. As a result, perturbation of energy levels of the spin system occurs and so slight but definite changes take place in the appearance of the signals which are coupled to the signal being tickled. Let us now consider the [1]NHMR spectrum of furfural as shown below:

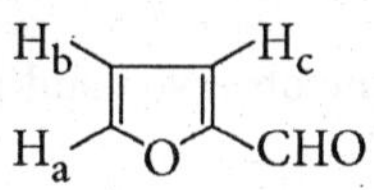

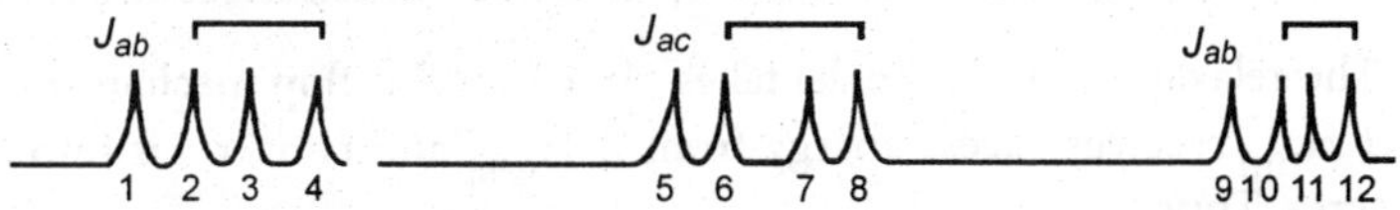

Protons H_a, H_b, and H_c are adjacent to each other and are non-equivalent. So proton H_a couples with H_b giving rise to a doublet but the proton H_c also couples with H_a, so that each line of H_a is further split into four lines in all (1, 2, 3, and 4) for H_a. Similarly, four lines result for each for H_b (9, 10, 11 and 12) and H_c (5, 6, 7 and 8). Hence the spectrum of furfural consists of 12 lines (three multiplets) for three protons H_a, H_b and H_c. Each multiplet displays two different values of J, i.e., (J_{ab}, J_{ac}), (J_{ab}, J_{bc}), (J_{ac}, J_{bc}). Now, if spin tickling is carried out by weakly irradiating one line (say, line 1) of H_a, it is seen that the line 7, 8 of H_c signal and 11, 12 of H_b signal are affected as is evident from the following figure.

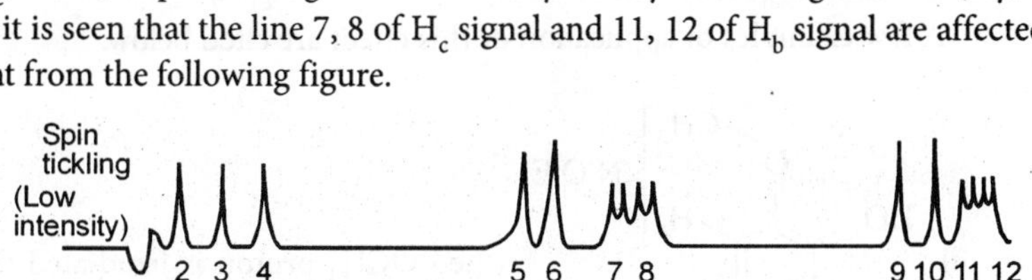

Since spin tickling on line 1 in the spectrum of H_a disturbs the lines 7, 8, and 11, 12, so we can say that all these lines originate from the same state of H_a. Hence it is clear that the signs of coupling constants J_{ac} and J_{ab} must be of the same type and is opposite to that of J_{bc}. Thus this technique is of immense importance in ascertaining the absolute sign of J value.

(h) *Nuclear Overhauser effect*: It is an important type of double resonance technique and is of great value in determining the molecular geometry of the compound. By this technique it may be possible to ascertain whether the two protons (which are not directly coupled) are in close proximity within the molecule or not—

This technique is based on the fact that the rate of relaxation and hence the intensity of the signal of a particular nucleus (observed) can be increased by irradiating another nucleus which is close to it in space but not coupled to each other with the required radio frequency. Let us assume that the nucleus $H_{(a)}$ is being observed and $H_{(b)}$ is another nucleus which is close to $H_{(a)}$ in space.

When $H_{(b)}$ is irradiated its rate of interconversion between the spin states will be increased and hence its rate of relaxation is also increased.The relaxation of this irradiated proton takes place by the transfer of spin energy to the observed proton and so the rate of relaxation of the observed proton $H_{(a)}$ and also its intensity is increased. This effect was first observed by Overhauser and as such it is known as *nuclear Overhauser effect* (N.O.E.). For successful operation of this effect following two conditions must be maintained:

(i) The distance between the observed and irradiated proton must be very short (< 3.5 Å).

(ii) The magnitude of N.O.E. depends upon the strength of both the nuclei.

The relaxation of two nuclei takes place by a flip-flop mechanism. This means that α-$H_{(Irr)}$ receives excess energy from β-$H_{(obs)}$ and converts it into α-$H_{(obs)}$ and vice versa. Thus,

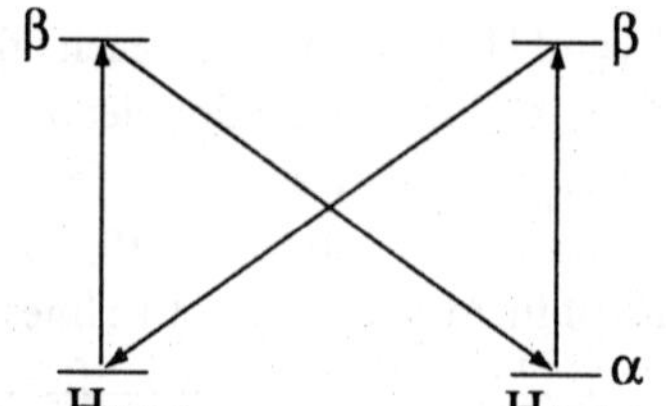

A few examples of application of this effect are cited below.

(i) 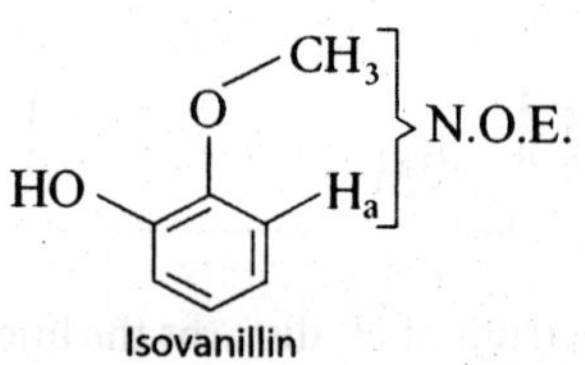

Isovanillin

when OCH_3 proton is irradiated the intensity of ortho-proton marked H_a increases..

(ii) 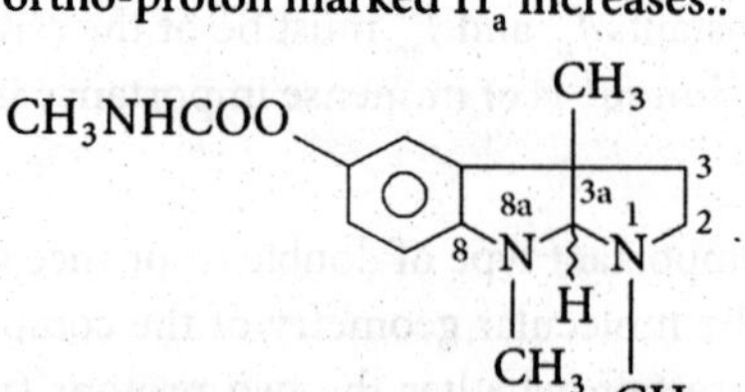

Signal irradiated	Change in intensity of the observed proton 8a
(i) 3a — CH_3	15%
(ii) 8N — CH_3	15%
(iii) 1N — CH_3	0%

Thus 3(a) $-CH_3$ and 8N$-CH_3$ are nearer to 8(a) proton i.e the ring juncture is *cis*.

(i) *Two dimensional NMR spectroscopy, 2D-NMR*: Two dimensional NMR spectroscopy or 2D-NMR spectroscopy is an important and useful technique to identify the coupling relationship in a molecule which gives rise to complex spectrum.

This technique is based on the fact that the intensity of a particular signal is proportional to the difference in population of nuclei in two spin states between which transition takes place. It has been observed that, in general, irradiation of one nucleus modulates the populations of the nuclei which are involved in couplings and hence intensities of the signals of the coupled nuclei will also be affected. This is due to fact that during irradiation transfer of population (*polarisation transfer*) takes place.

The technique as a whole is simple but very long. The sample is subjected to a broad strong pulse of radio frequency and the nuclei are allowed to relax for a short time (say, t_D). The nuclei relax to their original equilibrium state and emit radio frequency at the same frequency and intensity as the absorbed radio frequency. These emitted intensities are recorded and continued till relaxation is complete and are fed into computer. The experiment is repeated several times with same t_D (delay time), readings are collected and added. The time domain data thus obtained is subjected to Fourier transform (FT) operation to convert them into frequency domain data through computer. The frequencies are plotted along both *x*- and *y*- axes. The data are subjected to second FT operation to obtain intensity along the third axis (*z*). The spectrum is presented as a three-dimensional photograph like geographical models of plains and mountains which gives a look of the mountain range viewed from sky. These mountain like spectra are known as *stacked plot.* But these types of spectra are not useful for deduction of essential parameters. So it is usually presented as a two-dimensional spectrum like a geographical *contour plot*. This is more useful for studying the compound. In these contour plot, parts of each mountain are presented by contour circle, i.e., large dot.

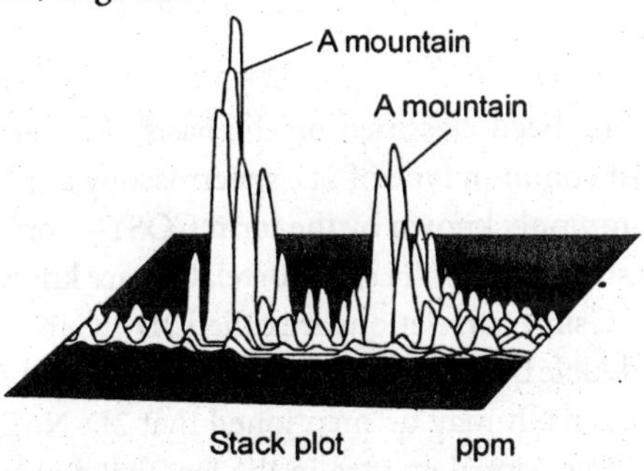

Stack plot

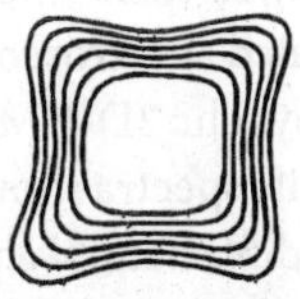
Contour plot

It may be remembered that since the positions of the signals are plotted with frequencies in both the axes (*x* and *y*), they appear at different places on the graph paper. When they are connected to each other by an imaginary line, most of the signals (contour) lie on the diagonal passing through the origin, while a few signals remain off the diagonal. These signals are known as *off-diagonal* or *cross-peaks*. Such types of spectrum is known as *2D-NMR spectrum*. In the spectrum, the conventional ordinary one-dimensional spectrum is shown on both *x*-(horizontal) and *y*-(vertical) axes to assist to identify which contour indicates which nucleus. It takes approximately 3–3.5 hours to record this spectrum. The contour on the diagnonal corresponds to normal signal while cross-peaks indicate the nuclei geminally or vicinally coupled.

To identify which nucleus is coupled to which is indicated by drawing an imaginary horizontal line from a cross-peak to intercept a contour on the diagonal and a vertical line from the same cross-peak to intercept another contour on the diagonal. The nuclei (contour) connected by horizontal and vertical lines through the same cross-peak are mutually connected to each other. A general look of 2D-NMR spectrum is shown in the figure below:

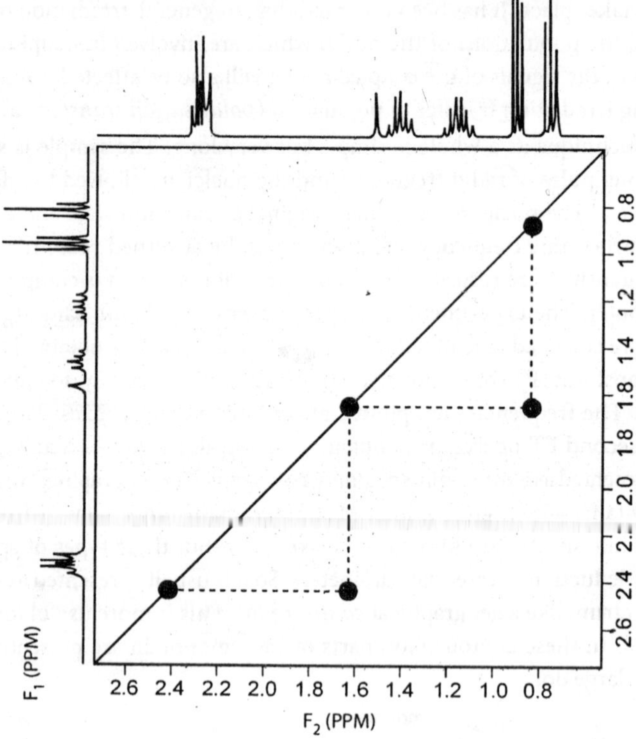

2D-NMR spectroscopy has been classified on the basis of nature of nuclei coupling to each other. The most common type of 2D-spectroscopy are ^1H–^1H shift correlated spectroscopy. It is commonly known by the term COSY—correlated Spectroscopy. The 2D-NMR spectra showing ^{13}C–^1H shift correlation are known as HECTOR while spectra showing ^{13}C–^{13}C shift correlations are called 2D ^{13}C INADEQUATE (incredible natural abundance double quantum transfer experiment) and are used to identify directly bonded cabon atoms. It may be mentioned that 2D-NMR spectra differ from the conventional one-dimensional spectra in the fact that a conventional NMR spectrum is a plot of frequency versus intensity, i.e., it has one frequency axis and one intensity axis while 2D-spectrum has two frequency axes and one intensity axis.

An example illustrating a COSY spectrum may be cited in this connection. In the diagonal of COSY spectrum of 3-hetptanone

$$\left(\overset{7}{CH_3}-\overset{6}{CH_2}-\overset{5}{CH_2}-\overset{4}{CH_2}-\underset{\underset{O}{\|}}{\overset{3}{C}}-\overset{2}{CH_2}-\overset{1}{CH_3}\right)$$

In the 2D spectrum given below five groups of signals are seen at δ 0.9, 1.00, 1.3, 1.55 and at 2.40.

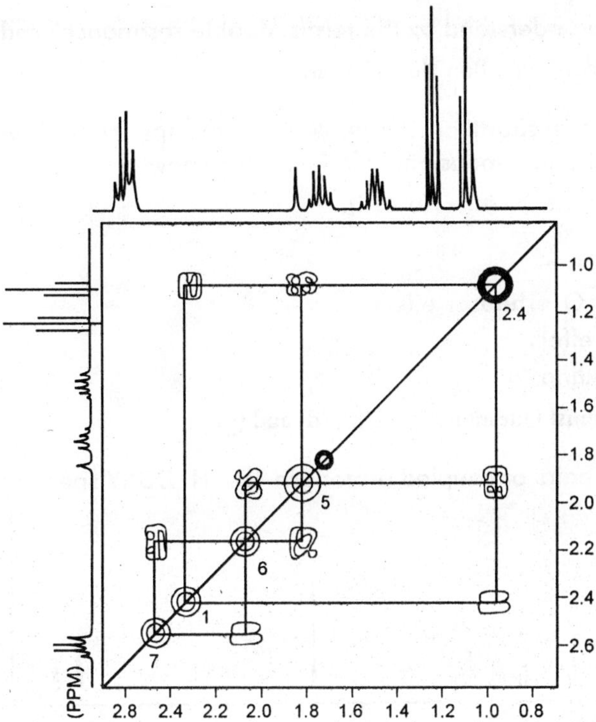

In 3-heptanone there are six groups of non-equivalent protons. But the diagonal shows only five groups of signals. Hence one of the signals must be due to two groups of non-equivalent protons which have very close chemical shifts. Scrutiny of the structure of 3-heptanone suggests signal at δ 2.40 should be due to CH_2 groups at 2 and 4 positions $\left(-\overset{4}{C}H_2-\underset{\underset{O}{\|}}{C}-\overset{2}{C}H_2-\right)$. This signal has two cross-peaks at δ1.55 and δ1.00.

Hence it is logical to assume that the signal at δ2.40 is coupled to signals at δ1.55 and at δ1.00. The signal at δ1.00 is most probably due to CH_3 at the position 1, while signal δ 1.55 is due to CH_2 at the position 5. Again, signal δ 1.30 has a cross-peak at δ 0.9 which is due to CH_3 at the position 7. The signal at δ 1.30 is due to CH_2 at C_6. Thus the final assignment of different signals of 3-heptanone may be shown as follows:

$$\left(\underset{\delta\,0.9\,(t)}{\overset{7}{C}H_3}-\underset{1.3\,(m)}{\overset{6}{C}H_2}-\underset{1.55\,(q)}{\overset{5}{C}H_2}-\underset{2.40\,(t)}{\overset{4}{C}H_2}-\underset{\underset{O}{\|}}{\overset{3}{C}}-\underset{2.40\,(q)}{\overset{2}{C}H_2}-\overset{1}{C}H_3\right)$$

Thus 2D-NMR spectra may be considered as an aid to identify the interacting nuclei in the complex spectra which are difficult to analyse by conventional methods.

42. What do you mean by shift reagent in NMR study ? Mention the names of some important shift reagents and their utility in NMR study.

➜ See answer against Question No. 41(b).

43. Write a short note on chemical exchange as observed in NMR spectroscopy.

➜ See answer against Question No. 41(c).

44. What do you understand by the terms 'double resonance' and 'spin-tickling'.

➜ See answer against Question Nos. 41(f) and (g).

45. Give a short account on 2D-NMR spectroscopy. How does it differ-from conventional one dimensional NMR spectroscopy?

➜ See answer against Question No. 41(i).

46. Write notes on—

(i) Nuclear Overhauser effect

(ii) Solvent effect

(iii) Spin-tickling

➜ See answer against Question No. 41(h), (d) and (g).

47. Identify the pairs of coupled protons in ^1H–^1H COSY spectrum given below:

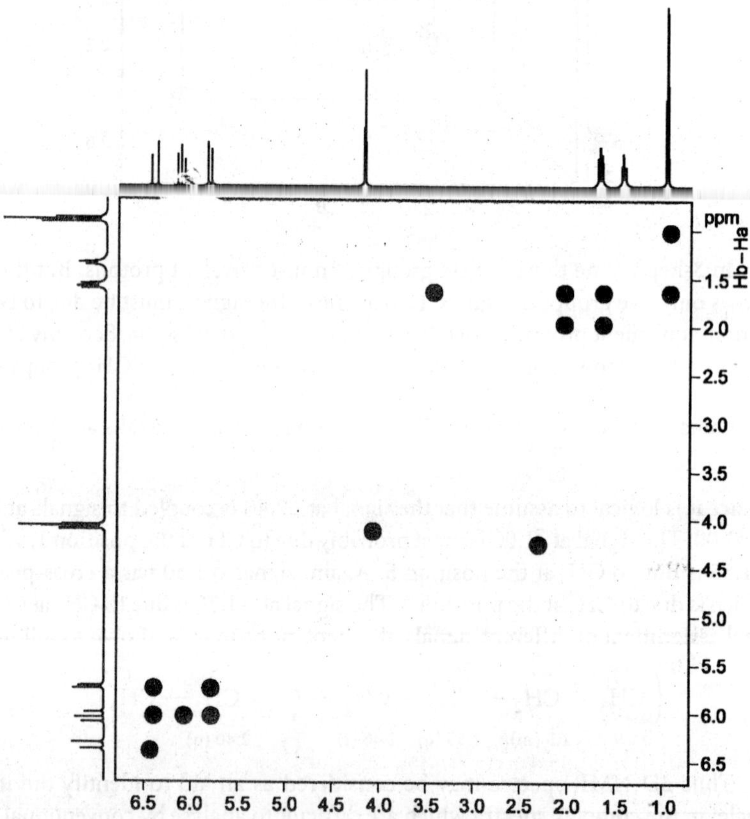

➜ As required for the analysis of 2D-NMR spectrum, a diagonal is drawn passing through the origin when the spectrum looks as shown below:

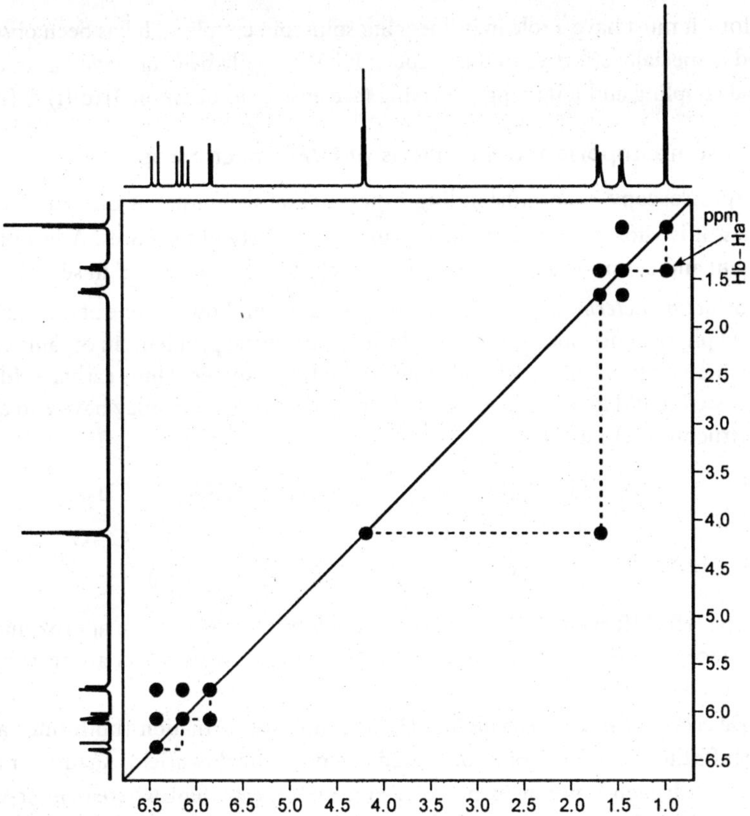

From the spectrum, pairs of coupled protons can be easily identified.

48. A natural product citral, $(CH_3)_2C{=}CH{-}CH_2{-}CH_2{-}\underset{\displaystyle CH_3}{\underset{|}{C}}{=}CH{-}CHO$ exhibits only two geometric isomers. Assign E and Z nomenclature to the isomers. On irradiation of C_3-methyl group of one isomer intensity of signal for C_2—H is increased by 20% but no such increase in intensity of the signal in the other isomer was observed. Suggest the stereochemical nature of this stereo-isomer.

➜ Citral has two double bonds but one of the double bonds cannot show geometric isomerism because the two groups at one end are not different. From the nuclear Overhauser effect the structure of the isomer comes out to be *Z* because in *Z*-isomer required spatial closeness of the involved groups is available.

49. Explain why quaternary carbon shows small nuclear Overhauser effect.

➜ We know that the magnitude of N.O.E. depends upon the distance from the irradiated proton to the observed centre. The quaternary carbon being always placed deeply inside the structure shows small N.O.E.

50. Write a brief note on the contribution of π(pai) bond on coupling of protons

➜ π-bond has immense contribution on the coupling of protons. It is obvious that any bond is a region of electron density. So bonding electrons can create a secondary magnetic field.

Therefore, it must have a role in influencing spin-spin coupling. It has been observed that π-bond is specially effective in this respect. It has actually been observed as in the case of geminal coupling and long range coupling (see answer to question 31(c)(i) & (ii).

51. Discuss some important applications of NMR spectroscopy.

➜ NMR spectroscopy is an outstanding experimental technique at the hands of organic chemists. It is widely applied by the chemists for variety of purposes. A few of the most important and common applications of this technique are discussed in sequel.

(a) *Structure determination*: This technique has been shown as one of the most successful physical methods for solving difficult structural problems in organic che-mistry where classical approach fails to give the final answer. Thus, earlier evidence suggested that diketene obtained by the pyrolysis of acetone may have two alternative structures (I) and (II).

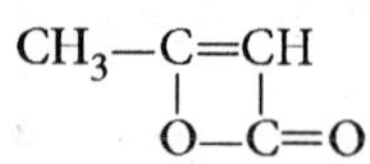

(I)

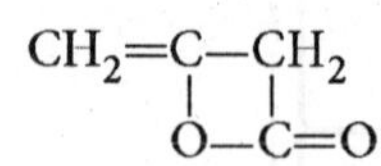

(II)

The ^{1}HNMR spectrum of the diketene exhibits two singlets of approximately equal intensity, and the chemical shift values of these two singlets are in favour of the structure (II).

(b) *Detection of hydrogen bonding*: Hydrogen bond formation (both inter and intra) shifts the absorption for a concerned proton to the downfield. Again, intra-molecular hydrogen bonding is not concentration dependent whereas inter-molecular hydrogen bonding varies with concentration and temperature and so these two types of hydrogen bonding can be distinguished.

(c) *Detection of aromaticity*: Protons attached to aromatic rings are highly deshielded due to ring current effect and so aromatic protons appear at a very low field. From this, aromaticity of the compound can be detected.

(d) *Detection of cis-trans isomers*: The *cis* and *trans* isomers of a compound can be easily distinguished from the coupling constant values. Thus,

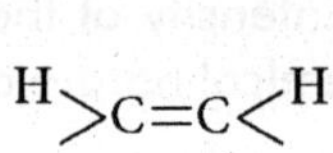

cis

$J = 7–10$ Hz

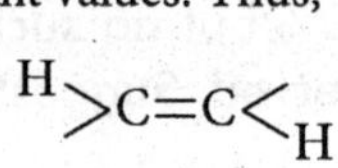

trans

$J = 8–18$ Hz

(e) *Detection of functional groups*: NMR spectroscopy is well suited for detecting certain functional groups in a molecule from their characteristic chemical shift values. Thus,

$C–CH_3$	$O–CH_3$	$N–CH_3$
δ 0.9–1.25	δ 3.5–4.5	δ 2–2.25

(f) *Determination of relative amount of tautomers in a tautomeric mixture*: Tautomeric equilibrium mixture can be analysed by this technique. Thus acetyl acetone exists as

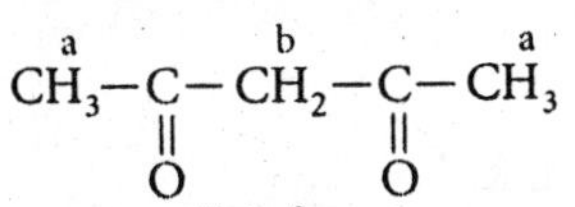

Keto form

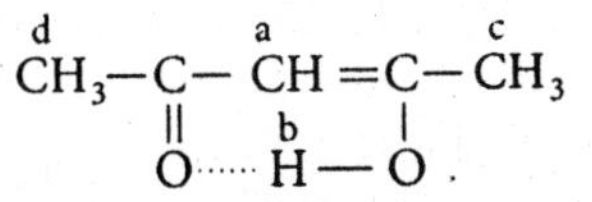

Enol form

The ^{1}HNMR spectrum of these two tautomers exihibits different chemical shift values.

Keto form	a =	2.15 δ
	b =	3.65 δ
Enol form	a =	2.00 δ
	b =	~14.00 δ
	c =	5.5 δ
	d =	2.15 δ

(g) *Determination of conformation*: NMR spectroscopy has been studied in determining the conformational isomers. Thus in cyclohexane there are two types of hydrogens—axial and equatorial. These two types of protons have different chemical shifts and give rise to broadbands. It has been found that euqatorial proton appears at lower field than axial proton.

(h) *Magnetic Resonance Imaging* (MRI): Application of NMR is not confined to chemical sciences but it has been applied in biological sciences too. Magnetic resonance imaging (MRI) is, in fact, an application of NMR that makes it possible to examine the inside of the human body using radio frequency radiation and is less damaging than X-rays as it contains lower energy. It does not require any imaging or contrast agents. It is a sophisticated technique and it does what other NMR spectrometers do. It detects proton and is essentially sensitive to biological materials such as water and lipids which are rich in hydrogen. The regions of the image are lighter or darker depending on the relative concentration of protons and their enviroments. It is known that in NMR, there must be continuous absorption of energy for which relaxation of excited protons must occur. The rate of relaxation depends upon the environments. In human body hydrogens are attached to the surface of the biological materials and their relaxation rate is high under normal condition. But slight change in the environment (tissue structure) may shorten the relaxation time. Such difference in environment is used to image the organs of human body by MRI as seen in the figure.

A normal brain

The dark shadow for tumour

CHAPTER 5

CARBON-13 NMR (CMR) SPECTROSCOPY

1. (a) What do you mean by Carbon-13 NMR spectroscopy?
(b) How it differs from ^{1}HNMR spectroscopy?
(c) How Carbon-13 spectra are measured?

➡ (a) Carbon-13 NMR technique is one of the most recent development in NMR spectroscopy. This technique has been developed in the years since 1970 and today this instrumental method is used routinely to complement proton NMR spectroscopy. Although ^{12}C is non-magnetic but the isotope ^{13}C is magnetic $\left(I=\frac{1}{2}\right)$. So in principle, it is not difficult to observe ^{13}C NMR. Consequently, this technique is based on the same principle as in PMR and the general procedure is also the same. Solvent used is commonly $CDCl_3$, D_2O, etc. and the internal reference used is also tetramethyl silane (TMS).

(b) This technique differs from ^{1}HNMR (PMR) in respect to the following characteristic features.

(i) The natural abundancy of ^{13}C is only 1.1%.

(ii) Its gyromagnetic ratio, i.e., precessional frequency per second is $\frac{1}{4}$ th of ^1H and its magnetic moment is $\frac{1}{4}$ th of ^1H. Hence, signals are very weak and very low energy radio frequency is to be applied. Generally, different carbons (^{13}C) resonate in the range of 12–220 ppm or δ.

(iii) The sensitivity of ^{13}C magnetic moment is very poor due to low abundancy and low precessional frequency. This creates extreme difficulty in recording the spectra. Difficulty primarily arises from the noise. It is known that noise originates from the fact that in atmosphere, there are a large number of random radiations which are picked up by the recorder. As a result base line of NMR spectrum is not exactly linear but consists of infinite number of signals of very low intensity. So confusion arises in identifying ^{13}C signals. Thus the low intensity of signals had been a major obstacle for the advent of ^{13}C NMR as the common procedure for recording ^{1}HNMR spectra. That is why this technique could not be fruitfully applied for obtaining ^{13}C NMR spectra for a long time.

(c) Different methods are now available to record ^{13}C spectra avoiding the difficulty arising out due to loss of signals in noise. These methods are discussed below:

(i) One of the most important methods to avoid the problem arising out from the noise is to use *repeated scanning.* The intensity data of each scan is fed into computer for addition. Many such scanning (at least 100) are carried out and sum is presented by the computer as ^{13}C spectrum. The addition necessary for obtaining a ^{13}C spectrum of the compound is based on the fact that noise is random, i.e., it occur in positive and negative directions and so are cancelled out statistically in totalling, but the signals due to ^{13}C are always unidirectional, i.e., positive and hence are added up. Further improvement of signal intensities are made by irradiating the sample with a strong broad range of radio frequency of the appropriate proton frequencies. This results in decoupling of ^{13}C signals from protons and so prevents splitting and only identifiable ^{13}C signals are obtained. This technique is known as computer averaging of transient abbreviated as CAT process or signal averaging process.

However, CAT process suffers from some drawbacks. The main disadvantage of this process is the time factor. The scanning is slow and it requires 250 seconds per sweep and one scan takes at least five minutes. So 100 scans require 8–9 hours while increase in intensity is only 10 times.

(ii) *Pulse Technique*: The disadvantage of time factor in CAT process has been removed by applying pulse technique. In this technique, the sample is subjected to a strong broad radio frequency band with frequency greater than ^{13}C frequency range, called *pulse* for a short time (10 – 100 μ seconds). The radiation is stopped for a short period. The nuclei (^{13}C) relax to their original equilibrium state and emit radio frequency at the same intensities and frequency as the absorbed radio frequencies. These intensities are detected and fed into a computer. The pulse is applied again for a short time, stopped and the process is repeated several times. The data are presented as intensity versus time graph. This graph is known as *time domain spectrum*.

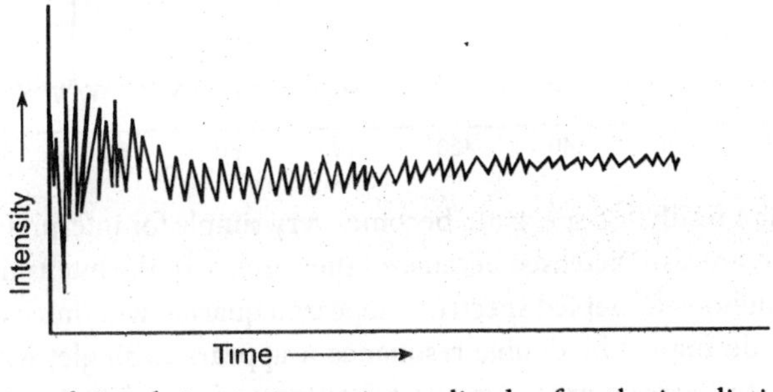

The intensity of signals is maximum immediately after the irradiation is stopped and decreases rapidly with time.

The time domain spectrum is of no use. It is now transformed to operationally simple procedure, termed as *Fourier Transform* (FT). It is a mathematical process carried out by a computer. The different frequencies and intensities are separated and plotted in the familiar form—frequency versus intensity spectrum. These are sometimes referred as *frequency domain spectra*. This pulse technique along with FT operation requires only 15–20 minutes for recording a spectrum.

However, pulse technique along with Fourier transform operation is not the final answer. Complications may arise from multiplicity of signals. Both ^{13}C and ^{1}H have

$I = \frac{1}{2}$ and are magnetically active but non-equivalent. So it may be expected that in ^{13}C spectrum coupling between (i) $^{13}C-^{13}C$ and (ii) $^{13}C-^{1}H$ may be observed. But since the abundance of ^{13}C is very low the probability of simultaneous presence of two ^{13}C at the adjacent positions is very remote. So, in practice, $^{13}C-^{13}C$ coupling does not occur. But coupling between ^{13}C and ^{1}H may occur and can split the ^{13}C signals into singlet, doublet, triplet, quartet, etc. depending upon the number of hydrogen atoms attached to ^{13}C. Coupling between ^{13}C and hydrogen is strong having J_{CH} in the range of 120 Hz and occurs even when they are separated by several intervening bonds. Consequently the spectra, thus obtained, are often quite complex and difficult to interpret.

Such coupling ($^{13}C—^{1}H$) has also been removed by a *double resonance technique* usually termed as broadband decoupling technique. In this process, as carbon (^{13}C) spectrum is being taken, the sample is simultaneously irradiated with a band of radio frequency radiation that excites all the hydrogens. This causes each of the hydrogens to flip rapidly between spin states, so its two magnetic field orientations average to zero. Hence, no coupling occurs and so in normal cases ^{13}C spectra consists of only singlet signals for different groups of equivalent ^{13}C as seen from the figure below:

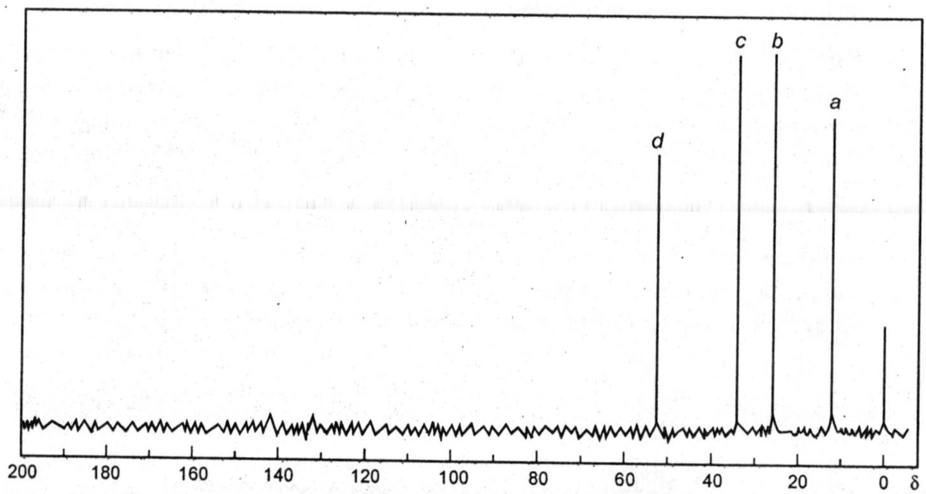

As a result, ^{13}C spectrum becomes very simple for interpretation and intensities of signals are also increased because of the absence of the intensities of ^{1}H. For example, CH_3 in non-decoupled spectrum appears a quartet with intensity ratio 1 : 3 : 3 : 1 but when decoupled by double resonance it appears as singlet with increased intensity. These two aspects are the major *advantages* of complete decoupled technique. However, this complete decoupling technique suffers from a serious *disadvantage*. This complete decoupling technique does not provide the information of exact nature of carbon. In any compound carbon may be present either as $-CH_3$, $-\underset{|}{C}H_2$, $-CH$ or as $-\overset{|}{\underset{|}{C}}-$. But such information cannot be obtained from fully decoupled technique because in this technique all carbons appear as singlet.

In order to get such information reagarding the nature of carbon several techniques are available of which *off-resonance technique* is one of the important technique and is usually applied.

In off-resonance technique, proton decoupling is carried out with radio frequency for all the protons except those attached to a particular carbon atom, i.e., the couplings between ^{13}C and protons attached to it are retained and ^{13}C spectrum is recorded as usual. It may be remembered that in order to carry out this procedure, the position of the particular proton signal should be first determined by recording ^{1}HNMR spectrum. In fact, the result of *off-resonance* is an incomplete collapse of the multiplicity—^{13}C are split only by the protons directly bonded to them. As a result, multiplets become narrow and are not removed completely as in fully decoupled spectra. So the useful information regarding the nature of carbon may be easily obtained without becoming complicated due to overlapping. In short, off resonance decoupling simplifies the spectrum by allowing some of the splitting information to be retained. These spectra are represented as $^{13}C - \{^{1}H\}$ spectra to indicate multiplicity as *s* (singlet for $-\overset{|}{\underset{|}{C}}-$), *d* (doublet for —CH), *t* (triplet for $-CH_2$ *q* and, (quartet for CH_3). The figure below gives a general view of a typical proton coupled spectrum.

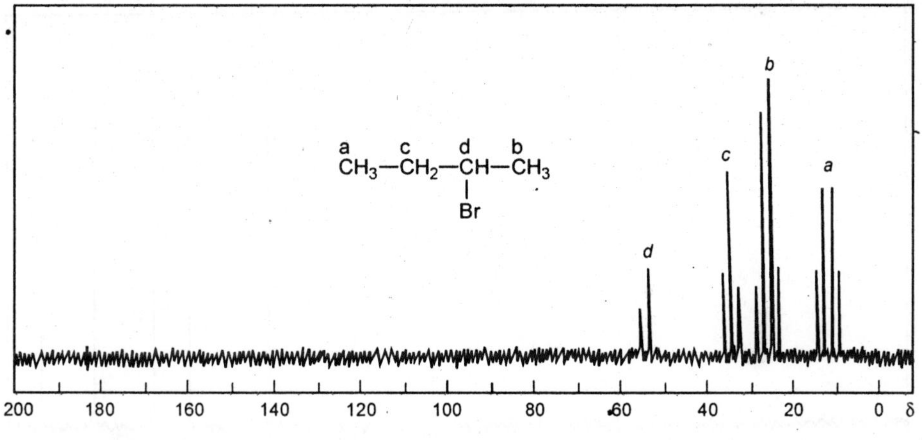

For routine work, the spectrum of the compound is run thrice: first to know the position of a particular proton or protons ^{1}HNMR spectrum should be recorded, next broadband proton decoupled spectrum to be recorded for indicating the number of non-equivalent carbon atoms and their chemical shifts. Finally, off-resonance decoupled spectrum is taken to indicate the number of hydrogen atoms bound to each carbon atom.

2. (a) What are characteristic features of chemical shifts in ^{13}C NMR?

(b) What are the factors which control the chemical shifts of ^{13}C?

(c) Give a chart of chemical shift of different types of carbon.

➜ (a) The chemical shifts of ^{13}C nuclei like ^{1}HNMR are expressed in units of parts per million (ppm) downfield from TMS. However, the range of ^{13}C chemical shifts is much larger than ^{1}H which is 0–220 δ for different carbon atoms. Due to this widespread of signals it is unlikely that any two ^{13}C nuclei will have identical chemical shifts unless they are equivalent. For example, every individual carbon atom in secondary butyl bromide can be observed in ^{13}C spectrum as is evident from the figures (a and b).

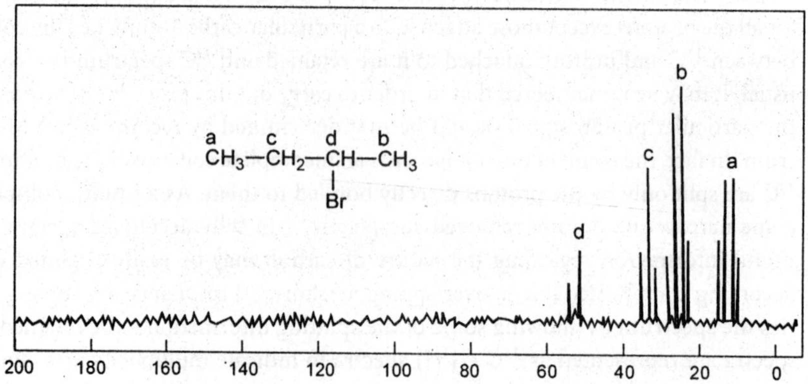

Fig. (a): Proton coupled ^{13}C-NMR spectrum of sec-Butyl bromide.

Scrutiny of the structure of the compound reveals that all the four carbon atoms are non-equivalent and so the spectrum exhibits four signals, one for each.

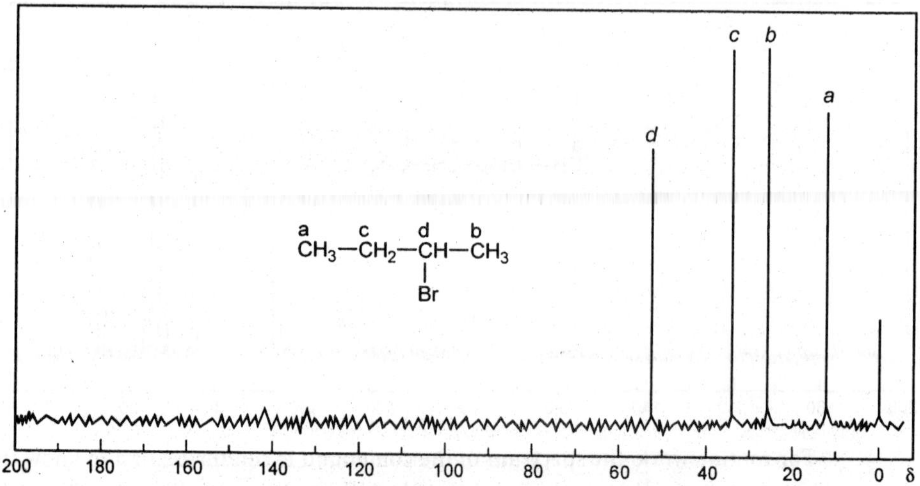

Fig. (b): Proton decoupted ^{13}C-NMR spectrum of sec-Butyl bormide.

Following characteristic features have been observed in the chemical shifts of ^{13}C nuclei.

(i) Alkane generally absorbs in the most upfield region.

(ii) Increasing alkylation generally moves the carbon resonance downfield.

(iii) The values of chemical shifts indicate the nature of hybridisation (sp^3, sp^2, sp) at each carbon.

(iv) Carbons of both aromatic ring and alkenes resonate in the same region. This makes ^{1}HNMR useful to distinguish between two types of compounds.

(v) Carbon of carbonyl group absorbs far downfield usually around 200 δ.

(vi) In proton-decoupled ^{13}C NMR spectra the number of signals exhibits how many non-equivalent carbons are present in the molecules.

(vii) In ^{13}C proton off-resonance decoupled NMR spectrum, the splitting of signals indicates the number of hydrogen atoms attached to the particular carbon. Thus,

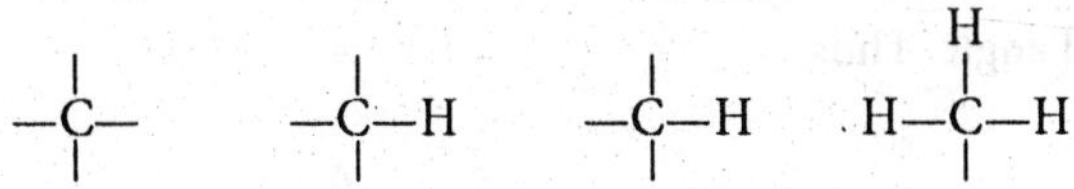

no proton (singlet) | one proton (doublet) | two protons (triplet) | three protons (quartlet)

(viii) In ^{13}C NMR spectrum, the peak areas are not necessarily peoportional to the number of identical ^{13}C nuclei.

(b) The most dominating factor which determines the chemical shifts of different carbon atoms is the state of hybridisation. sp^3 carbons absorb.at higher field while sp^2 carbons at lowest field and sp carbons appear in the region in between sp^3 and sp, i.e., $sp^3 > sp > sp^2$.

The ranges at which these types of carbon appear are:

$$sp^3 = 12 - 50\ \delta$$
$$sp = 65 - 95\ \delta$$
$$sp^2 = 110 - 160\ \delta$$

Aromatic carbons also appear in sp^2 region.

It may be mentioned that carbonyl carbon atom is mostly investigated and generally appears in the lowest field in the region 180–220 δ. Within this range chemical shift varies depending upon the electron density surrounding a particular carbon as has been observed in ^1H. Besides the nature of hybridisation next important factor which controls the chemical shift, is the substituent and its position. It has been observed that the substituent at α and β positions on sp^3 carbon causes downfield shifts with shift being greater at α position than at β position but the substituent at γ position causes upfield shift. Thus,

$$\underset{13.7}{CH_3}-\underset{22.6}{CH_2}-\overset{\gamma}{\underset{34.5}{CH_2}}-\overset{\beta}{\underset{22.6}{CH_2}}-\overset{\alpha}{\underset{13.7}{CH_2}}-H$$

$$CH_3-CH_2-CH_2-CH_2-CH_2-Cl$$
$$\delta \rightarrow \quad 13.6 \quad 22.2 \quad 29.2 \quad 32.5 \quad 44.2$$

These effects are known as α-, β- and γ-effects. The magnitude of shifts varies with the nature of substituent. Thus the α-effects caused by various substituents are:

Substituent	*Shift at α position δ*
—Cl	+30.6
—Br	+19.3
—F	+70.1
—OH	+48.3
$-NH_2$	+29.7
$-NO_2$	+64.5
$-CH_3$	+9

Similar effect of downfield shift has been observed at β position but with lesser extent. For chlorine the magnitude of this shift is ≈ +10 δ; other substituents at β position follow the same pattern of shift as has been seen for α-substituents. However, substituent at γ-position causes though smaller but upfield shift. This anomalous observation has been explained on the basis of *gauche* interaction, i.e., steric compression which depends upon the dihedral angle. Thus

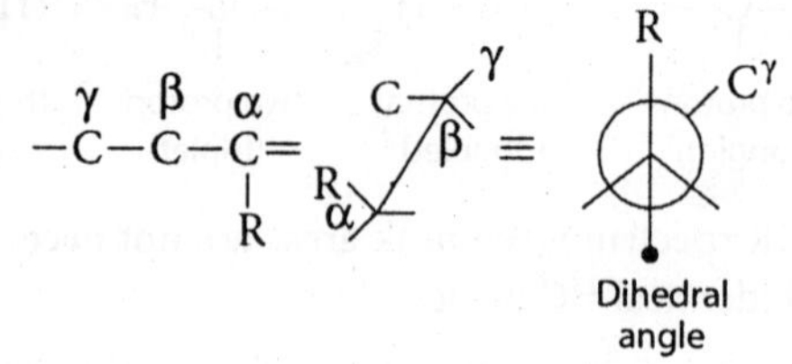

In sp^2 and aromatic system substituents' effects are different. In sp^2 system, geometric factor is also important. Thus,

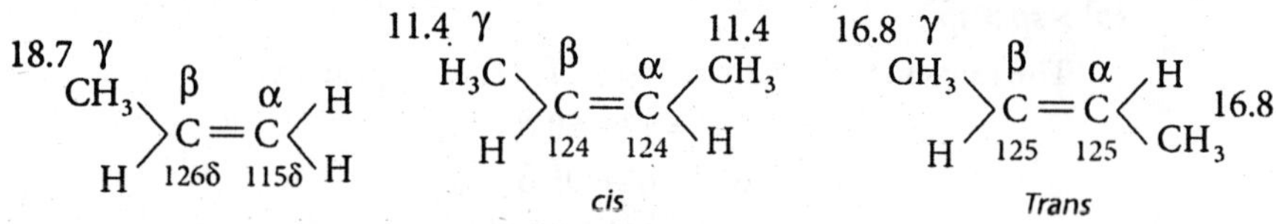

The γ-effects are upfield (–7.3 ppm) for *cis*-isomer and –1.9 ppm for *trans*-isomer, i.e. the γ-effect is stronger for *cis*-isomer.

It may be mentioned that ^{13}C chemical shifts of olefinic and aromatic carbons are very useful to identify the polar substituents. Unlike saturated carbon atoms, the polar effects of substituents on unsaturated carbons are longer-ranging due to resonance in addition to inductive effects. Thus

(i) $CH_2{=}CH{-}\ddot{O}{-}CH_3 \longleftrightarrow \bar{C}H_2{-}CH{=}\overset{+}{O}{-}CH_3$

Vinyl ether

(ii) 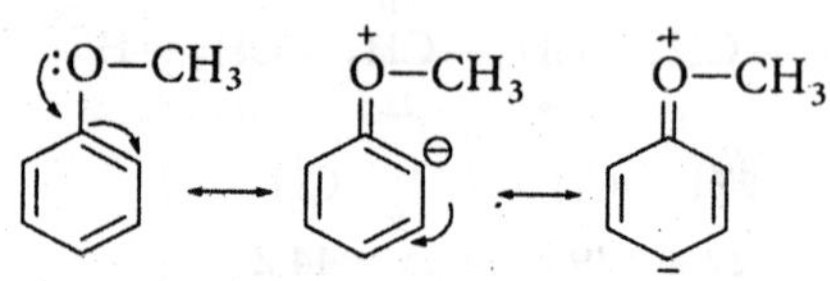

In this case -OCH_3 group deshields the carbon to which it is attached but *ortho* and *para* carbons are shielded due to the electron releasing effect of oxygen. Similar effect is also observed in α, β-unsaturated carbonyl compounds.

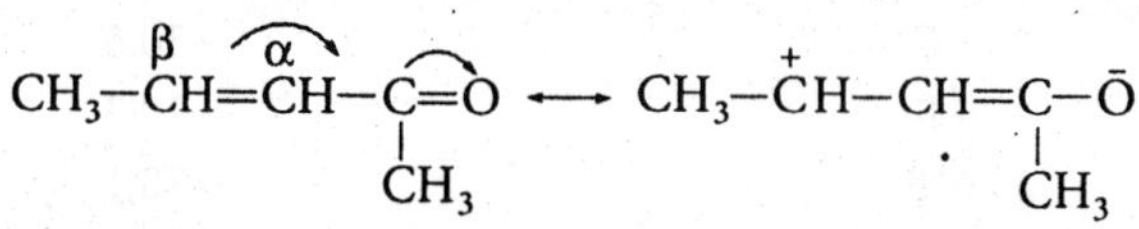

Here α-carbon atom is shielded but β-carbon is deshielded.

The chemical shifts of carbonyl carbons are also affected by the substituent attached to the carbonyl carbon atom.

(c) A chart of chemical shifts of common type of carbon atoms may be presented as below:

Type of carbon	Chemical shift (δ)
Alkane (sp^3)	12–50
Alkene (sp^2)	110–160

Alkyne (*sp*)	65–95
Aromatic (*sp*^{2})	110–140
Alcohol or ether	50–90
Amide; Carboxylic acid	150–185
Aldehydes	185–205
Alkyl halide	10–65
$R-\underset{\underset{O}{\|}}{C}-OR$	160–175
$R-\underset{\underset{O}{\|}}{C}-OH$	160–185
Ketone	190–220

3. Discuss with appropriate examples how you will deduce the structure of organic compound with the help of ^{13}C NMR spectrum.

➻ For predicting the structure of organic compound from ^{13}C spectrum following steps are to be followed:

(i) Number of signals in the spectrum is to be counted. This gives the number of non-equivalent carbon atoms in the molecule.

(ii) The intensities of the signals are noted; non-proton-bearing carbons give lower intensity signals while groups of two or more equivalent carbons give higher intensity signals.

(iii) Multiplicity of signals, i.e., *s, d, t* or *q* are noted which gives the information regarding primary, secondary, tertiary or quaternary nature of carbon.

(iv) Finally, the chemical shifts values of different carbon atoms are compared with the standard values from table to identify different carbon atoms.

Let us now consider the following examples for the elaboration of the above steps.

(a) A compound, $C_9H_{10}O$ exihibits the following signals in ^{13}C spectrum.

δ	8.2	31.6	128.3	128.6	432.8	137.2	200.0
	q	*t*	*d*	*d*	*d*	*s*	*s*

Predict the structure.

➻ From the chemical shift values and their multiplicities following conclusions may be drawn:

(i) The signal at 8.2 δ is due to $-CH_3$

(ii) The signal at 31.6 δ is due to $-CH_2$

(iii) Signals at 128.3, 128.6 and 132.8 d are due to C–H carbon atoms of aromatic system

(iv) Signal at 137.2 δ may be assigned to $-\overset{|}{\underset{|}{C}}-$ atomatic and

(v) Signals at 200 δ is for Ketocarbonyl $-\underset{\underset{O}{\|}}{C}-$ of aromatic system.

On the basis of above the part structure of the compound stands as

$$CH_3,\ -CH_2-\underset{\underset{O}{\|}}{C}-,\ C_6H_5$$

Presence of signals in the sp^2 (aromatic) region suggests C_6H_5 portion is a benzene with substituents of other carbons at one position. The most probable structure, thus, may be

$$C_6H_5-\underset{\underset{O}{\|}}{C}-CH_2-CH_3$$

(b) An organic compound, $C_9H_{10}O_3$ displays in its ^{13}CNMR spectrum the following signals

δ	56.00	56.10	109.0	110.7	126.5	130.3	149.8	154.6	190.0
	q	*q*	*d*	*d*	*d*	*s*	*s*	*s*	*d*

Suggest the probable structure of the compound.

➜ The positions and multiplicities of the signals suggest the following features of the compound.

(i) Signals at δ 56.00 (q) and 56.10 (q) are due to $-OCH_3$ groups.

(ii) Doublets at δ 109.0, 110.7 and at 126.5 δ are assignable to aromatic (sp^2) unsaturation of the type (HC = CH)

(iii) Singlets at δ 130.3, 149.8 and at δ 154.6 are due to aromatic quarternary carbon.

(iv) Doublet at δ 190.00 reveals the presence of an aldehydic function $H-\underset{\underset{O}{\|}}{C}-$.

The appearance of three aromatic signlets is a clear indication in benzene rings three groups $-OCH_3$, OCH_3 and the CHO groups are connected, i.e., the part structure of the compound stands as

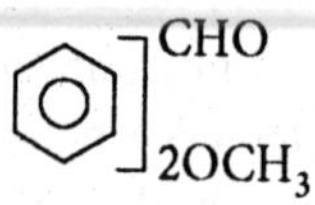

Complete analysis of the δ values of different signals suggests that the probable structure of the compound may be written as

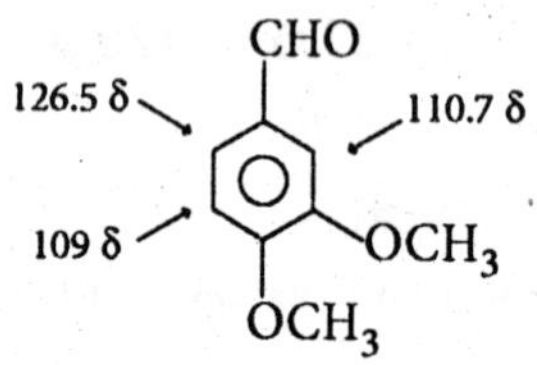

(c) Predict the chemical shift positions for different carbon atoms in butyl ethyl ketone.

➜ In order to predict the chemical shift of different carbons in butyl ethyl ketone we must know the δ values of butane and ethane. Let us write structural formulae of these compounds.

$$\overset{7}{C}H_3-\overset{6}{C}H_2-\overset{5}{C}H_2-\overset{4}{C}H_2-\underset{\underset{O}{\|}}{\overset{3}{C}}-\overset{2}{C}H_2-\overset{1}{C}H_3$$

Butyl ethyl ketone

$CH_3-CH_2-CH_2-CH_3$

Butane

CH_3-CH_3

Ethane

From the table δ value for CH_3 of ethane is 5.7 δ. As there is carbonyl at β position, the δ will be increased (i.e., resonance occurs at lower field) slightly (normally by 2 ppm), So C_I in the ketone will appear at 7-8 δ as triplet with respect to C_2. Carbonyl group is at α-position. So C_2 will be shifted downfield to a larger extent approximately by +35 ppm. Hence, C_2 may be around 40 δ.

C_4 will have chemical shift similar to C_2 and its signal position may be at ~40δ. C_5 will appear at slightly higher field than C_4 around 30δ while C_6 will appear at higher field than C_5, and C_7 may have normal δ value. Thus the predicted δ values of different carbons of butyl ethyl ketone is as shown.

$$\overset{15}{CH_3}-\overset{25}{CH_2}-\overset{30}{CH_2}-\overset{40}{CH_2}-\overset{210}{\underset{\underset{O}{\|}}{C}}-\overset{40}{CH_2}-\overset{8.0\ \delta}{CH_2}$$

(d) Predict the δ values for carbons of *p*-hydroxyacetophenone.

➜ The structural formula of *p*-hydroxyacetophenone is

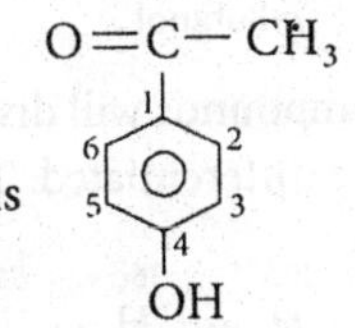

The expected δ values for different carbon atoms are as under:

C_5 → 130 δ, C_2 and C_6 → 130 δ, C_3 and C_5 → 115 δ and C_4 → 160 δ, Carbonyl carbon → 200 δ and Keto methyl carbon around 25 δ.

3. Explain the followings.

(a) α-, β- and γ- effects in ^{13}C NMR spectroscopy.

➜ See answers against Question No. 2(b).

(b) ^{13}C–^{13}C coupling is not a factor in ^{13}C NMR spectroscopy?

➜ The natural abundance of ^{13}C is so low that it is practically impossible that a particular ^{13}C– nucleus in a molecule will have another ^{13}C– nucleus as an immediate neighbour, so ^{13}C– ^{13}C coupling in ^{13}C– spectroscopy is very much unlikely.

4. How will you differentiate between the followings from ^{13}C NMR spectroscopy?

(a) *cis* and *trans* 2 butene.

➜ The structural formula for *cis*- and *trans*- 2-butene may be written as

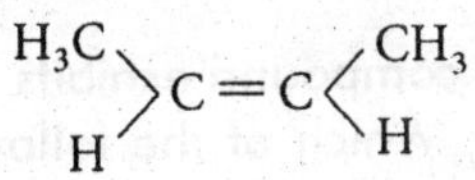

cis

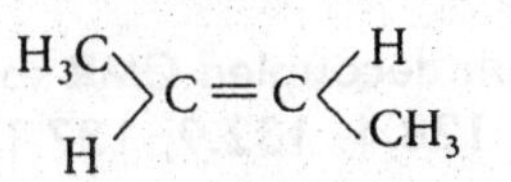

trans

From chemical shift values, these two isomers can be differentiated. γ-effect is the deciding factor. Due to γ-effect carbon atoms of CH_3 group in *cis* will have lower δ value than its *trans* counterpart. Thus *cis* methyl carbons will resonate at around 11.4 δ while *trans* methyl carbons will give signal at 16.8 δ.

(b) $CH_3-\underset{\underset{O}{\|}}{C}-CH_2-CH_3$, $CH_3-CH_2-CH_2-CHO$ and $(CH_3)_2CH-CHO$

(I) (II) (III)

➜ Compound (I) is a ketone while (II) and (III) are aldehydes. Hence multiplicity of carbonyl carbon in (I) can be differentiated from that of (II) & (III). Carbon in keto carbonyl will appear as singlet but in (I) and (II) carbons of carbonyl will appear as doublet.

Again, the aldehyde $CH_3-CH_2-CH_2-CHO$ will give two triplets for two CH_2 and one quartet for CH_3 but the compound (III) will give beside aldehyde carbon another doublet for CH and one quartet.

(c) $\overset{4}{CH_3}-\overset{3}{CH_3}-\overset{2}{CH_2}-\overset{1}{CH_2}-OH$ $CH_3-CH_2-\underset{\underset{OH}{|}}{CH}-CH_3$

n-Butanol 2-Butanol

➜ Both these compounds will display four signals each but the chemical shifts are different, so they can be differentiated. Their chemical shifts are as under:

$\overset{13\delta}{CH_3}-\overset{20\delta}{CH_2}-\overset{36\delta}{CH_2}-\overset{51.4\delta}{CH_2}-OH$ $\overset{10\delta}{CH_3}-\overset{32\delta}{CH_2}-\underset{\underset{OH}{|}}{\overset{59\delta}{CH}}-\overset{22\delta}{CH_3}$

5. A tribromo benzene derivative exhibits only two signals. Suggest the probable structure.

➜ The tribromo benzene may be represented by three structures.

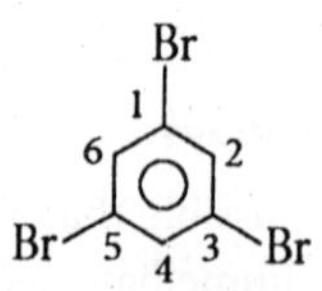

1,3,5-Tribromo benzene

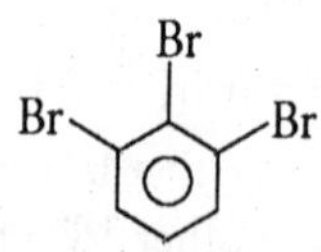

(I)

1,2,6-Tribromo benzene

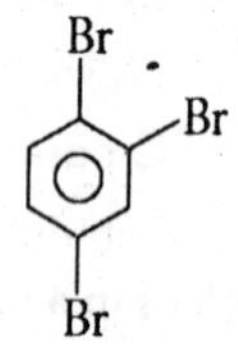

(II) (III)

1,2,4-Tribromo benzene

Out of these three isomer, 1,3,5 tribromo benzene is symmetrical and has two different types of carbons (2, 6) and 4. So only two signals will appear in its ^{13}C NMR spectrum. Hence, the compound is 1,3,5 – tribromo benzene (I).

6. A proton decoupled CMR spectrum of a compound exhibits signals at δ 26.3, 128.2, 128.4, 132.9, 137.1 and 197.6. Which of the following structures is consistent with the data?

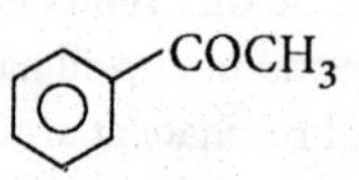

(I)

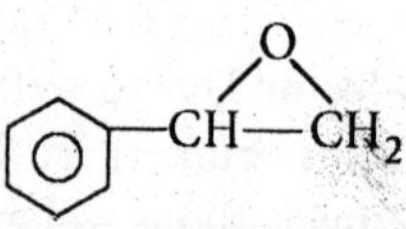

(II)

➜ The appearance of a signal at δ 197.6 is a clear indication of the presence of a keto carbonyl carbon. Hence the data are more consistent with (I). The assignment of different signals are shown as below.

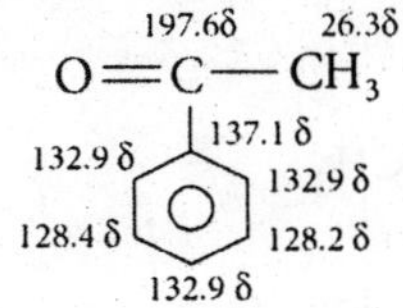

7. How many signals are expected in ^{13}C NMR spectrum of *o*-, *m*- and *p*-xylenes.

➜ The structural formulae of these three isomers are as under:

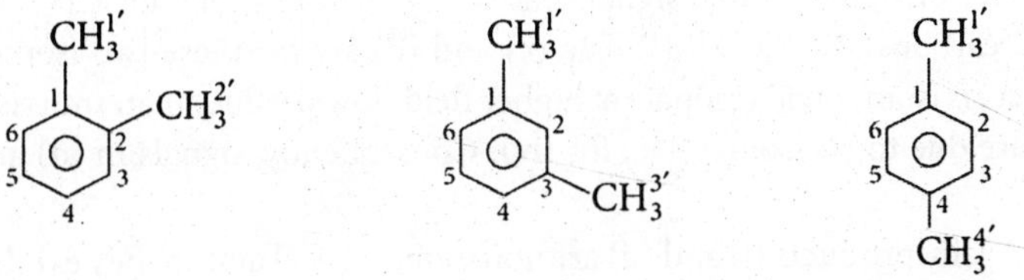

o-Xylene   *m*-Xylene   *p*-Xylene

Orthoxylene will show four signals for (i) CH_3 (1′, 2′); (ii) — C — CH_3 (1, 2); (iii) Ortho carbons (3, 6) and (iv) para carbons (4, 5); metaisomer will exhibit five signals for (i) CH_3 (1′, 3′); (ii) meta carbons (1, 3); (iii) (4, 6) carbons, (iv) (5) carbon and (2) carbon but para isomer will display three signals for (i) CH_3 (1′, 4′); (ii) para carbons (1, 4) and (iii) (2, 3, 5, 6) carbons.

8. 3-Hexanol when dehydrated forms a mixture of four unsaturated compounds. Their ^{13}C NMR spectra exhibit following signals:

(a) δ 12.30, 13.50, 23.00, 29.30, 129.70 and 130.70

(b) δ 13.50, 17.20, 22.9, 35.00, 125.00 and 132.00

(c) δ 14.30, 21.00 and 130.00

(d) δ 14.00, 26.00 and 131.00

Analyse the data and assign the structures of four compounds.

➜ When 3-hexanol is subjected to dehydration experiment, following products may be obtained

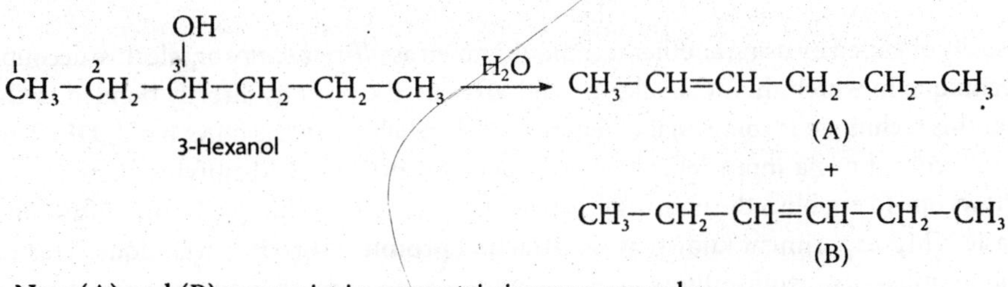

3-Hexanol

Now (A) and (B) may exist in geometric isomers as under:

(i) *trans*   (ii) *cis*

Similarly,

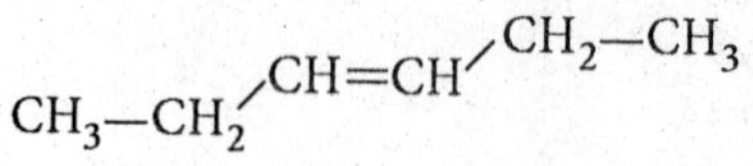

(iii)
trans

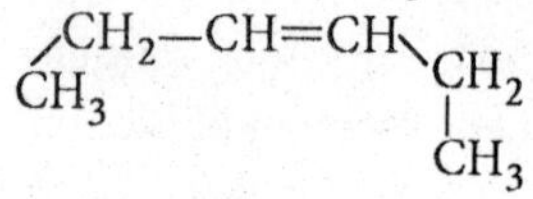

(iv)
cis

Now the isomers (iii) and (iv) are symmetric in nature. Hence they will exhibit less number of signals (three signals one for sp^2 carbons, one for CH_2 carbons and one for CH_3 carbons). So ^{13}C NMR data (c) and (d) are for these two isomers. Now due to γ-effect *cis* isomer will resonate at higher field (low δ value) than *trans* isomer. So signals in (c) are due to *cis* isomer, i.e., for (iv). Consequently, signals in (d) are for *trans* isomer (iv).

Again, products (i) and (ii) are non-symmetrical and so they exhibit larger number of signals. Hence, signals (a) and (b) are for (i) and (ii). The product (ii) is *cis* and (i) is *trans* isomer. *cis* isomer will have some upfield signals (lower δ values) than *trans* isomer. Hence, signals (a) are for (ii), i.e., *cis* isomer and signals (b) are due to *trans* isomer (I).

9. Write a brief note on off-resonance technique. What are its advantages over double resonance?

➜ See answer against Question No. 2(c).

10. What do you mean by distortionsless enhanced polarisation transfer (DEPT) spectra of ^{13}C? Explain it with appropriate examples

➜ Distortionless enhancement by polarisation transfer (DEPT) is a recently developed technique used as a means of enhancing intensity of signals and editing spectra. It is known that broadband proton decoupling technique removes multiplicity in carbon resonance and provides right number of singlet signals at proper chemical shift values for the carbon atoms present in the molecule under consideration. However, some important and valuable information regarding the nature of carbon, viz., primary (CH), secondary (CH_2), tertiary (CH_3) and quaternary ($-\overset{|}{\underset{|}{C}}-$) are lost in such technique. Such informations are usually obtained by using another technique known as *off-resonance* or selective decoupling technique in which the ^{13}C nuclei are split only by the protons directly bonded to them. But this technique is somewhat older and not so reliable. On the other hand, **DEPT** technique constitutes a more reliable and convenient method of identifying these sites and allows one to establish the nature of carbon correctly. This technique is a modification of a basic NMR experiment known as the **attached proton test (APT)** technique. This technique utilises two transmitters,—one operating at the proton resonance frequency and other the at ^{13}C resonance frequency. The proton transmitter serves as proton decoupler; it can be switched on and off at appropriate intervals during the pulse sequence. The ^{13}C transmitter provides different pulses along *x*-axis, usually 90° and 135° pulses. Thus, in this technique three spectra are obtained— one is normal broadband (BB) decoupled spectrum, another is DEPT-90° spectrum while the third one is DEPT–135° spectrum. The procedure adopted in this technique is as deplicted below:

1. $\frac{\pi}{2}(90^\circ)$ or $\frac{3\pi}{4}(135^\circ)$ pulse as the case may be is applied for a short period (t_p) at ^{13}C frequency. Pulse is usually described in terms of the angle through which M [net magnetic moment of assembly of different non-equivalent nuclei (^{13}C)] is displaced from the z-axis (z-axis is the axis on which the applied field acts). Again, the magnitude of this angle depends upon the time (t_p) through which the pulse is passed as shown in the figure below:

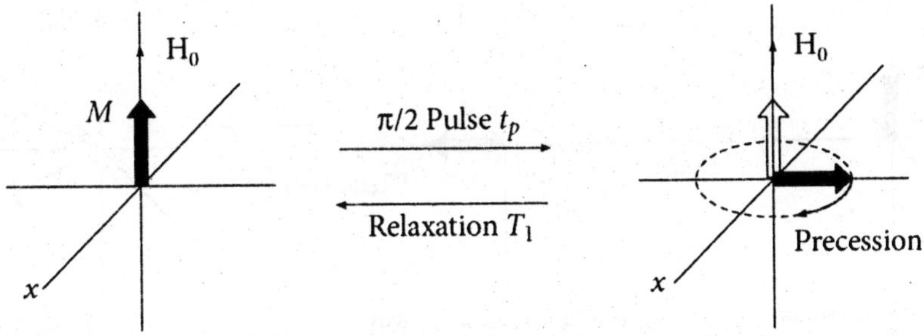

2. The pulse is stopped and allowed to relax for a short period, t_D (delay time). During the delay time signals are splitted in different number depending upon the number of protons attached to the carbons and precess in opposite directions due to spin orientation. The rate of precession depends upon J-value. It may be mentioned that the delay time may be suitably manipulated to produce periodic functions in the signal intensity.
3. Next π-pulse is applied when the spins of the signals are placed in the inverted (mirror) or echo region and is allowed to relax for a short time t_D.
4. After the π-pulse deoupling radio frequency is applied for H when protons are decoupled and as a result splitted signals are collapsed to a singlet and precess in y-axis.
5. ^{13}C signals are collected.

The above steps are repeated several times, added together through computer and is subjected to Fourier transform to get usual ^{13}C frequency domain spectrum. The signal intensities may be determined as described below:

A. **Pulse applied 90°**

(i) **Methine carbons (CH)**

In this system only one proton is attached with ^{13}C. So ^{13}C will appear as a doublet with coupling constant J, i.e.,

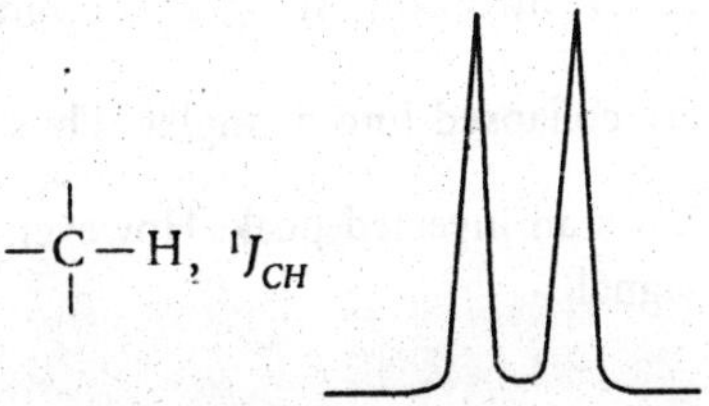

when a 90° $(\pi/2)$ pulse is applied the net magnetic moment vector, M, is directed along y-axis—the receiver coil (chemical shift axis) as a single line since there is only one vector component. However, in this case the attached hydrogen splits this resonance signal into a doublet. One of the component vectors rotates at $J/2$ Hz per sec in clockwise direction while the other rotates at $J/2$ Hz per sec in anti-clockwise direction as shown in the figure.

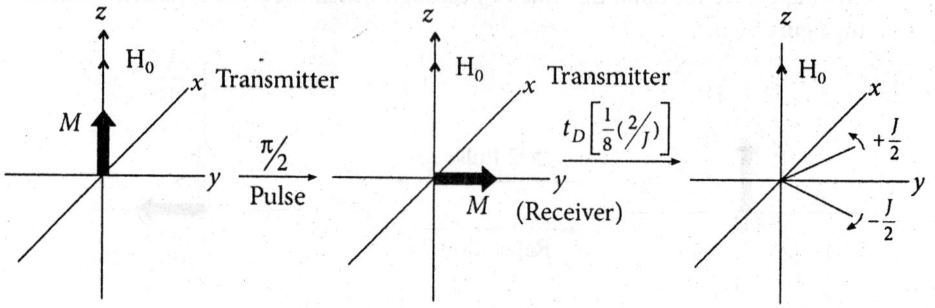

At thermal equilibrium (a) | *M* is directed along y-axis (b) | (c)

The component vectors are rotating at a rate of $\frac{J}{2}$ revolutions per second but in opposite direction. The time required for one revolution is, therefore, the inverse of this rate, i.e., $\frac{2}{J}$ seconds. Now at time, $t_D, \frac{1}{8}\left(2/J\right) = \frac{1}{4J}$, both the component vectors will make one-eight, of a resolution, i.e., 45° from *y*-axis and will be opposite to each other (Fig. c). Now a π-pulse is applied when the spin signals are placed in the inverted, i.e., mirror region (Fig. d).

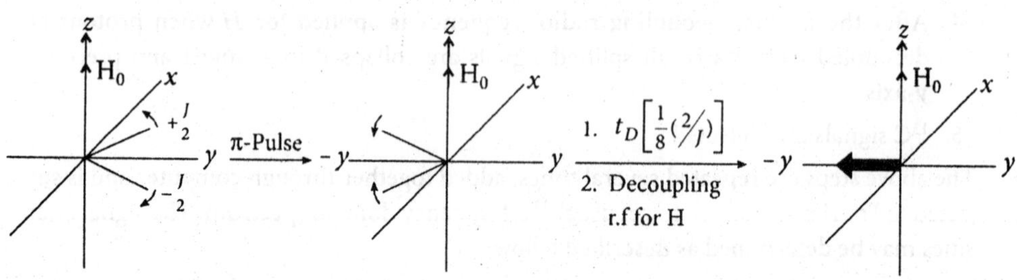

(c) (d) (e)

Now allowing the situation (Fig. d) to relax for $t_D[\frac{1}{8}(2/J)]$, the vectors realigned along *y*-axis but in the negative direction. At this decoupling radio frequency for *H* is applied when the signals are collapsed into a singlet. Thus at the time $t_D = \frac{1}{8}\left(2/J\right)$ second, methine carbon shows an inverted peak. However, during editing it is displayed as normal (positive) signal.

(ii) Methylene carbons (CH_2)

In the case of methylene carbons two protons are attached with ^{13}C, i.e.,

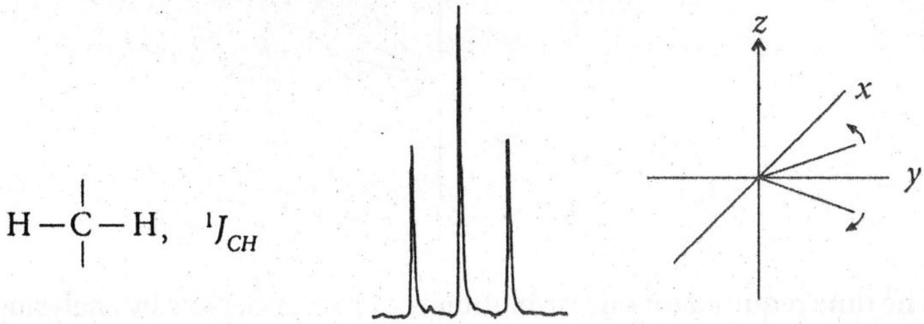

So after $\pi/2$ pulse there are three vectors for ^{13}C nucleus because the two protons attached with it split the ^{13}C resonance signal into a triplet. One of these vectors remains stationary while other two move apart in opposite direction with a speed of J revolutions per second (where J is the coupling constant) during revolution period (t_D), The time required for one revolution is, therefore, $1/J$ second. Now at time $t_D = \frac{1}{4J}$, the vectors will move 90° from y-axis and will be in opposite directions along the x-axis. At this time no signal is detected by the receiver because there is no component of M along y-axis (the resultant of these two vectors is zero). So, applying π-pulse and decoupling radio frequency for hydrogen the situation will be the same, i.e., no signal will appear.

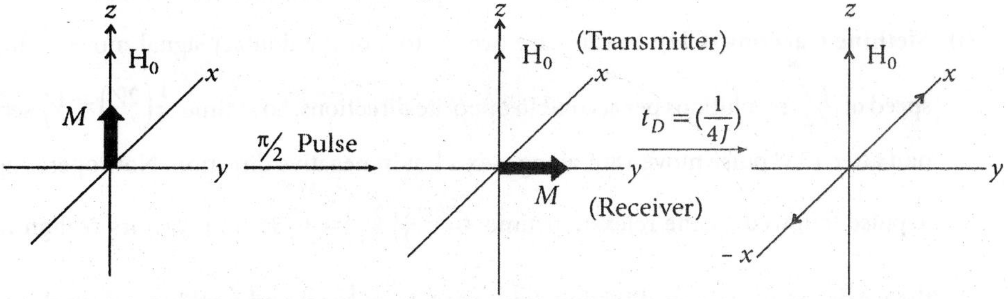

(iii) Methyl Carbons (CH_3): In methyl carbon, three protons are attached with ^{13}C.

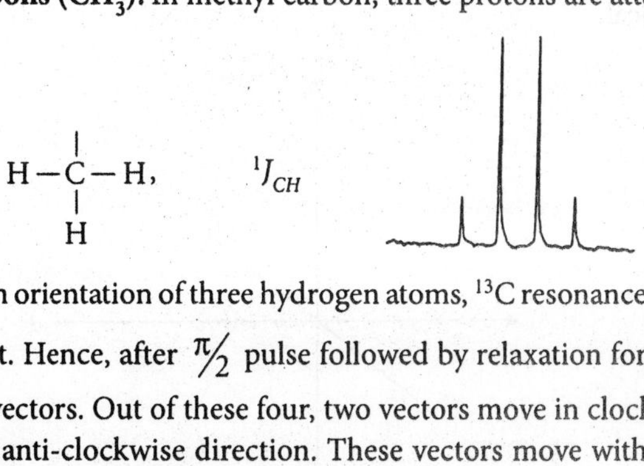

So due to spin orientation of three hydrogen atoms, ^{13}C resonance signal will be splitted into a quartet. Hence, after $\pi/2$ pulse followed by relaxation for a short period there will be four vectors. Out of these four, two vectors move in clockwise direction while other two in anti-clockwise direction. These vectors move with a speed of J revolutions per second as shown in the figure below:

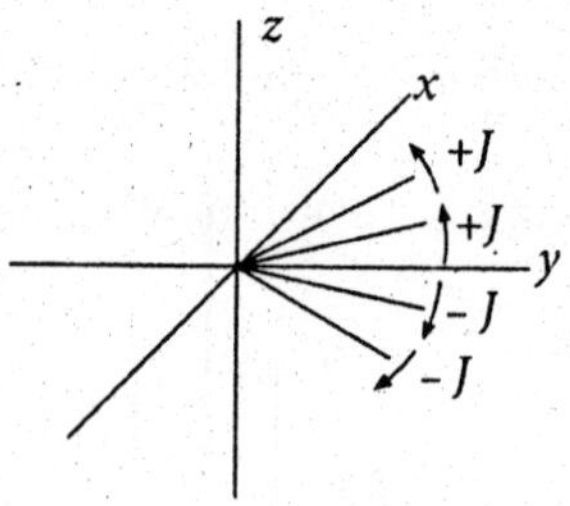

The time required for one revolution is $1/J$ second. Now by analysing the precessional frequencies of four vectors as in the case of methylene carbons it is evident that after time $t_D = \frac{1}{4J}$, the vectors move 90° from y-axis and will be in opposite directions along x-axis. At this time there will be no component of M along y-axis. So no signal will be detected by the receiver even after applying π-pulse and decoupling radio frequency for hydrogen.

(iv) **Quaternary carbons (C):** In the case of quaternary carbon no hydrogen atom is attached with ^{13}C. Since DEPT technique is based on the polarisation transfer, i.e., the transfer of proton magnetisation into the directly bonded carbon, such carbon (quaternary carbon) are not seen in DEPT spectra.

B. Pulse applied 135°

Same arguments may be applied for the analysis of the nature of carbon during 135° pulse as in the case of 90° pulse and the results may be summarised as below:

(i) **Methine Carbons (CH):** In this case two vectors of the doublet signal move with a speed of $J/2$ revolutions per second in opposite directions. So at time $\frac{1}{2}(2/J) = 1/J$ second after 135° pulse move 180° along y-axis but in negative direction. Now operating π-pulse followed by the relexation time $t_D = \frac{1}{2}(2/J) = 1/J$ second, vectors realign in the y-axis in the negative direction. So at time $t_D = 1/J$ second methine carbon shows an inverted peak.

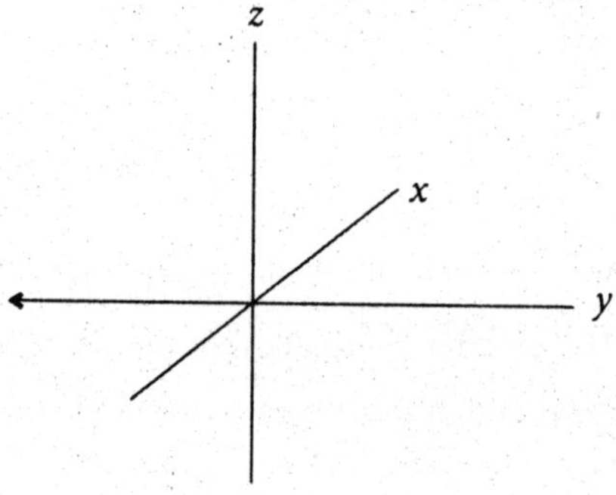

At time $t_D = 1/J$

(ii) **Methylene Carbons (CH_2):** In methylene carbons, one of the vectors of the triplet remains stationary but other two move in opposite direction at the rate of *J* revolutions per second. So at time $t_D = 1/J$ second they will realign along the positive *y*-axis. Thus, the vectors will form normal (positive) peak if they are detected at time $t_D = 1/J$.

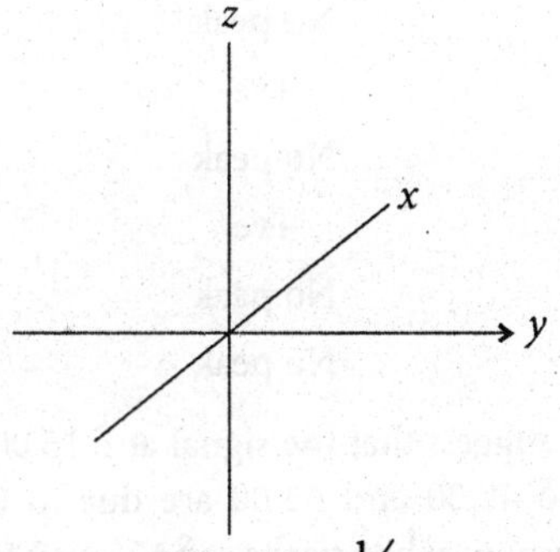

Time $t_D = 1/J$

(iii) **Methyl Carbons (CH_3):** An analysis of the precessional frequencies of the four vectors of the quartet reveals that at time $t_D = 1/J$ second the methyl carbon exhibits an inverted peak because at time $t_D = 1/J$ second the vectors realign along the negative *y*-axis.

C. Conclusion

From the above discussion it is evident that in the DEPT experiment it is possible to get separate spectra for methyl, methylene, and methine carbons by varying the pulse-width and delay time. It has been observed that for every methylene carbon a normal (positive) peak and an inverted (negative) peak for each methine and methyl carbon are obtained. Thus we can tell whether the number of hydrogen atoms attached to the carbon is even or odd.

It may be mentioned that in many cases during the presentation of DEPT spectra, the intensities are presented in the reverse manner of what has been stated above through editing, i.e., carbon atoms that bear odd numbers of hydrogens (methine, methyl) appear as positive peaks and carbon atoms that bear even numbers of hydrogens (methylene) appear as negative peaks as summarised below:

Spectrun	**Signal intensity**		
	CH	CH_2	CH_3
DEPT—90°	+ve	Zero	Zero
DEPT—135°	+ve	–ve	+ve

It may further be methioned that a variety of pulse angles and delay time may be incorporated into the pulse sequence.

D. Application

Suggest the tentative structure of a compound having molecular formula, $C_{10}H_{12}O_3$, which exhibits the following signals in different spectral measurements.

^{1}HNMR : δ 1.4(t), 4.00 (q), 6.80(d) and 7.20(d).

^{13}C (BB) δ	DEPT—90°	DEPT—135°
15.00	No peak	+ve
40.00	No peak	–ve
63.00	No peak	–ve
115.00	+ve	+ve
125.00	No peak	No peak
130.00	+ve	+ve
158.00	No peak	No peak
179.00	No peak	No peak

➜ The DEPT spectral results suggest that the signal at δ 15.00 in normal ^{13}C spectrum is a CH_3 group and peaks at δ 40.00 and 63.00 are due to CH_2 functionality. Signals at δ 115 and 130 are for CH groups while peaks at δ 125 and 158 may be ascribed to quaternary carbons. The peak at δ 179.00 in the ^{13}C spectrum is a $>C{=}O$ group and the chemical shift value is suggestive of an ester or carboxylic acid. The chemical shift value of methylene carbon at δ 63.00 suggests $—O—CH_2$ grouping which is supported by the apprearance of a signal at δ 4.00 in ^{1}HNMR spectrum and the presence of a triplet at δ 1.40 for a CH_3 group in ^{1}HNMR spectrum indicates the presence of an ethoxy grouping ($—OCH_2—CH_3$). The appearance of two signals at δ 680 (d) and 720(d) in ^{1}HNMR spectrum indicates a para disubstituted aromatic ring system. This is also supported by the presence of two CH groups at δ 115 and 130 and two quaternary carbons at δ 125.0 and 158.00. The remaining methylene group at δ 40.00 in the (^{13}C) carbon spectrum has no adjacent protons. The tentative structure of the compound is as below:

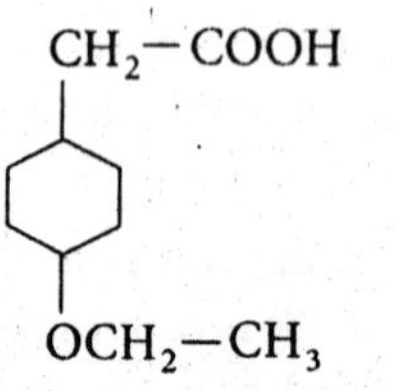

4-Ethoxy phenyl acetic acid

2. Acetaldehyde on reaction with dimethyl malonate in basic medium furnishes a compound having molecular formula, $C_7H_{10}O_4$. The ^{1}HNMR spectrum shows signals at δ 1.95(d), 3.85(s), 3.75(s) and 7.15(q) and its normal ^{13}C carbon and the DEPT spectral results are as below:

^{13}C Spectrum δ	DEPT—135°	DEPT—90°
16	+ve	No peak
52.2	+ve	No peak
52.3	+ve	No peak
129	No peak	No peak

146	+ve	+ve
164	No peak	No peak
166	No peak	No peak

Suggest the probable structure of the product formed.

➜ The reaction between acetaldehyde and dimethyl malonate in basic condition may be outlined as below:

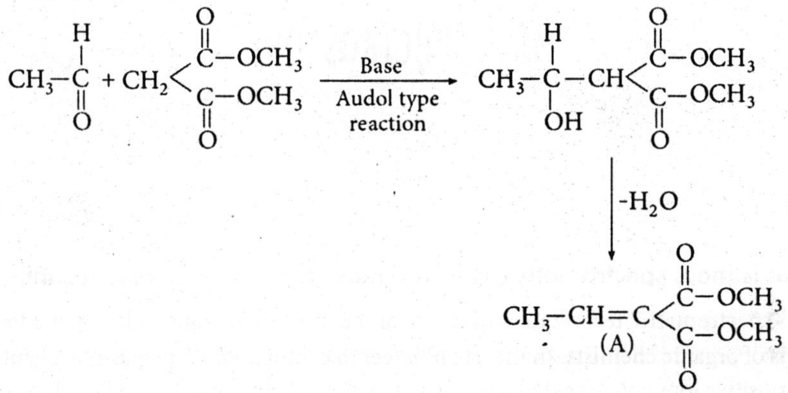

The doublet at δ 1.99 in ^{1}HNMR spectrum may be due to CH_3 group of CH_3—CH— while signals at δ 3.75 and 3.85 are due to —OCH_3 groups of ester functions and the quartet at δ 7.15 may be asigned to methine proton (CH_3—CH=). Again the DEPT spectra reveals that the signals at δ 52.2 and 52.3 may be ascribed to two CH_3 groups of two —CH_3 functionalities. Further, peak at δ 129 is due to quaternary carbon (=C) and peak at δ 146 may be asigned to CH carbon. The two peaks at δ 164 and 166 are due to C=O carbon of ester groups at 7. Thus the structure of the product is as shown above (A).

CHAPTER 6

MASS SPECTROSCOPY

1. What is mass spectrometry? How it differs from spectroscopic technique?

➜ Mass spectrometric technique is now one of the most significant and essential tools at the hands of organic chemists. In the late nineteenth century, W. Wein was successful to show that positive ions could be deflected by magnetic and electric fields and in the early twentieth century, J. J. Thomson was the first to record the mass spectra of molecule having low molecular weight. Originally, this instrumental technique was designed to determine the relative masses of atoms and molecules. Our knowledge on existence of isotopes and the stable elements has been derived from the evidence provided by this technique. Fundamentally, it is a micro-analytical technique and requires only a few milligrams of the sample for providing the valuable and characteristic information regarding the structure and molecular weight of the molecule under study. This technique involves the generation of positive ions in the gas phase and the separation of the ionised molecules and their decomposition products followed by the measurement of the relative abundance of different ions formed.

Mass spectrometry is, in principle, quite different from the usual spectroscopic techniques because in this technique absorption or emission of electromagnetic radiation is not involved in any way. This technique usually bombards molecules in the vapour phase with a high energy electron beam and records the results of electron impact as a spectrum of positive ions separated on the basis of mass/charge (m/z).

2. Discuss the basic principle of mass spectrometry.

➜ Primarily mass spectrometry is an electron impact phenomenon and is based on the same principle as applied in the case of separation of isotopes from its mixture. For the separaion, mixture of isotopes is normally bombarded with a beam of electron having sufficient energy. As a result of such bombardment one or more electrons may be abstracted from each of the isotopes forming uni- and multipositively charged ions. There is also a possibility of generation of negatively charged ions due to the absorption of one electron by the isotopes as depicted below:

$$M + e \text{ (Mixture of isotopes)} \longrightarrow \begin{cases} M^- \text{ (uninegative ions)} \\ M^+ \text{ (unipositive ions)} + 2e \\ M^{n+1} \text{ (Multipositive ions)} + 2e \end{cases}$$

Of these ions unipositively charged ions play the most important role in mass spectrometric technique. This is because of their easy formation and relative stabilities. It may be remembered that the stability of ions is the most important factor in mass spectrometry. After the formation of unipositively charged ions they are accelerated by a strong electrostatic force and are deflected to a magnetic field. As the ions are magnetic in nature they are attracted by the magnetic field and depending upon their masses, are deflected at different positions in the magnetic field—the heavier isotope at closer distance while lighter ones are deflected at a longer distance. In this way, different ions may be separated, collected and their masses may be determined. An obvious question—whether different collectors for different ions or a single collector to be used for the collection of different ions—may arise at this juncture. For the mixture of isotopes of unknown masses placement of different collectors at different positions in the magnetic field is not possible and so single collector is used for the collection of ions of different masses but different deflecting force is applied maintaining the same distance between the source of unipositive ions and the collector. As soon as the unipositive ion reaches the collector a signal is obtained, it is amplified by means of an amplifier and recorded. From the accelerating forces, masses of the different ions are determined. However, it may be mentioned in this connection that the accelerating force (deflecting force) applied does not give the *absolute mass* as is evidenced from the equation (I) deduced as in the page 176.

Let us assume that the unipositive ions of isotopes are accelerated (deflected) in an electrostatic field by the voltage V as shown in the figure below.

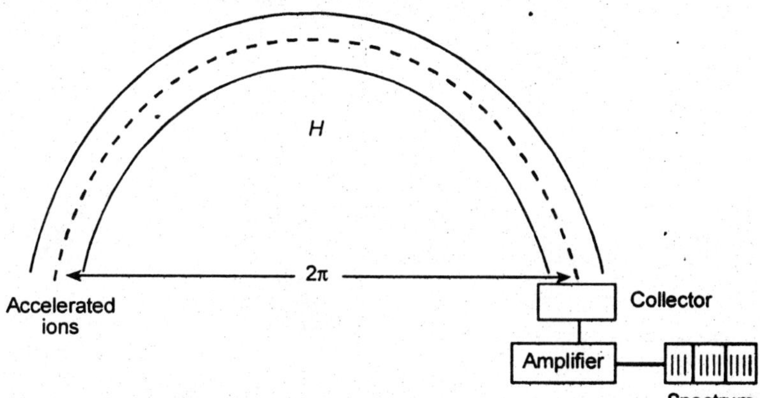

Under this condition the potential energy gained by each unipositive ion is zV where z is the charge of the unipositive ion. It may be mentioned that since all the ions have the same charge and the same voltage has been applied to all the ions they all will have same potential energy.

When the ions pass from accelerating zone (deflecting zone) towards the collector through the magnetic field, the potential energy gradually changes and the ions gain kinetic energy at the cost of potential energy and ultimately potential energy is fully converted to the kinetic energy. Thus, the kinetic energy of each ion becomes equal to zV. Again, we know that kinetic energy is expressed as $\frac{1}{2}mv^2$ where m is the mass and v is the velocity of the moving particle (ion).

Thus, the kinetic energy of the ion is

$$\frac{1}{2}mv^2 = zV \quad \text{(a)}$$

$$\text{or,} \quad v = \sqrt{\frac{2zV}{m}} \quad \text{(b)}$$

From the equation (b) it is evident that the different ions move with different velocities depending upon their masses (z is same for all the ions). Further, it is known that positive ions when pass through magnetic field are forced to move in a circular path through 180° and they are attracted by a centripetal force, Hzv, where H is the strength of the magnetic field through which the ions pass. It produces an acceleration of $\frac{v^2}{r}$ in a circular path of radius r. This centripetal force is balanced by the centrifugal force which is in accordance with Newton's second law of motion ($p = mf$) is $\frac{mv^2}{r}$.

$$\text{Thus} \quad Hzv = \frac{mv^2}{r} \quad \text{(c)}$$

$$\text{or,} \quad Hz = \frac{mv}{r}$$

$$\text{or,} \quad v = \frac{Hzr}{m} \quad \text{(d)}$$

Now putting this value of v in (b),

$$\frac{Hzr}{m} = \sqrt{\frac{2zV}{m}}.$$

Squaring both the sides,

$$\frac{H^2z^2r^2}{m^2} = \frac{2zV}{m}$$

$$\text{or,} \quad \frac{H^2zr^2}{m} = 2V$$

$$\text{or,} \quad \frac{z}{m} = \frac{2V}{H^2r^2}$$

$$\text{or,} \quad \frac{m}{z} = \frac{H^2r^2}{2V}. \quad \text{(I)}$$

Thus from equation (I), $\frac{m}{z}$ ratio can be determined by varying anyone and keeping other two constant. For convenience in almost all the mass spectrometer r is kept constant and $\frac{m}{z}$ ratio can be determined in two ways—

(i) by varying H keeping V constant—this is known as *magnetic scanning.*

(ii) by varying V keeping H constant—this is termed as *electric voltage scanning.*

From the above equation (I), following conclusion can be drawn:

Mass spectrometer does not give the absolute mass but it gives the value of ratio of mass and charge. But as the ions are in general unipositive (i.e., $z = 1$), from equation (I), mass value can be determined .

From the above discusson it is also evident that the major functions of mass spectrometer are: (i) generation of unipositive ions in vapour phase; (ii) their separation and (iii) determination of the masses of the unipositive ions formed. Consequently, the basic principle of any mass spectrometry that may be stated is to form unipositive ions and to determine their masses as regards to their $\frac{m}{z}$ ratio after usual separation .

3. Why is the formation of unipositive ion considered to be the most significant feature of mass spectral studies?

The formation of unipositive ion is considered to be the most significant feature of mass spectral studies because these ions are easily formed and are relatively stable in nature than the other ions that may be generated.

4. Explain the significance of mass spectral technique in organic chemistry.

It is known that the basic principle of mass spectral technique is centered around the formation of unipositive ions and determination of masses of the unipositive ions formed. From this principle it may obviously be concluded that if in any way from any compound formation of unipositive ions are possible then their masses can be determined by mass spectrometry.

It has been found that organic compounds can be ionised by electron bombardment. It has also been observed that if electron beam of sufficient power is used during bombardment organic molecules undergo cleavage with the formation of a large number of unipositive ions (carbocations). Hence, in such a situation an organic molecule may be regarded as a mixture of several unipositive ions. So knowing the masses of each unipositive ion generated from the molecule and computing the masses of different unipositive ions in a proper way by the application of fundamental concepts of organic chemistry, an idea regarding the structure of the molecule may be obtained.

The basic aspect of organic mass-spectrometry, thus consists of bombarding the vapour of an organic compound with high energy electrons, and separating, collecting and analysing the unipositive ions (fragments) formed as regards to their $\frac{m}{z}$ ratio (m = mass, z = charge primarily it is one) and abundance. To illustrate this let us consider an arbitrary organic compound ABCDE containing several chemical bonds. It is subjected to electron bombardment. As a result, ionisation of the organic compound occurs. According to probability factor during ionisation only a portion of the sample undergoes ionisation while a large amount remains as residual portion. Thus, in the first step a single electron is expelled from the molecule with the formation of unipositive ion of the molecule and in the next step this ion undergoes bond cleavage forming a large number of other ions. Thus,

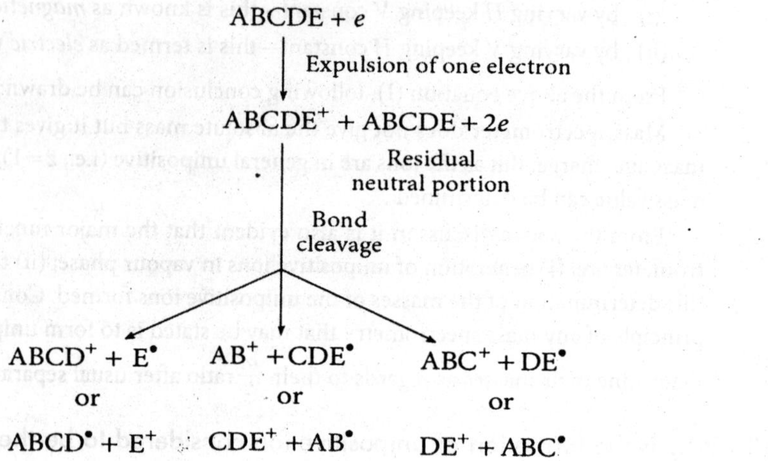

The first step is the one electron abstraction process and the ion that is formed is the radical ion of the molecule but simply it is called molecular ion. In the next step heterolysis, i.e., transfer of pair electrons occurs as shown.

$$\mathrm{ABC{-}\dot{D}\overset{+}{E}} \longrightarrow \mathrm{ABC^+ + DE^\bullet}$$

$DE^\bullet$ portion is short of one electron and so transfer of pair of electrons makes this portion a radical. Normally electron beam of 70 eV is applied. This is equivalent to 1600 kcal/mole which is much higher than ionisation energy (equivalent to 10–15 eV/mole) and fragmentation occurs. As a result several unipositive are formed as shown above. It may be mentioned in this connection that all the ions will not be formed in equal amounts. Hence, the intensity, i.e., abundances of different ions will be different. These different ions are presented by a graph. This graph is a plot of relative intensity versus $\frac{m}{z}$ of ions. In such a graph different ions appear as peaks. This graph is termed as *mass spectrum* of the compound under study as shown in the figure below:

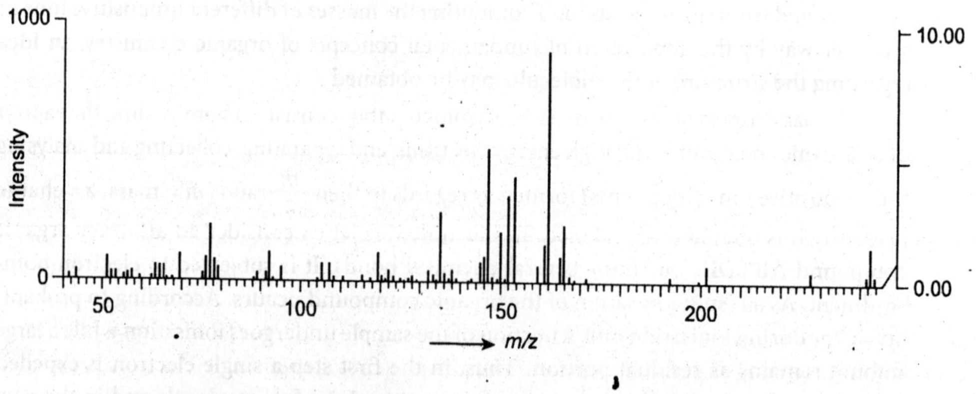

The $ABCDE^{\bullet +}$ ion is termed as molecular ion or parent ion and is generally indicated by M^+, or more precisely as M^+ (at $\frac{m}{z}$). The other ions are known as fragment ions reprsented by $\frac{m}{z}$.

It may be mentioned in this connection that the ratio $\frac{m}{z}$ is a dimensionless ratio of the mass number (m) of a given particle and the number (z) of the electrostatic unit (e) carried by the particle. Prior to 1980, the mass to charge ratio was incorrectly expressed as $\frac{m}{e}$ which was, in fact, the ratio of mass of a particle divided by the value of the electrostatic charge of an electron (1.6×10^{-19} coulombs).

5. (a) What do you mean by the following terms: (i) molecular ion/peak, (ii) quasi-molecular ion/peak, (iii) base peak, (iv) fragment ion peaks and (v) relative intensity? (b) What are factors that control the formation of molecular ion and fragment ions?

➜ (a) (i) *Molecular ion peak*: It is the peak in the mass spectrum due to ion formed from a molecule by the removal of one electron of low ionisation potential. In general, highest mass ion peak in the spectrum is the molecular ion peak. It is indicated by M^+ at $\frac{m}{z}$ and corresponds to the molecular weight of the molecule. But sometimes extreme difficulties may appear to identify the molecular ion peak. This is due to the collision of molecular ion formed initially with the huge amount of residual portion giving rise to some ions having masses more then original molecule which also appear in the mass spectrum. Hence confusion may arise in identifying the actual molecular ion peak. In such situation some special techniques are applied for the recognition of the molecular ion.

(ii) *Quasi-molecular ion/peak*: It has been observed that sometimes the molecular ion formed may undergo collision with the large amount of residual neutral portion. As a result some portion from neutral part may be detached and that part may be attached to the molecular ion formed forming ions having masses more than the parent molecule. These ions are known as quasi-molecular ion and the corresponding peaks in the mass spectrum are termed as the *quasi-molecular ion peaks*.

(iii) *Base peak*: It is obvious that all the ions may not be formed in equal amounts because the stabilities of all the unipositive ions may not be the same. So the abundances and hence the intensities of different ions in the mass spectrum will be different. The most intense peak in the mass spectrum is termed as *base peak*. It is assigned a relative intensity of 100% and the intensities of other peaks including molecular ion peak are expressed as percentage of base peak.

(iv) *Fragment ions*: The ions formed from the molecular ion by the cleavage of chemical bonds are known as fragment ions. They are indicated by $\frac{m}{z}$ and have lower masses. They are, in fact, used as building blocks to reconstruct the molecular structure.

(v) *Relative intensity*: The relative intensity of a particular peak indicates its intensity relative to that of base peak which is the most intense peak in the mass spectrum. The relative intensity of base peak is arbitrarily assigned as 100%. By convention, the relative intensity of any peak in the mass spectrum is indicated by the ordinate on the left side of the graph.

5. (b) Factors controlling the formation of molecular and fragment ions.

➜ It has been observed that in some cases during electron bombardment, the molecular ion formed may collide with the vast amount of residual neutral portion with the generations of ions having masses more than the mass of the molecule creating a lot of confusions in identifying the actual molecular ion. But if we go into details of the process of electron bombardment it reveals that such situation arises (collision between molecular ion and residual portion) only when the molecular ion initially formed is *very unstable* because we know that the stability of the unipositive ion is the most important factor in mass spectrometry. So in order to achieve the desired stability, molecular ion formed undergoes collision with residual neutral portion for abstracting a portion from the residual neutral portion of the molecule and attaching that portion to itself. But in that case the most important factor is that the neutral portion must be *less stable* so that the abstraction of a certain part from it may be easy. Again, which portion of the neutral portion will be detached that depends upon the strength of its different bonds. Preferentially less stable bond in neutral portion will undergo cleavage. It may be mentioned that normally in organic compounds containing heteroatom, particularly nitrogen atom, formation of such ions having masses more than molecular ion is observed. It may further be mentioned that formation of (M + 1) ion is generally common because the condition under which spectral technique is conducted only a very small portion, i.e., H atom can be abstracted from the neutral portion and may be attached to the molecular ion giving rise to (M+1) ion.

Once again, hydrogen atom may be present in the molecule at several positions. Which hydrogen will be removed depends upon the geometry of the molecule.

Thus from the above discussion it is evident that the following four factors are responsible for the formation of molecular and fragment ion.

(i) Stability of unipositive ions including molecular and fragment ions.
(ii) Stability of residual neutral portion of the molecule.
(iii) Stability of the bonds and
(iv) Geometry of the molecule.

6. Describe in detail the instrumental aspects for scanning the mass spectrum of an organic compound.

→ In general, the major functions that any mass spectrometer have to perform are:

(i) Vaporisation of the sample and formation of unipositive ions.
(ii) Separation of unipositive ions.
(iii) Determination of masses of unipositive ions as regards to their $\frac{m}{z}$ ratio (where m and z are the mass and charge of the ions) and their relative abundances. Therefore, the mass spectrometer to be used should consist of the following parts as shown in the figure.
    (i) The inlet system or sample handling
    (ii) Ionising chamber or ion source
    (iii) The accelerating system
    (iv) Ion-separation (Magnetic field)
    (v) Detector (collector)
    (vi) Amplifier and recorder.

(i) *Inlet system (Handling of the sample)*: Before ionisation, sample must be in vapour phase. So the type of inlet system to be used depends upon the nature of the sample. For solid and liquid samples there is a heating device where the sample is heated at a temperature of 350–400 °C in such a way that the vapour of the sample can easily enter into the ionising chamber. For gaseous sample no such heating arrangement is required. It involves simply the transfer of gaseous sample into the ionising chamber. In case of highly volatile liquids cold inlet system is used.

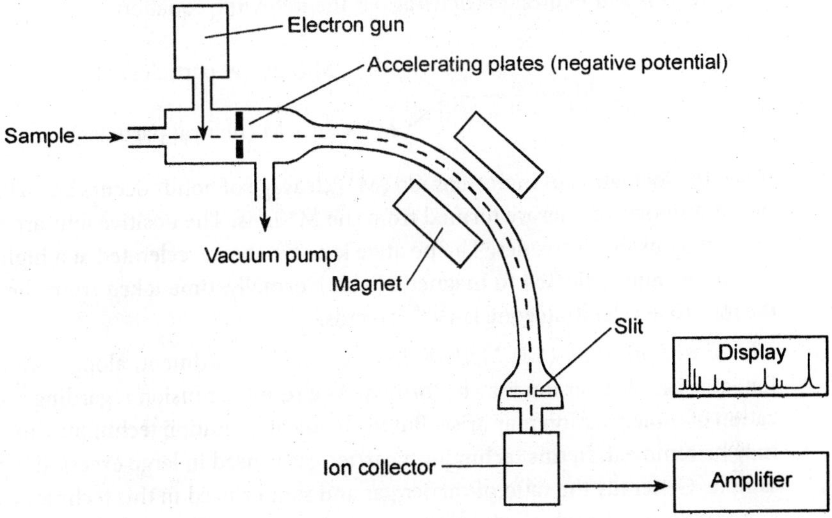

It may be mentioned that only a few nanomole of the sample is required of which only a few percentage of this actually enter into ionisation chamber and only about 1% of them is ionised.

(ii) *Ionisation chamber* (*Generation of ions*): From the inlet system, the sample in vapour phase is introduced into the ionising chamber. For the generation of ions there are various methods, a short account of which are furnished below:

A. *For volatile materials*:

(a) *Electron Ionisation* (*E I*) *Method*: This is routinely used for volatile sample. In this method, the sample is first volatilised at a pressure of 10^{-6} mm. The vapour is allowed to pass through a slit into the ionisation chamber and the vapour is bombarded by a

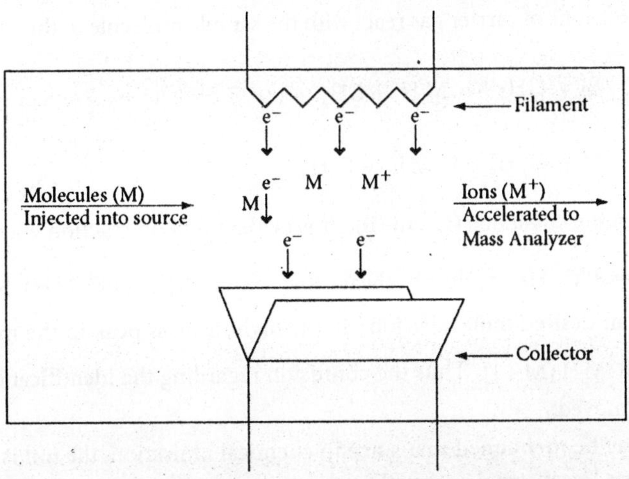

stream of electrons. Electrons required for bombardment are obtained from electrically heated tungsten or rhenium filament. On the side of the filament there is a plate known as electron trap as shown in the figure below. From the trap electrons are allowed to bombard the sample. The bombarded electrons are of high energy (70 eV) and after the impact on the sample, form ions, mainly positive ions, along with small number of negative ions and neutral particles as per the following equation:

$$M + e \longrightarrow \begin{cases} M^- + M \text{ (neutral particles)} \\ \dot{M}^+ + 2e + M \text{ (neutral particles)} \end{cases}$$

After the formation of molecular ion (M^+) cleavage of bonds occurs and a large number of unipositive ions are formed from the M^+ ions. The positive ions are selectively pushed by using electrodes. The positive ions are then accelerated at a high potential (1–10 kV) and deflected to magnetic field. Normally, time taken from the source of the ions to accelerating zone is 10^{-6} seconds.

(b) *Chemical Ionisation (CI) Method*: In electron bombardment, along with molecular ion, quasi-molecular ion may be formed. As a result confusion regarding the identification of molecular ion may arise. But in chemical ionisation technique this confusion may be removed. In this technique a carrier gas is used in large excess along with the sample. Generally the ratio of carrier gas and sample used in this technique is 100 : 1. Commonly used such carrier gases are methane, isobutane and ammonia of which most commonly used is methane gas. Now this mixture of the carrier gas and the sample is subjected to electron bombardment as in the case of electron ionisation method. Since carrier gas is present in large excess of the sample the beam of electrons primarily attacks this carrier gas when the following reactions occur.

$$CH_4 + e \rightarrow \overset{\bullet\,+}{CH_4} + 2e$$

$$\overset{\bullet\,+}{CH_4} + CH_4 \text{ (residual portion)} \rightarrow \overset{+}{CH_5} + \overset{\bullet}{CH_3}$$

$$\overset{\bullet\,+}{CH_4} + \overset{\bullet}{CH_3} \rightarrow \overset{+}{C_2H_5} + H_2\uparrow$$

Now these ions of carrier gas react with the sample molecule in the following manner.

$$M + \overset{+}{CH_5} \rightarrow \overset{+}{MH} + CH_4 \qquad \text{(i)}$$

$$M + \overset{+}{C_2H_5} \rightarrow \overset{+}{MH} + C_2H_4 \qquad \text{(ii)}$$

From the equations (i) and (ii), it is evident that on reaction with secondary ions ($\overset{+}{CH_5}$ and $\overset{+}{C_2H_5}$) molecule always form unipositive ions ($\overset{+}{MH}$) having one mass more than desired molecular ion. So the highest mass peak in the mass spectrum is always at $\overset{+}{MH}$ (M+1). Thus the confusion regarding the identification of molecular ion is removed.

It may be mentioned that since in chemical ionisation, the ion is formed chemically it has less internal energy than the unipositive ion formed by electron ionisation method. So this ion is stable and gives more intense (M + 1) ion peak. However, since the ion is stable it has less tendency to undergo fragmentation. So the number of peaks in the spectrum obtained by this technique is less than the number of peaks obtained by electron ionisation method.

Chemical ionisation is a process of acid-base reaction

$$\text{i.e., } \underset{\text{(Base)}}{M} + \underset{\text{(Acid)}}{\overset{+}{CH_5}} \rightleftharpoons \overset{+}{MH} + CH_4$$

So the formation of $\overset{+}{MH}$ depends upon the proton affinity of the two species. Lesser the proton affinity of acid (secondary ion) and greater the proton affinity of the base (sample), more is the probability of formation of $\overset{+}{MH}$ ion.

It may be mentioned that the tendency of losing proton is very high for $\overset{+}{CH_5}$ than the corresponding ions ($\overset{+}{N}H_4, C_4\overset{+}{H_9}$) of the other carrier gases. That is why methane is commonly used as carrier gas in chemical ionisation.

B. *For non-volatile materials*: There are a large number of organic compounds which are not volatile, so electron ionisation and chemical ionisation techniques cannot be applied. For the ionisation of such molecules desorption ionisation technique may be adopted. This technique is applicable to carbohydrates, proteins, peptides, etc. which do not exhibit molecular ion peaks on electron impact or chemical ionisation technique. There is a variety of desorption ionisation (DI) technique of which following are most important.

(a) *Field desorption ionisation* (FDI) *method*: In this method a thin wire with sharp needles is taken. This wire is supported between two rods as shown below:

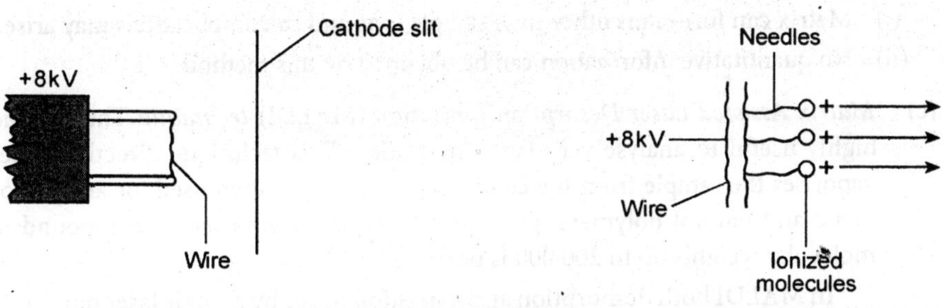

A solution of a small amount of the sample (1 μg) is deposited on the wire. The wire is kept at +8 kV and is heated. Since the needles are very sharp, the field at their tips may be high (10^8 V/cm) which causes the discharge of an electron from the sample into the vacant orbitals of the metal from which the wire is made. As a result positive ions are formed on the wire which is desorbed by coulombic repulsion often with the help of controlled heating of the wire. In this way positive ions are thrown into the gas phase without thermal decomposition.

(b) *Fast atomic bombardment* (FAB) *technique*: It is the most popular method of ionisation and is now frequently used. In this method the sample is dissolved in a viscous liquid. The commonly used such liquid is glycerol. This liquid solvent is termed as *matrix material*. The solution of the sample in viscous liquid is coated on a thin metal plate. This thin metal plate coated with the sample solution is known as *sample matrix*. The sample matrix is bombarded by a beam of fast moving neutral atom. Commonly used such neutral atoms are usually inert gases either xenon or organ. It is prepared by ionising first a small portion of inert gas in a separate chamber and then accelerated with an energy of 6-9 keV to inert gas chamber where the previously ionised inert gas reacts with the rest of the inert gas atoms when the transfer of charge occurs as under:

$$Xe \rightleftharpoons \overrightarrow{\dot{X}e^{+}}$$

$$\overrightarrow{\dot{X}e^{+}} + Xe \rightleftharpoons \overrightarrow{Xe} + \dot{X}e^{+}$$

Fast moving atom

Now the fast moving atom $\overrightarrow{Xe}$, thus formed, impacts into the soluton of the sample in the matrix as shown below.

$$M + \overrightarrow{Xe} \rightarrow \dot{M}^{+} + Xe$$

Now the sample desorbed often as an ion by a momentum transfer. The beam of sample ion is then analysed as usual.

This method has following advantages and disadvantages.

Advantages

(i) It can be used for non-volatile high molecular weight sample of biological importance.

(ii) Abundant (M+1) ion are formed and hence molecular ion can be determined without any confusion.

(iii) It can be used for mass analysis of salts of the sample.

Disadvantages

(i) Matrix can form ions other than sample ions and so complications may arise.

(ii) No quantitative information can be obtained by this method.

(c) *Matrix Assisted Laser Desorption Ionisation* (MALDI) *technique*: This technique is highly useful to analyse very large molecules. This technique directly ionises and vaporises the sample from the condensed phase. It is often used for analysis of synthetic and natural polymers, proteins and peptides. Analysis of compound having molecular weights up to 200,000 is possible.

In MALDI both desporption and ionisation occur by a single laser pulse as shown in the figure below.

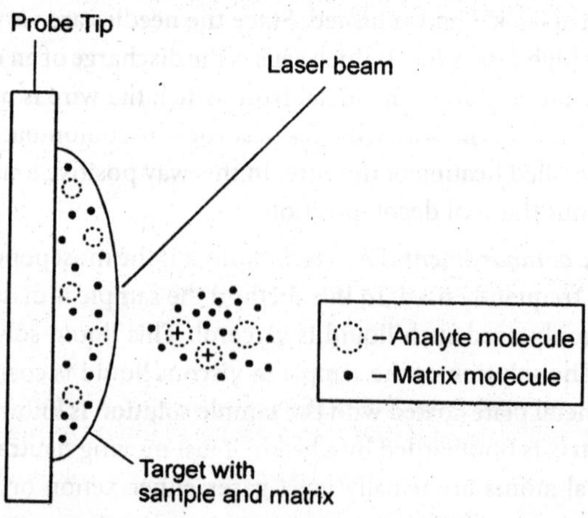

The sample is prepared by mixing the moecule under consideration with a matrix compound chosen to absorb the laser wavelength. This sample is placed on the tip of the probe and dried. A vacuum lock is used to insert the probe into the source region of the mass spectrometer. A laser beam is allowed to pass on the dried mixture and the energy from the laser pulse is absorbed by the matrix. This energy ejects ions of the molecule from the surface so that a mass spectrum is obtained for each laser pulse. This technique is more general than other laser ionisation processes because matrix absorbs laser pulse whereas in other laser ionisation techniques, the molecules under consideration absorb laser wavelength.

(iii) *Accelerating system*: The unipositive ions formed in the ionisation chamber are withdrawn by a strong electrostatic field of 400–4000 V which accelerates the ions to the ion separator, i.e., magnetic field.

(iv) *Ion separation (Magnetic field)*: The separation of ions, i.e., resolution is an important aspect of mass spectrometer on which the identification of ions depends. In this part of the mass spectrometer, ions are separated according to their mass and charge ratios. An analyser (separator) must possess the following characteristics:

(a) It should have a high resolution capacity.

(b) It must have a high rate of transmission of ions.

Unfortunately, above two characteristics are incompatible. When slits are made narrow, the resolution increases but transmission of ions decreases. For this reason a compromise has been reached. There are different kinds of mass spectrometers which differ mainly in the way they sort out the ions. Some of these are as described below.

(a) *Single focussing magnetic deflection*: This is the most common type of analyser. It has low power of resolution.

(b) *Double focussing magnetic deflection*: This type of analyser has higher power of resolution than the single focussing instruments. However, this type of deflectors have low rate of transmission of ions. This type of instrument is generally used when highest resolution is required and normally is applied when precise mass of the molecule is to be determined.

(c) *Time-of flight* (TOF) *mass spectrometer*: The *time of flight* (TOF) technique is based on the fact that the ions produced in the source leave the accelerating zone with different velocities depending on their masses. If these ions are allowed to travel along a straight line through a *magnetic field-free region*, they will take different time to travel a given distance. The measurement of 'this line of flight' forms the basis for this non-magnetic separator as shown in the figure below:

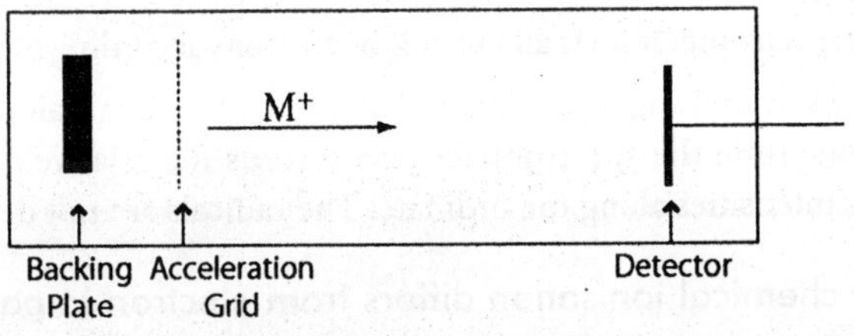

In the source of TOF analyser, a packet of ions is formed by a very fast ionisation pulse. These ions are accelerated into the flight tube by an electric field

($\approx$ 2–2.5 kV) applied between the backing plate and acceleration grid. Since all the ions are accelerated across the same distance by the same force, they have the same kinetic energy ($\frac{1}{2}mv^2 = zeV$, where v is the velocity of the ion, V is the accelerating voltage, m is the mass of the ion and *e.z.* is the charge of the ion). Now during acceleration the ions will begin to separate according to their masses as they drift along the flight tube and the ions of different mass arrive at the detector at different time t as per the following equation,

$$t = \frac{L}{\sqrt{2V \times e}} \times \sqrt{\frac{m}{z}},$$

where L is the length of drift region,

The mass spectrum is obtained by measuring the detector signal as a function of time for each pulse of ions produced in the source.

This is a very simple mass spectrometer that uses fixed voltage and does not require a magnetic field. But TOF instruments have poor mass resolution and low detection limits. However, recently higher resolution technique has been introduced in this process.

(v) *Ion detector*: The ions which are separated by the analyser are detected and measured electrically or photographically. The photographic plate is generally used with radio frequency spark. The ions pass through the collecting slit one after another and fall on the detector. The spectrum is scanned by going up the scale. The spectrum is recorded by using a fast scanning oscillograph

7. In many compounds molecular ions are very weak or completely absent in the mass spectrum. Explain how to overcome this limitation.

➧ The presence of weak molecular ions or their complete absence means that the definite information regarding the molecular weight is very difficult to obtain under high energy electron ionisation. This problem may be solved by lowering the electron energy but this may be at the expense of absolute intensity. The most preferred method is to adopt soft ionisation process like chemical ionisation.

8. How the molecular ion peak can be accurately identified?

➧ See chemical ionisation part in answer against Question No. 6(ii) A(b).

9. After ionisation and fragmentation what function a mass spectrometer performs to produce mass spectrum?

➧ The mass spectrometer sorts all the unipositive ions according to their $\frac{m}{z}$ values and records these values decreasing from right to left as the line signals along the abscissa of the chart. At the same time the spectrometer also records the relative abundance as signal height plotted as intensities along the ordinate. The radicals are not detected.

10. (a) How chemical ionisation differs from electron impact?

(b) How does it help in determining the accurate molecular ion peak?

➧ (a) The chemical ionisation technique differs from electron impact phenomenon in the following respects.

(i) Since in chemical ionisation the ion is formed chemically it has less internal energy than the ion formed by electron ionisation. Hence, this ion is stable and displays intense (M + 1) ion peak.

(ii) Since the ion is stable it has less tendency to undergo fragmentation. So the number of peaks in the spectrum obtained by this technique is less than the number of peaks obtained by electron impact method.

(iii) Chemical ionisation is an acid -ase reaction, i.e.,

$$M + \overset{+}{C}H_5 \rightleftharpoons M\overset{+}{H} + CH_4$$

So the formation of $M\overset{+}{H}$ ion depends upon the proton affinity of the sample—greater the proton affinity of the sample more is the probability of formation of $M\overset{+}{H}$ ions.

(b) See answer against Question No. 6(ii) A(b).

11. What is the mass spectrum? Write what you know about the formation and stability of the molecular ion.

➻ See answer against Question No. 5.

12. Name the five basic sections of a mass spectrometer. What is the function of each?

➻ See answer against question No. 6.

13. Discuss in short how ions are generated from volatile organic compounds.

➻ See answer against Question No. 6(ii) A.

14. Write a short note on fast atomic bombardment (FAB) technique for ionisation of organic molecules. What are its advantages and disadvantages.

➻ See answer against Question No. 6(ii) B(b).

15. Discuss how the exact formula of an organic compound can be determined from the accurate mass obtained from the mass spectrum?

or

How will you proceed to determine molecular formula of an organic compound using mass spectrometer?

➻ The mass spectrum of an organic compound records accurately the mass of the compound up to six decimal places. From this accurate mass the exact molecular formula of the compound can be deduced without the help of estimation of carbon, hydrogen and other elements that may be present. Several observations serve as the guiding factors in deduction of molecular formula. Such deductions may be noted as under:

(i) It has been found that organic compounds generally contain only a fewer elements and the vast majority of organic compounds are commonly made up of carbon, hydrogen, oxygen and nitrogen and only in fewer cases halogen, sulphur and phosphorus are present. Because of this observation along with the consideration of the valencies of these elements *the number of possible molecular formulae for a given integral molecular weight* is finite and relatively smaller and can be written down systematically. Application of another important generalisation in this context becomes highly useful. This rule is known as *Nitrogen rule. This rule states that an organic compound containing odd number of nitrogen will have odd molecular mass while an organic compound having even numbered molecular mass must contain either no nitrogen or even number of nitrogen atoms.* This rule is applicable to all compounds containing carbon, hydrogen, oxygen, nitrogen, sulphur and halogens. Thus by applying this rule some of the possible structural formulae from the smaller number may be ruled out as the exact molecular formula. After writing down all the possible moelcular formulae the exact molecular weight of each is calculated by using the accurate atomic weights. From the accurate molecular weight the correct molecular formula from the list may be easily ascertained.

An example may be cited in this connection to explain the above process.

An organic compound records its mass as 100.0635 in its mass spectrum. What should be its exact molecular formula?

The integral part of the molecular mass is 100. On the basis of this mass number following molecular formulae may be listed.

C_8H_4	$C_4H_4O_3$	$C_3H_8N_4$
C_7H_{16}	$C_6H_{12}O$	$C_5H_{12}N_2$
	$C_5H_8O_2$	$C_4H_8ON_2$

Now taking the atomic weights of the following elements as

H = 1.0078	N = 14.0030
C = 12.0000	O = 15.9949

The exact molecular weights of each of the molecular formulae may be calculated as below.

Molecular Formula	**Molecular Weight**
C_8H_4	100.0313
C_7H_{16}	100.1251
$C_6H_{12}O$	100.0887
$C_5H_8O_2$	100.0522
$C_4H_4O_3$	100.0161
$C_5H_{12}N_2$	100.0004
$C_4H_8ON_2$	100.0637
$C_3H_8N_4$	100.0748

Now comparing these calculated values with the experimental one it is seen that only the molecular formula, $C_4H_8ON_2$ has molecular weight very close to the experimental value. So the molecular formula of the compound is $C_4H_8N_2O$.

16. What is nitrogen rule? Explain with suitable example its importance in mass spectral analysis.

➜ Nitrogen rule is an important guiding factor in deciding the molecular ion of an organic compound. In a mass spectrum there may be a large number of ion peaks. Many of these peaks may be ruled on the grounds of reasonable structural requirement. *Nitrogen rule* is often helpful in this regard. This rule states that *a molecule with even numbered molecular mass must contain either no nitrogen or even number of nitrogen atoms* and a *molecule having odd numbered molecular mass contains an odd number of nitrogen atoms.* This rule has been found to be applicable for all organic compounds containing carbon, hydrogen, oxygen, nitrogen, halogens and sulphur.

An important corollary of this rule states that the fragmentation at a single bond gives an odd numbered fragment ion from an even number fragment ion. Similarly, an even numbered fragment ion results from an odd numbered molecular ion.

An example may be cited in this connection to explain this rule. The organic compound nitro-benzene, $C_5H_5NO_2$ records its molecular ion at $\frac{m}{z}121$. This mass is an odd number. Hence the molecule must contain odd number of nitrogen which is actually found to be so. Again, nitrobenzene exhibits two important fragment ions at $\frac{m}{z}46$ and 30. As required by this rule both of these fragment ions have even number of masses. Similarly, the mass spectrum of 2,4–dinitrophenol records molecular ion peak at $\frac{m}{z}184$ along with fragment ions at $\frac{m}{z}183$ and 155. Since dinitro-benzene has two nitrogen atoms, its molecular ion appaears at even number of mass. The two fragment ions containing both the nitrogen atoms ($M^+ - H$ at $\frac{m}{z}183$) and ($M^+ - H - CO$ at $\frac{m}{z}155$) appear at odd mass number jusifying thereby the validity of nitrogen rule.

17. What are metastable ions? How are they recognised? Explain their significance in mass spectral analysis.

It is obvious that all the unipositive ions formed in the ion chamber will not have equal stability—different ions formed will have different stabilities. So unipositive ions of different stabilities will have different internal energies and hence their lifetime will be different. It has been observed that any unipositive ion formed at the source takes 10^{-6} seconds to reach the accelerating zone from source and requires 10^{-5} second to reach the collector. So it is logical to assume that any unipositive ion to be fully accelerated to the collector must have half-life period more than 10^{-5} second. The ions formed in the ionising chamber may have either of the following properties:

(i) Ions may have lifetime less than 10^{-6} second and so they may decompose at the source and will not reach the collector.

(ii) The ions formed may have half-life period longer than 10^{-5} second. These ions may survive long enough to reach the collector. However, it is logical to assume that there may be a third alternative in between these two options. This third option is that there may be formation of some unipositive ions having half-life period longer than 10^{-6} second but less than 10^{-5} seconds. This means such ions will leave the ionisation chamber intact but decompose en route to the collector. It may be mentioned that when just accelerated from the accelerating zone, say, by a voltage V, all the ions will have their potential energy equal to zV (where z is the charge) which will gradually change to kinetic energy during the passage because the ions now start moving. Let us suppose an unipositive ion m_1^+ formed in the source has lifetime in the range 10^{-6} – 10^{-5} second, i.e., long enough to reach the accelerating zone but not long enough to become fully accelerated to the collector. So such ion decomposes somewhere in the passage to the collector. Let us consider it decomposes to m_2^+ and a neutral portion B as under:

$$m_1^+ \rightarrow m_2^+ + B$$

The kinetic energy which is equal to zV of the parent ion $\left(m_1^+\right)$ will now be divided between the daughter ion $\left(m_2^+\right)$ and the neutral portion B in proportion to their masses (conservation of momentum). So the daughter ion $\left(m_2^+\right)$ must have lower kinetic energy than the parent ion $\left(m_1^+\right)$. The kinetic energy will be reduced by the factor $\frac{m_2}{m_1}$. But if m_2^+ would have been formed in the source its kinetic energy would have been zV (all ions will have same energy). It may be mentioned in this connection that when the ions are deflected from the accelerating zone to the magnetic analyser, the attraction by the magnetic field depends upon its energy which is finally now kinetic energy—greater the kinetic energy of the ion, higher will be the $\frac{m}{z}$ value. Since m_2^+ has lower kinetic energy it will not appear at m_2 mass value on mass scale but will appear at a lower value being lowered by a factor $\frac{m_2}{m_1}$. This lower mass value is usually indicated by m^*. Thus,

$$m^{\bullet} = m_2 \times \frac{m_2}{m_1} = \frac{m_2^2}{m_1} \tag{i}$$

The ion of the mass value m^* is known as *metastable ion* and the corresponding ion peak at $\frac{m}{z}(m^{\bullet})$ in the mass spectrum is known as *metastable ion peak*.

The metastable ions and corresponding peaks in the mass spectrum can be easily identified from their following characteristic features:

(i) They have very low intensity.

(ii) They appear at non-integral value and

(iii) It has been oberved that the formation of metastable ion is possible for those compounds for which there are formation of both parent and daughter ions in the chamber and hence always there are appearance of three peaks i.e., for parent, daughter and metastable ions in the mass spectrum. This is due to fact that since the parent ion forms daughter ion in the passage to the collector due to its decomposition, it is reasonable to assume that the parent ion may decompose in the ion chamber and there is also the formation of daughter ion and may be collected and recorded at the mass value of daughter ion.

Significances of metastable ions/peaks: The formation of metastable ions and their appearance in the mass spectrum are highly significant in the interpretation of the mass spectral data. The formation of ions suggests that the parent ion decompose to daughter ion by a single step and hence helps in interpreting the process of fragmentation.

If there is an ion peak of lower intensity at non-integral mass value and if such peak is related to two other peaks (parent and daughter ion peaks) by the equation (I) mentioned earlier, it is reasonable to assume that the daughter ion has been formed from the parent ion by a single step and hence is helpful in the mass spectral fragmentation. Following examples may be cited to explain the significance.

(i) Acetophenone, $C_6H_5COCH_3$ in its mass spectrum records the following ion peaks at $\frac{m}{z}$ 120, 91.88 and 105. The fragmentation may be explained as under:

The peak at $\frac{m}{z}$ 91.88 is the metastable ion peak in between the parent ion at $\frac{m}{z}$ 120 and daugther ion at $\frac{m}{z}$ 105 as it is related by the equation (I).

$$91.99(m^{**}) = \frac{m_2^2}{m_1} = \frac{m_2^2}{120}$$

or, $$m_2^2 = 91.88 \times 120$$

or, $$m_2 = \sqrt{91.88 \times 120} = 105.$$

Thus the ion peak at $\frac{m}{z}$ 105 has been generated from the parent ion at $\frac{m}{z}$ 120 directly as shown below.

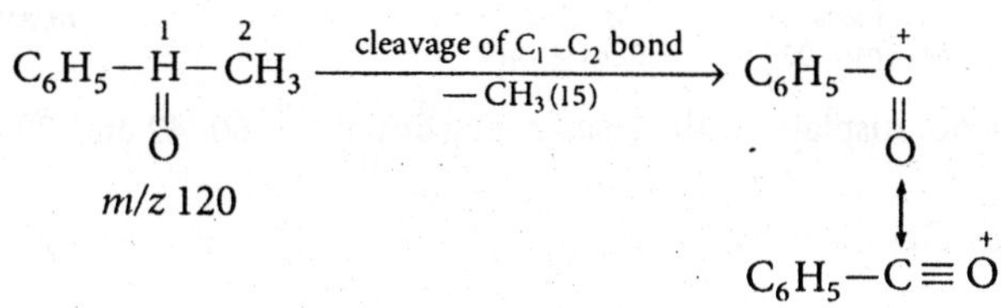

(ii) Anisidine exhibits following ion peaks in the mass spectrum at $\frac{m}{z}$ 123, 108, 80, 94.8 and at 59.2.

Now peak at $\frac{m}{z}$ 94.8 is the metastable peak in between the peaks at $\frac{m}{z}$ 123 and 108 because equation (I) is satisfied, i.e.,

$$m^* = \frac{m_2^2}{m_1} = \frac{(108)^2}{123} = 94.8.$$

Again, the ion at non-integral value 59.2 is the metastable peak in between the peaks at 108 and 80, i.e.,

$$m^* = \frac{m_2^2}{m_1} = \frac{(80)^2}{108} = 59.2.$$

Thus, it may be concluded that peak at $\frac{m}{z}$ 108 has been originated from the peak at $\frac{m}{z}$ 123 by single step. Similarly, peak at $\frac{m}{z}$ 80 has been directly formed from the peak at $\frac{m}{z}$ 108. Thus,

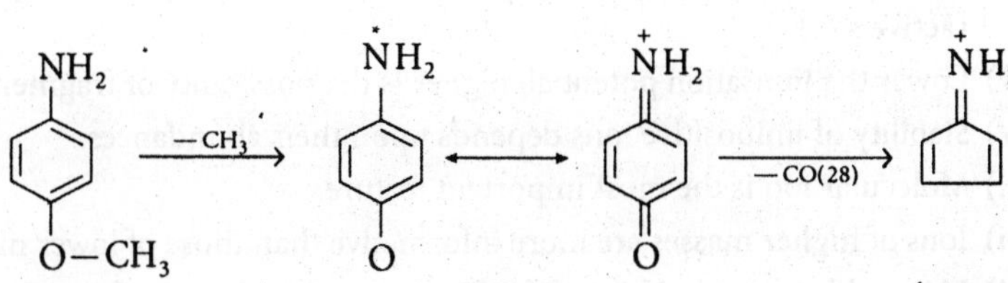

(Aniside) m/z 123 m/z 108 m/z 80

(iii) Toluene, $C_6H_5CH_3$, shows two strong peaks at $\frac{m}{z}$ 91 and 65 along with molecular ion peak at $\frac{m}{z}$ 92. It has been suggested that during fragmentation toluene loses a hydrogen atom and forms stable tropylium ion ($\frac{m}{z}$ 91) which then loses a molecule of acetylene (26 mass units) forming an ion $C_5\overset{+}{H}_5$ (at $\frac{m}{z}$ 65). This suggestion for explanation of fragment may be proved correct if there is a formation of a meta-stable ion peak at $\frac{m}{z}\frac{65^2}{91} = 46.4$. In actual practice a metastable peak at $\frac{m}{z}$ 46.4 appears in the mass spectrum of toluene. Thus

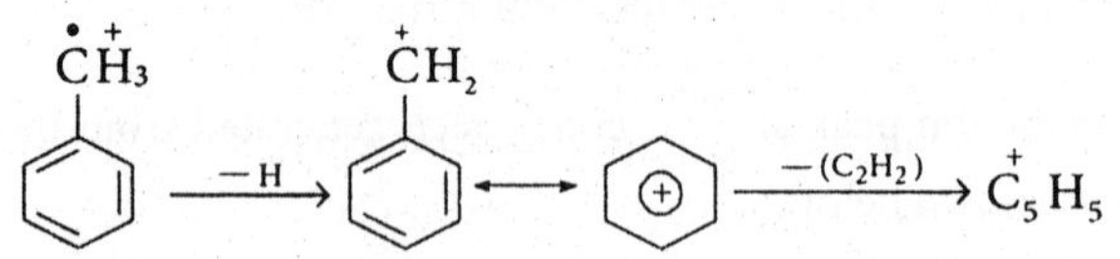

Toluene M⁺ (m/z 92) — m/z 91 Benzyl ion — m/z 91

(iv) n-Propanol displays in the mass spectrum at $\frac{m}{z}$ 60, 42 and 29.4. Now 29.4 mass units staisfies the equation (I), i.e., $\frac{42^2}{60} = 29.4$. Hence, the mass ion at $\frac{m}{z}$ 42 generates directly from the molecular ion at $\frac{m}{z}$ 60. Thus

$$\underset{\substack{n\text{-Propanol} \\ M^+ \text{ at } m/z\,60}}{CH_3{-}CH_2{-}CH_2{-}OH} \xrightarrow[\text{(18 mass units)}]{-H_2O} \underset{m/z\,42}{CH_3{-}CH{=}CH_2}$$

18. Write in detail the general modes of fragmentation processes in mass spectrometry.

➻ The primary aim of organic chemists is to interpret properly the mass spectral data, i.e., to find out the origin of different ion peaks in the mass spectrum in terms of correct molecular structure. For such purposes organic chemists use an important empirical approach involving the study of formation of different ions in terms of entire molecule. This empirical approach is commonly termed as *Quasi Equilibrium Theory* (QET). This theory was first proposed by Erying and is based on the following observations:

(i) Stability of unipositive ions is a most important factor.

(ii) On ionisation both ions and radicals may be formed but ions are stabler than radicals.

(iii) During ionisation normally that electron is expelled which requires least energy (active site).

(iv) Lower the ionisation potential, higher is the possibility of fragmentation.

(v) Stability of unipositive ions depends upon their abundances.

(vi) Molecular ion is the most important feature.

(vii) Ions of higher masses are more informative than those of lower masses.

(viii) Metastable ions provide useful information regarding the fragmentation of the molecule.

Based on such observations along with appropriate application of the concepts of carbocation chemistry and mechanistic paths of organic reactions fragmentation processes have been considered by several groups of chemists as indicated below.

The different fragmentation processes observed have been classified into two major groups: (A) simple cleavage and (B) Rearrangement process.

A. *Simple Cleavage*: This type of fragmentation involves the simple cleavage of sigma bonds without any kind of migration or rearrangement. It is known that the sigma bonds in organic compounds may be present in between carbon atoms or it may be present between a carbon atom and a hetero atom—consequently this type of cleavages have been considered under the following headings as mentioned below:

(i) *Cleavage of carbon-carbon sigma bond*:

(a) *Cleavage of sigma bond of saturated hydrocarbon*: In this process of fragmentation carbon-carbon sigma bond undergoes cleavage as shown below:

$$-\overset{|}{\underset{|}{C}}-\overset{|}{\underset{|}{C}}- \xrightarrow{-e} -\overset{|}{\underset{|}{C}}-\overset{|\cdot+}{\underset{|}{C}}- \longrightarrow -\overset{|+}{\underset{|}{C}} + \overset{|\cdot}{\underset{|}{C}}-$$

It may be mentioned that in organic compound there may be large number of carbon-carbon sigma bonds. Since the stability of unipositive ion is the prime consideration, normally that carbon-carbon bond undergoes cleavage by which comparatively stable unipositive ion (carbocation) is formed. Thus,

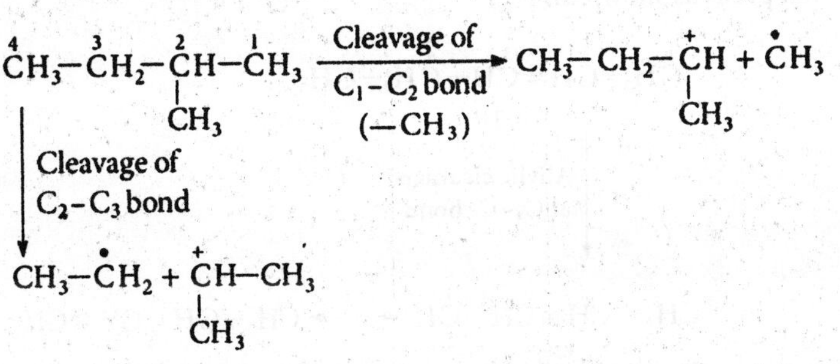

Such type of fragmentation frequently occurs in hydrocarbon class of organic compounds and is of highly helpful in differentiating isomeric hydrocarbons as shown below:

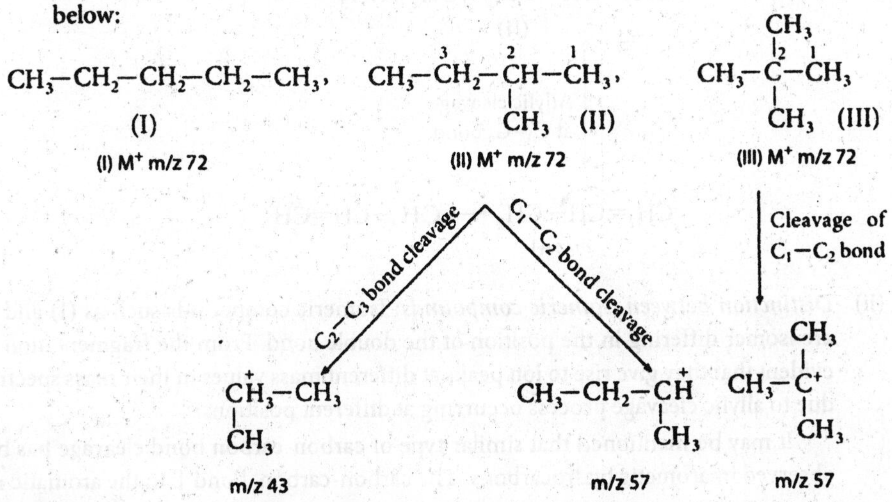

From the above mode of fragmentation it is thus evident that (I) exhibits only the molecular ion peak whereas (II) displays peaks at $\frac{m}{z}$ 57 and 43 along with molecular ion peak but the compound (III) in its mass spectrum shows intense peak at $\frac{m}{z}$ 57 along with M^+ ion and so they may easily be distinguished.

(b) ***Influence of π-bond in the cleavage of* C–C *sigma bond (Allylic Cleavage)*: The presence of π-bond in a compound greatly influences the fragmentation of the C–C sigma bond. In this situation carbon-carbon sigma bond, β to the double bond (*allylic bond*) preferably undergoes cleavage and is a favoured fragmentation mode in unsaturated compound. The resonance stabilisation of allyl cation results in the increased probability of the fragmentation of carbon-carbon bond β to a double bond. Thus,**

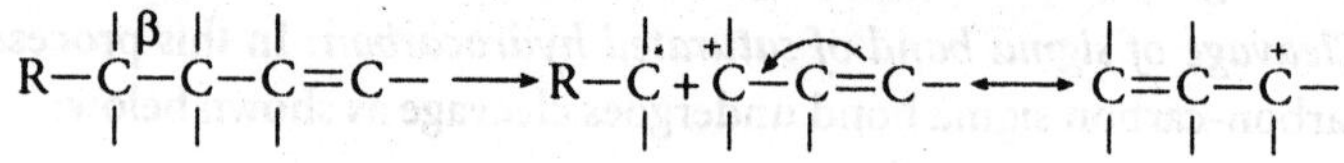

Allylic Cation

This type of fragmentation process is commonly known as *allylic cleavage* and is very useful for variety of purposes as illustrated below:

(i) *Location of double bond in unsaturated compound*: In unsaturated compond the location of the double bond may be ascertained from the mass value of the fragment ion peaks as shown under:

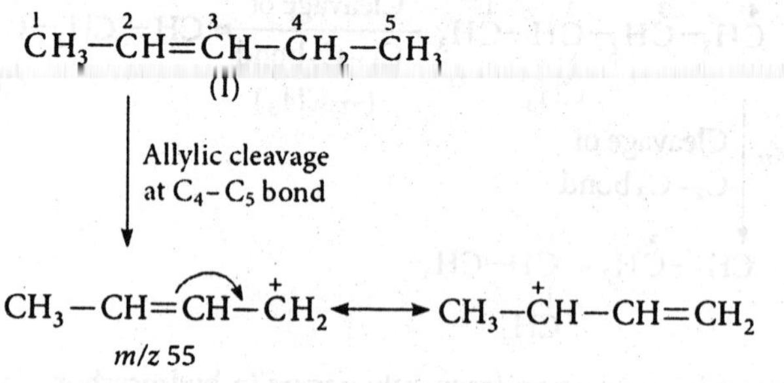

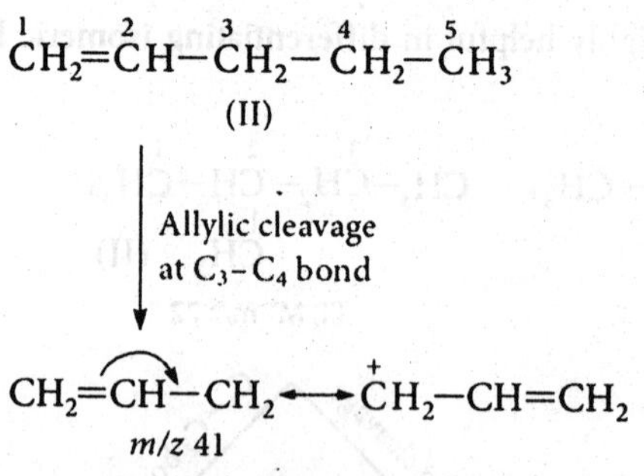

(ii) *Distinction between isomeric compounds*: Isomeric compounds such as (I) and (II) are isomer differing in the position of the double bond. From the fragmentation it is evident that they give rise to ion peaks at different mass values in their mass spectrum due to allylic cleavage process occurring at different positions.

It may be mentioned that similar type of carbon-carbon bond cleavage has been observed in aromatic hydrocarbons. The carbon-carbon bond β to the aromatic ring

(benzylic position) is an energetically preferred fragmentation mode. This type of fragmentation process may be looked upon as an extension of allylic cleavage and is usually termed as *benzylic cleavage*. This type of fragmentation process may be exemplified as under:

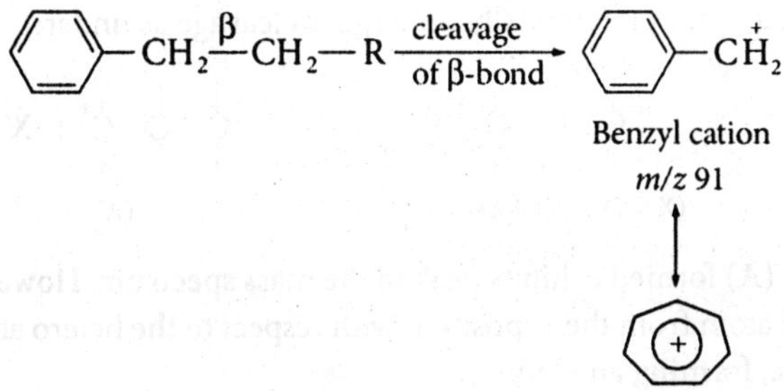

(iii) *retro-Diels-Alder Cleavage*: When the double bond is present in a cyclic system (Cycloalkene) β-bond is available in both the sides of the double bond. So double allylic cleavage can occur. As a result there are formation of two unsaturated fragments with the generation of two new π-bonds. Thus,

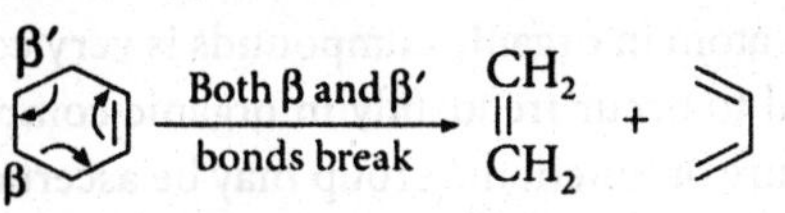

Since this cleavage process is a reverse process of a very well known chemical reaction—'Diels-Alder reaction' it is generally termed as *retro-Diels-Alder Cleavage* and is a characteristic fragmentation mode of cyclic olefins and frequently occurs in cycloalkene system.

This process of fragmentation is highly useful in the following cases.

(A) Location of substituent in the cycloalkene:

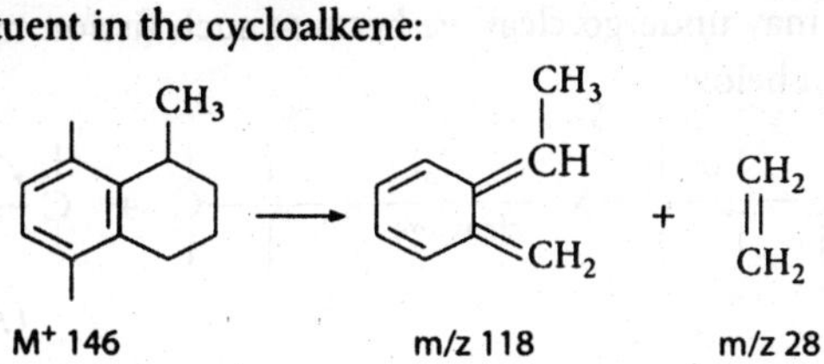

M^+ 146 m/z 118 m/z 28

α-Methyltetralin

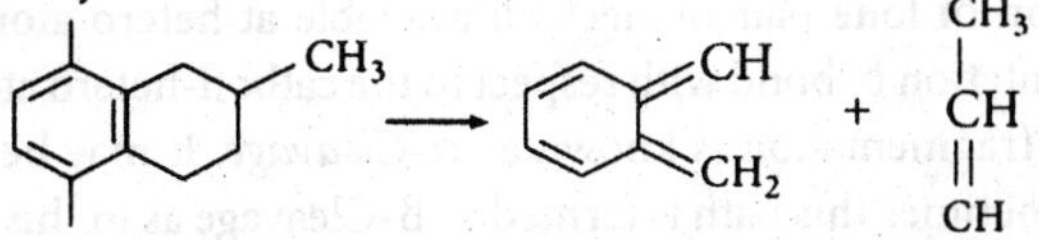

β-Methyltetralin m/z 104 m/z 42

(B) Location of double bond in cycloalkene system.

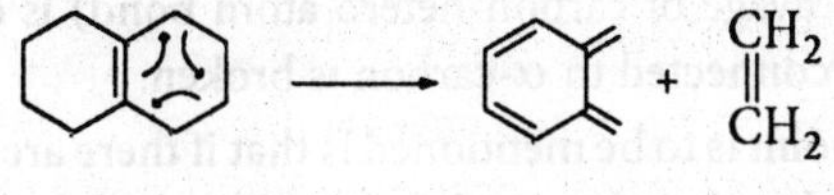

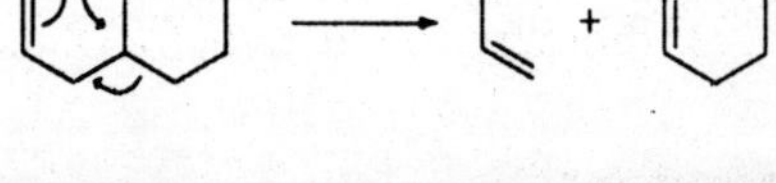

(C) *Differentiation of Isomeric compounds*: Isomeric compounds as in above example can be distinguished from the mass values of fragment ions.

(c) *Influence of heteroatom in the cleavage of sigma bonds*: When a hetero atom is attached to a carbon atom since the carbon-hetero atom bond is comparatively weak, normally that bond preferentially undergoes cleavage as under:

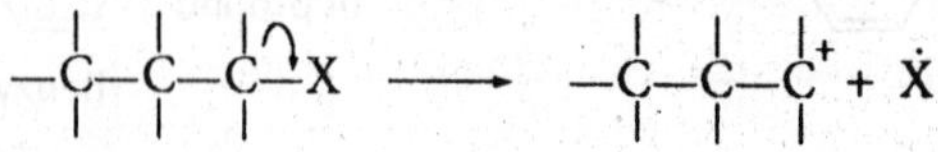

(X = O, N, Halogen, etc) (A)

The ion (A) formed exhibits peak in the mass spectrum. However, if the ion (A) is not stable H atom from the β-position with respect to the hetero atom is eliminated, if it is available, forming an olefin.

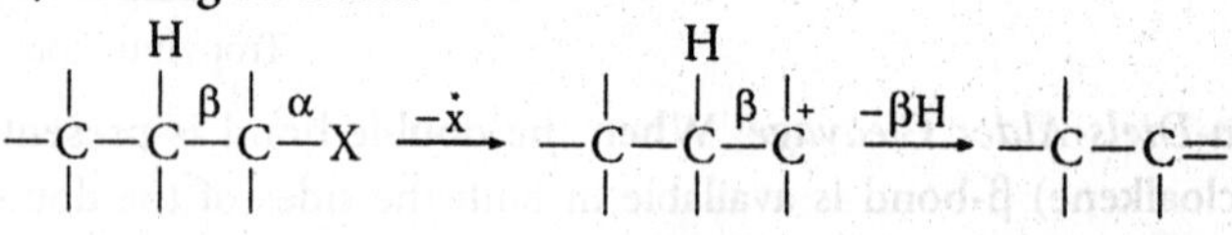

(A) Olefin

Thus, in this situation, the net elimination is HX from the parent ion. Since the presence of hetero atom in organic compounds is very common, this process of fragmentation is found to occur frequently in organic compound and by this process of fragmentation nature of functional group may be ascertained. Thus, an organic compound containing hydroxyl function (–OH) readily an ion peak at M–17 (M–OH) is formed in its mass spectrum and if there is any mass peak at M–18 the molecule must contain a H atom at β position along with OH function.

α-*Cleavage*: When a hetero atom is present in an organic molecule, in addition to the cleavage of carbon-hetero atom bond, the carbon-carbon bond α-to-carbon-hetero atom bond may undergo cleavage because such fission may form stable unipositive ion as shown below.

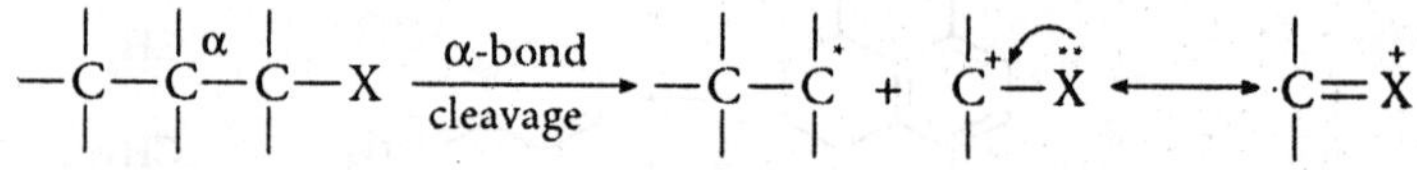

(A)

The initially formed unipositive ion (A) is stabilised by resonance due to the participation of lone pair of electron available at hetero atom. Since in this process of fragmentation α-bond with respect to the catbon-hetero atom undergoes cleavage this path of fragmentation is known as α-*Cleavage*. It may be mentioned in this context that sometimes this path is termed as β-Cleavage as in this case C–C bond attached to β-carbon with respect to hetero atom ($-\overset{\beta}{C}-\overset{\alpha}{C}-X$) undergoes cleavage and the previous path (cleavage of carbon-hetero atom bond) is obviously called to α-cleavage because bond connected to α-carbon is broken.

Another point is to be mentioned is that if there are more than one α-bond preferentially that α-bond undergoes breakage by which neutral portion of higher mass is eliminated. Thus,

$$CH_3 \overset{\alpha}{—} CH_2—O—CH_2 \overset{\alpha'}{—} CH_2—CH_3 \xrightarrow[\text{cleavage of a' bond}]{\text{preferential}} CH_3—CH_2—O—\overset{+}{C}H_2 \longleftrightarrow CH_3—CH_2—\overset{+}{O}=CH_2$$

Since the presence of hetero atom is extremely common in organic molecules this path of fragmentation is of frequent occurrence and on the basis of this fragmentation definite conclusion regarding the structure of the molecule may be obtained. Thus, with the help of this fragmentation process the structures of primary, secondary and tertiary alcohols may be ascertained or these three types of alcohols may be differentiated.

(i) 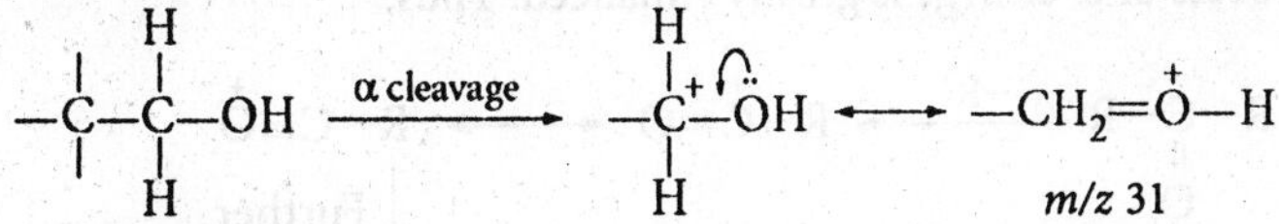

Primary alcohol

(ii) 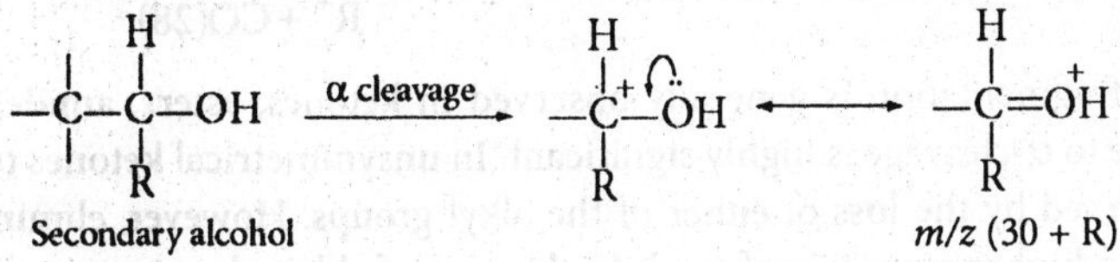

Secondary alcohol

(iii) 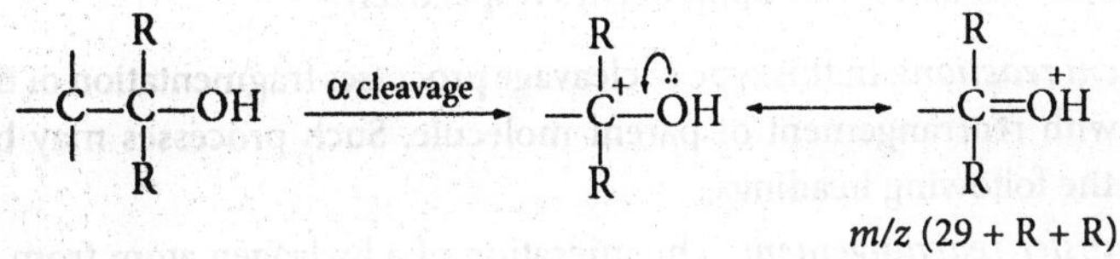

Similarly,

$$CH_3 \overset{\alpha}{—} CH_2—O—CH_2 \overset{\alpha'}{—} CH_2—CH_2—CH_3 \xrightarrow[\text{cleavage of } \alpha' \text{ bond}]{\text{preferential}} CH_3—CH_2—O—\overset{+}{C}H_2 \longleftrightarrow CH_3—CH_2—\overset{+}{O}=CH_2$$

$$CH_3—CH_2 \overset{\alpha}{—} CH_2—O—CH_2 \overset{\alpha}{—} CH_2—CH_3 \xrightarrow[\text{cleavage}]{\alpha} CH_3—CH_2—CH_2—O—\overset{+}{C}H_2 \longleftrightarrow CH_3—CH_2—CH_2—\overset{+}{O}=CH_2$$

The location of functional group may also be settled by interpreting the mass spectral data in terms of α-cleavage as indicated below:

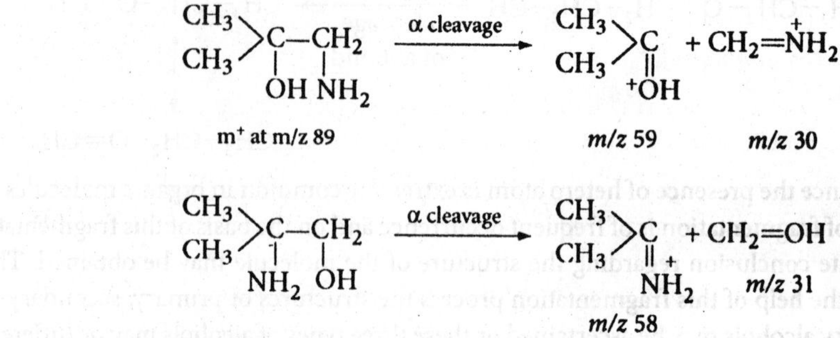

Further it may be mentioned that if the hetero atom is connected to a carbon atom by a π-bond the process of α-cleavge is greatly enhanced. Thus,

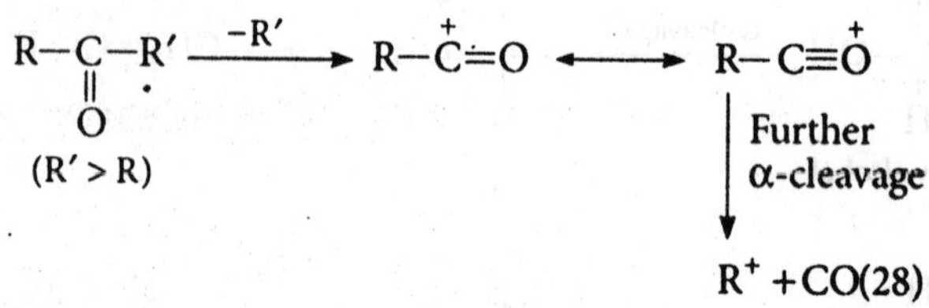

This type of fragmentation is generally observed in ketones, esters, amides, etc. In ketones peaks due to α-cleavage is highly significant. In unsymmetrical ketones two types of peaks are observed by the loss of either of the alkyl groups. However, elimination of alkyl group having higher mass is preferred. In the case of aldehydes, esters and amides breakage of C—H, C—O and C—N bond occurs respectively.

B. *Rearrangement reactions*: In this type of cleavage processes fragmentation of the molecule occurs with rearrangement of parent molecule. Such processes may be considered under the following headings.

(i) *Hydrogen transfer rearrangement*: The migration of a hydrogen atom from one carbon atom to another over a π-bond system during fragmentation is very common in mass spectrometry and has been established by deuterium labelling experiment. These rearrangements involve specific bonds and specific hydrogen atom in the molecule. In aliphatic hydrocarbons and aromatic compounds these processes occur with intramolecular hydrogen transfer rearrangements. Thus,

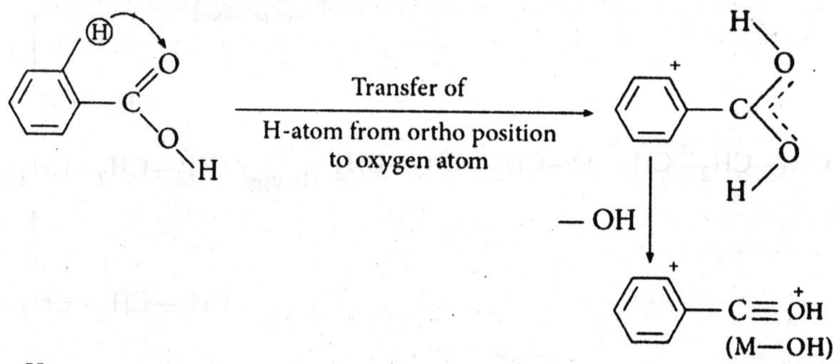

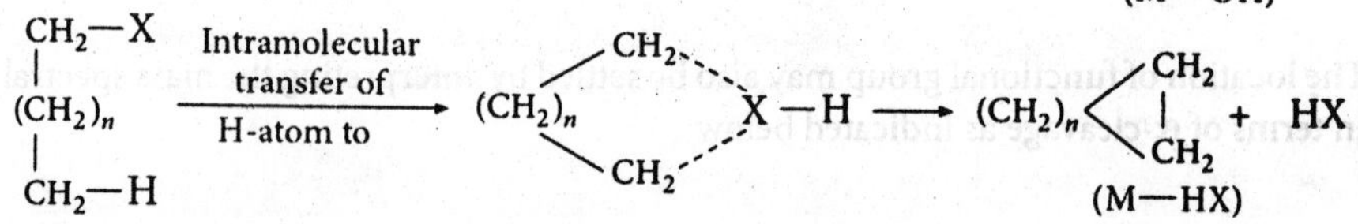

The mechanism of the formation of ions in this process is not known with certainty. The migration of H atom occurs disorderly; that is why this transfer is also termed as *scrambling.*

(ii) *Elimination*: In these fragmentation a neutral molecule is eliminated by the cleavage of two bonds followed by a formation of new bond. This type of fragmentation is generally observed in amides, alcohols, halides and esters. The elimination reaction occurs not only from the molecular ion but also from the fragmentions. Thus,

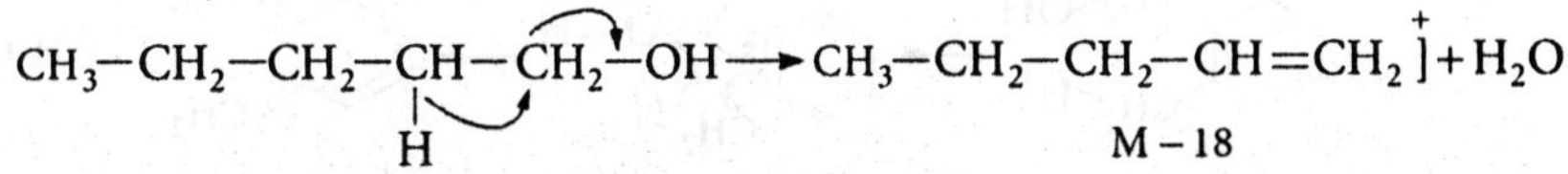

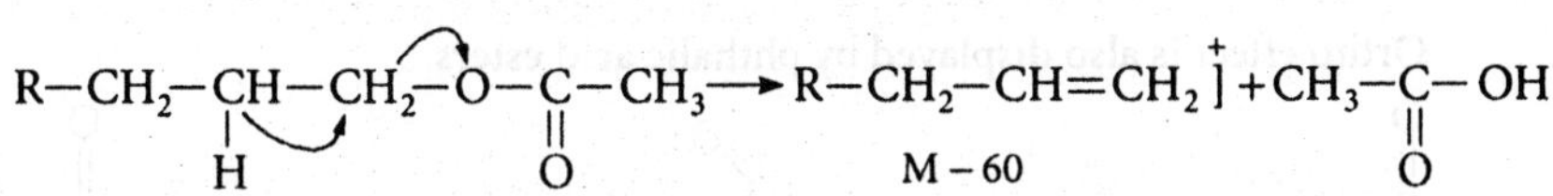

The elimination of ketene ($CH_2{=}C{=}O$) is a characteristic fragmentation mode of *N*-alkyl amides and *o*-acetates of phenols.

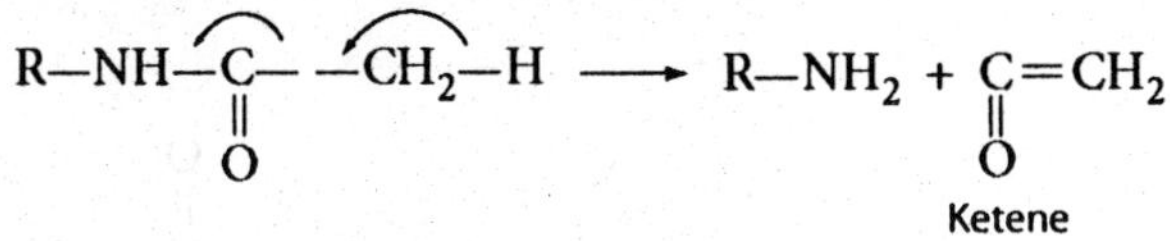

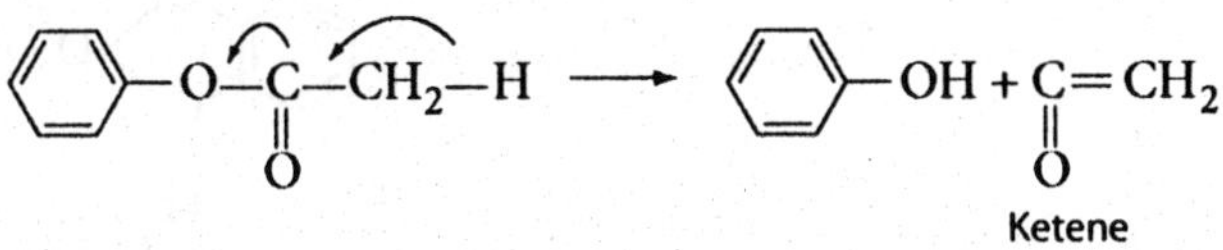

(iii) *Ortho effect*: In ortho substituted aromatic compounds or cis-olefines, the substituent and a hydrogen atom can come in close proximity so as to help the elimination of neutral molecule. This effect is known as *ortho effect.*

The elimination of methyl alcohol from *cis*-methyl crotonate may be cited as an example of ortho effect.

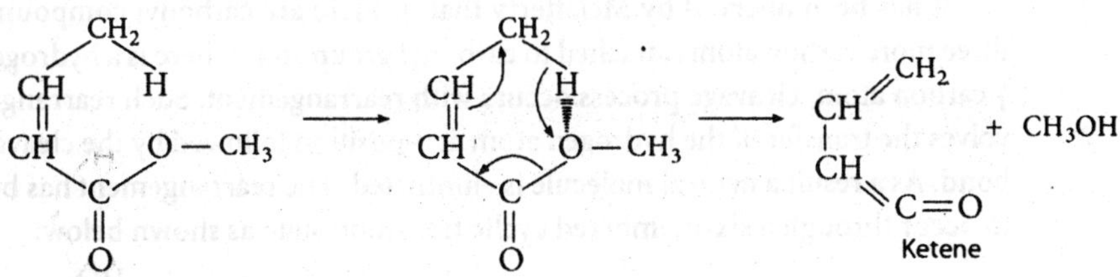

cis-Methyl Crotonate

When aromatic acids, esters and amides contain a hydrogen bearing substituent in the ortho position this effect operates. The general path of this rearrangement may be depicted as below.

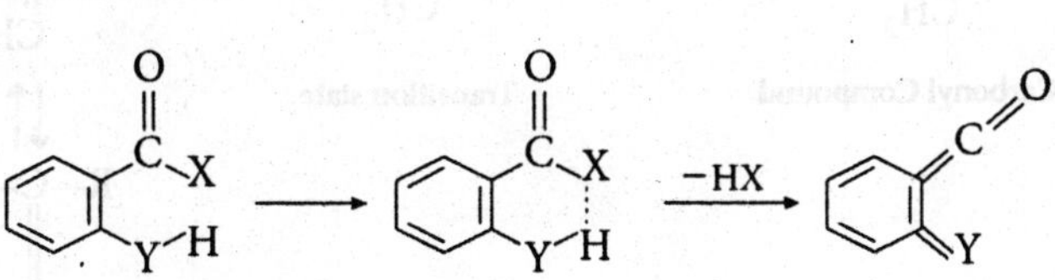

$X = OH, OR, NH_2, Y = CH_2, O, NH$

Examples cited in this connection are:

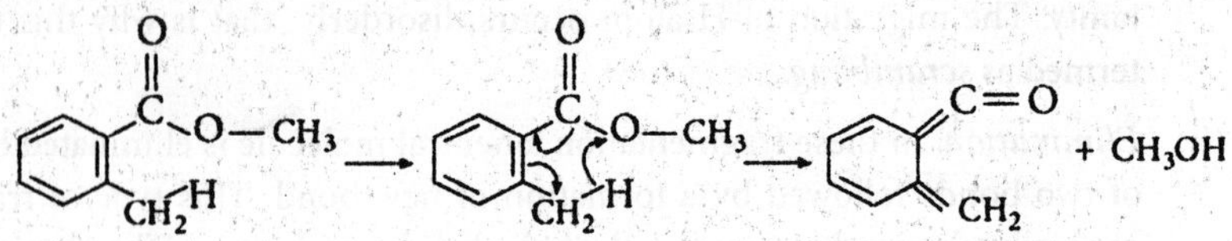

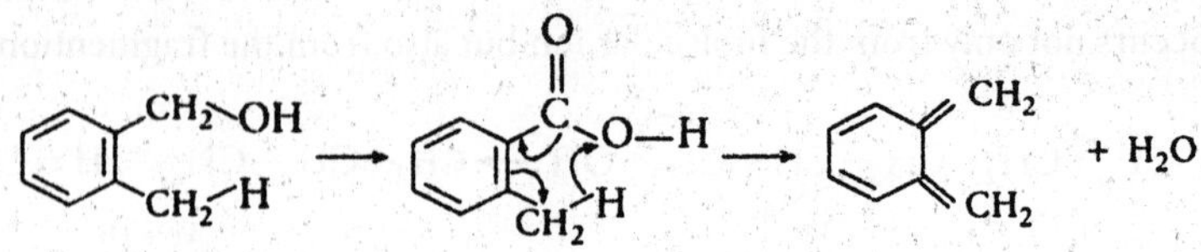

Ortho effect is also displayed by phthalic acid esters.

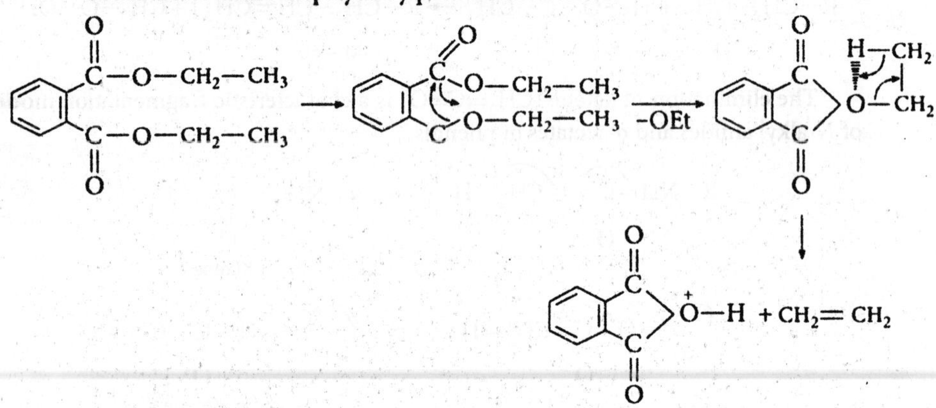

This type of fragmentation process may be applied to distinguish between *cis* and *trans* isomers and *ortho* from *para* and *meta* substituted isomers.

(iv) *Mclafferty rearrangement*: This rearrangement is an important mass spectral fragmentation process and has been proved to be very valuable in the structure deduction of organic molecules.

It has been oberved by Mclafferty that if there are carbonyl compounds having three more carbon atoms attached to carbonyl group and if there is a hydrogen atom at γ-carbon atom, cleavage process occurs with rearrangement. Such rearrangement involves the transfer of the hydrogen atom at γ-position followed by the cleavage of a β-bond. As a result a neutral molecule is eliminated. The rearrangement has been found to occur through a six-membered cyclic transition state as shown below:

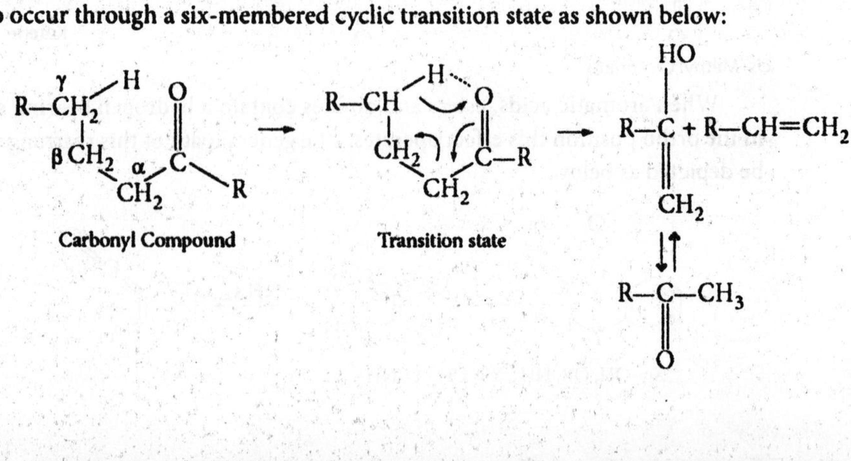

This mass spectral reaction has been well studied and found to have the following characteristics.

(a) **The cyclic transition state requires a double bond and a hydrogen atom at γ-carbon atom.**

(b) **The ring atom in the cyclic transition state may be C, O or N. This process can occur in compounds like alkene, aromatic compounds, aldehydes, ketones, esters, amines. etc., although ketones and aldehydes undergo this rearrangement as a major fragmentation pathway. The general mechanistic path of this fragmentation process may be depicted as below:**

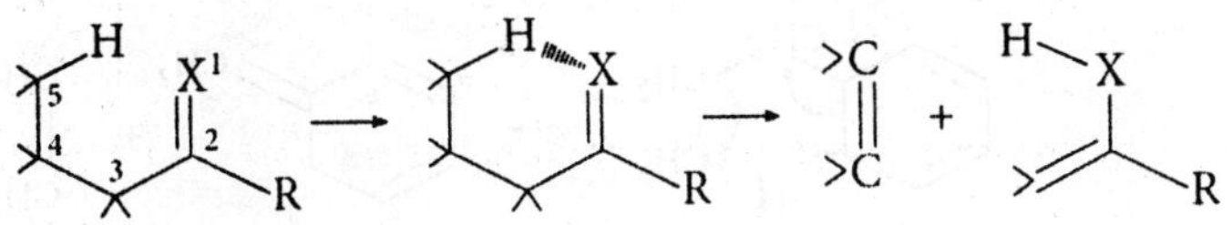

X=C,O,N

(c) **If more than one γ-H hydrogen is available, then comparatively more reactive γ-hydrogen is involved to a greater extent. The order of reactivity of γ-H is CH > CH_2 > CH_3.**

(d) **From mass values of the ion peaks, information about the substituent at 2, 3, 4 and 5 positions may be obtained. Thus, if there is a methyl group at 4 or 5 position, propene will be the product.**

(e) **Consecutive Mclafferty rearrangements have also been observed in some cases.**

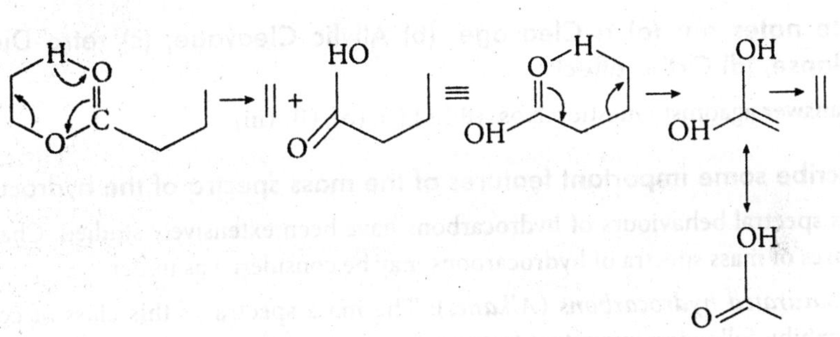

A few more examples of Mclafferty rearrangements are

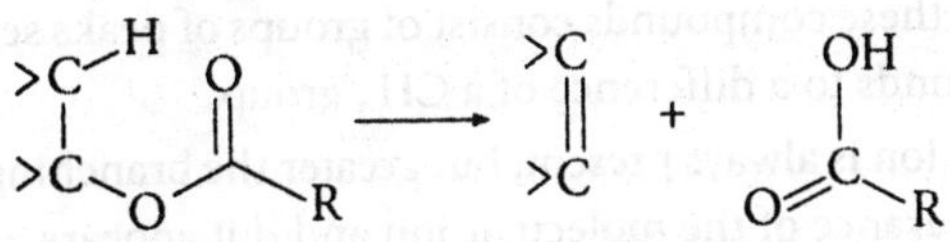

Ester

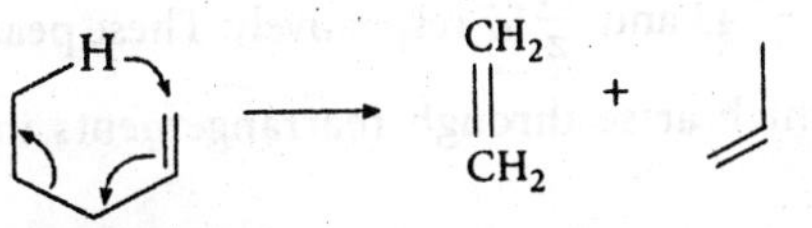

Hydrocarbon

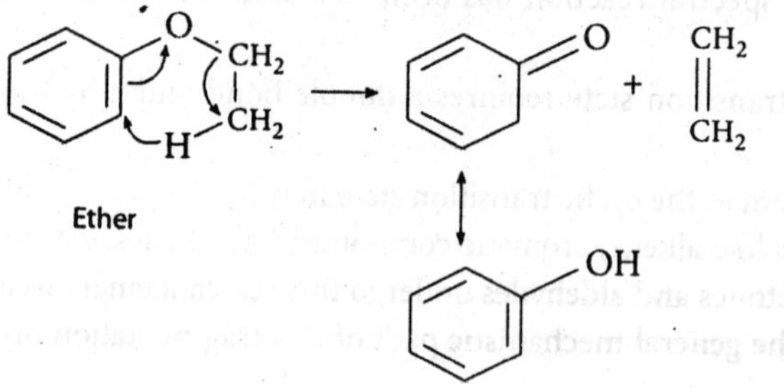

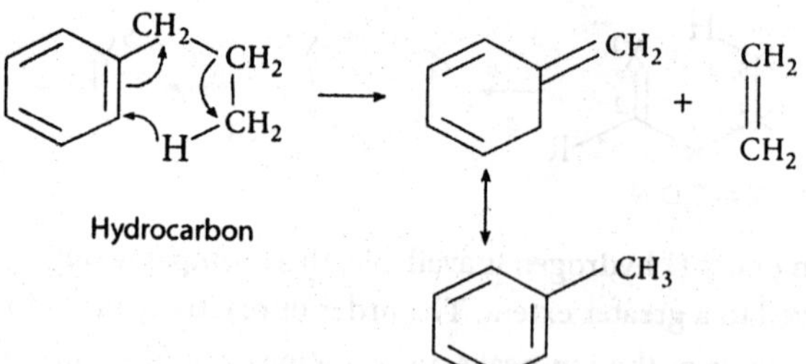

19. Write a note on Mclafferty rearrangement with suitable examples.

➧ See answer against Question No. 18 (B) (iii).

20. Write notes on: (a) α-Cleavage, (b) Allylic Cleavage, (c) retro-Diels-Alder Collapse, (d) Ortho effect.

➧ See answer agaomst Question Nos. 18 (A) (i), (b), (B) (iii).

21. Describe some important features of the mass spectra of the hydrocarbons.

➧ Mass spectral behaviours of hydrocarbons have been extensively studied. Characteristic features of mass spectra of hydrocarbons may be considered as under.

(a) *Saturated hydrocarbons* (*Alkanes*): The mass spectra of this class of compounds exhibit following important features:

(i) The intensity of molecular ion decreases with the increase in molecular mass.

(ii) The spectra of these compounds consist of groups of peaks separated by 14 mass units which corresponds to a difference of a CH_2 group..

(iii) The molecular ion is always present, but greater the branching, the less is the probability of the appearance of the molecular ion and if it appears, its intensity is often low.

(iv) The most intense peaks in the spectrum are due to ions containing three and four carbon atoms at $\frac{m}{z}$ 43 and $\frac{m}{z}$ 57 respectively. These peaks are primarily due to highly branced ions which arise through rearrangements in the case of straight chain alkanes.

(v) In the case of branched chain alkanes bond cleavage takes place preferably at the site of branching as a result more stable secondary or tertiary carbocations are formed. Generally, largest substituent at the branch is eliminated readly as a radical.

(vi) Greater number of fragments result from a branched chain compound compared to open chain compound.

Following examples may be considered in this connection:

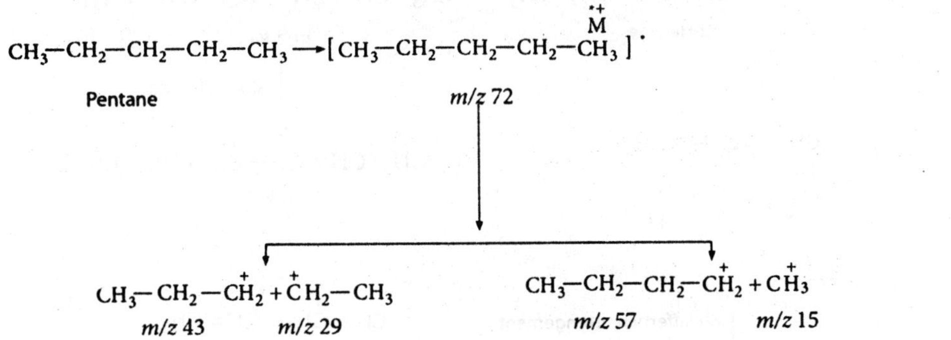

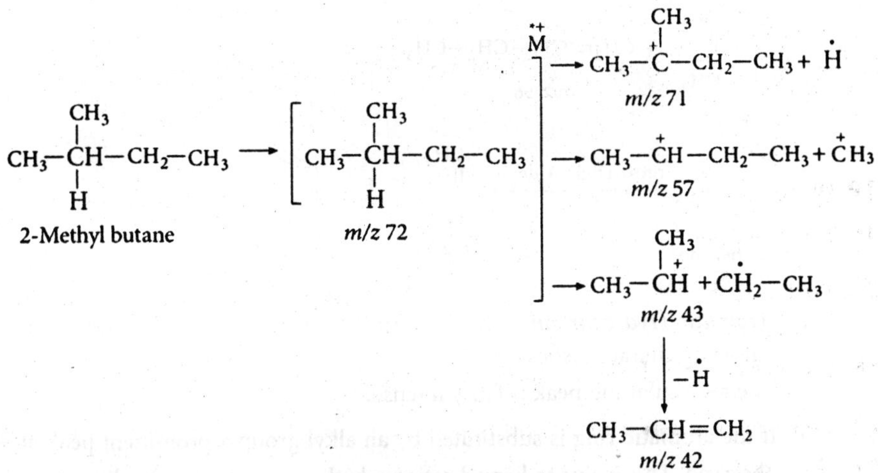

(b) *Unsaturated hydrocarbons* (*Alkenes and Alkynes*): Following important features have been observed in the mass spectra of alkenes and alkynes.

(i) Molecular ion peak in the spectra is more intense than the saturated analogues.

(ii) The relative abundance of molecular ion peak decreases with the increase in the molecular mass.

(iii) Groups of peaks appear at 14 mass units apart.

(iv) The general mode of fragmentation induced by the double bond in alkenes is the allylic cleavage.

(v) Ion peaks due to Mclafferty rearrangement are more intense.

(vi) The mass spectra of alkynes are similar to those of alkenes but the *m/z* values of the major peaks in the group of homologous series of ions have the general formula, C_nH_{2n-3}.

(vii) Cycloalkenes with the double bond in a six-membered ring undergoes retro-Diels-Alder collapse.

Thus,

(i) 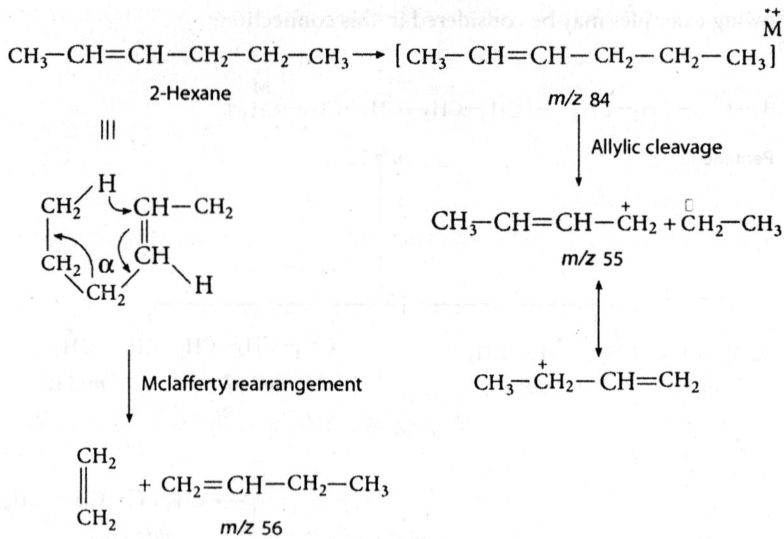

(ii) 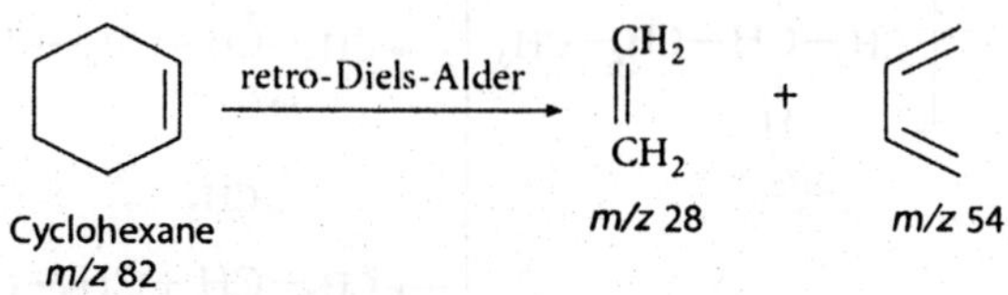

(c) *Aromatic Hydrocarbons*: The mass spectra of aromatic hydrocarbons display the following characteristics.

(i) The molecular ion peak is fairly intense.

(ii) If the aromatic ring is substituted by an alkyl group a prominent peak at $\frac{m}{z}$ 91 is observed. This is due to benzyl cation which rearranges to tropylium cation. Thus,

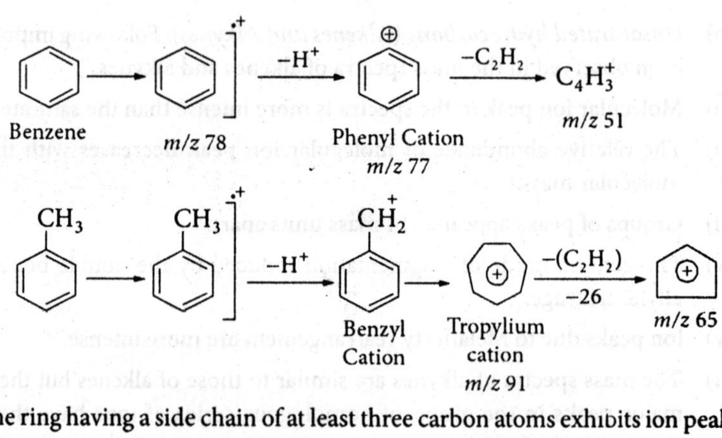

Benzene ring having a side chain of at least three carbon atoms exhibits ion peak due to Mclafferty rearrangement in addition to the ion peaks at $\frac{m}{z}$ 91 and 65.

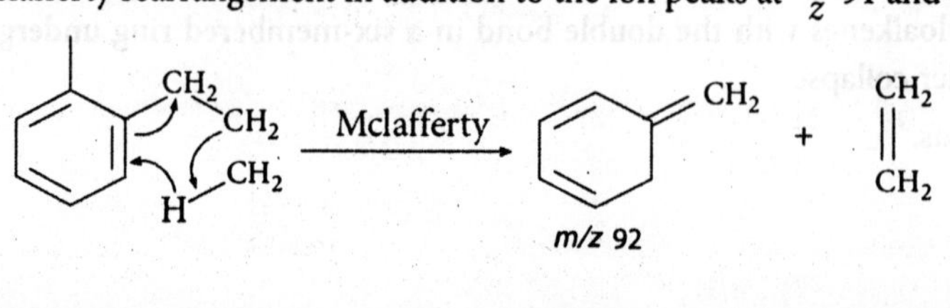

22. Discuss the characteristic features of mass spectra of hydroxy compounds.

➜ Hydroxy compounds has been subjected to mass spectral studies. The important features of their mass spectra may be discussed as below:

(a) *Aliphatic alcohols*: The mass spectra of alcohols have following characteristics:

(i) The molecular ion peaks of primary and secondary alcohols are very weak and in the case of tertiary alcohols molecular ion is either very weak or absent.

(ii) Number of fragmentation modes depends upon their nature (primary, secondary or tertiary).

(iii) The major fragmentation is the α-Cleavagele, i.e., cleavage of C–C bond next to C–O bond.

(iv) Primary alcohols show M–18 peak due to loss of H_2O forming alkene.

(v) The signal at $\frac{m}{z}$ 31 appears in large abundance in the mass spectrum of primary alcohols.

(vi) Primary alcohols having a chain of four or more carbon atoms shows a peak which corresponds to M–(18–C_nH_{2n}) as shown below:

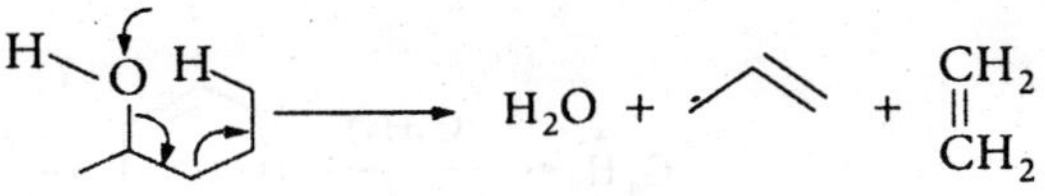

Let us consider the following examples in this context.

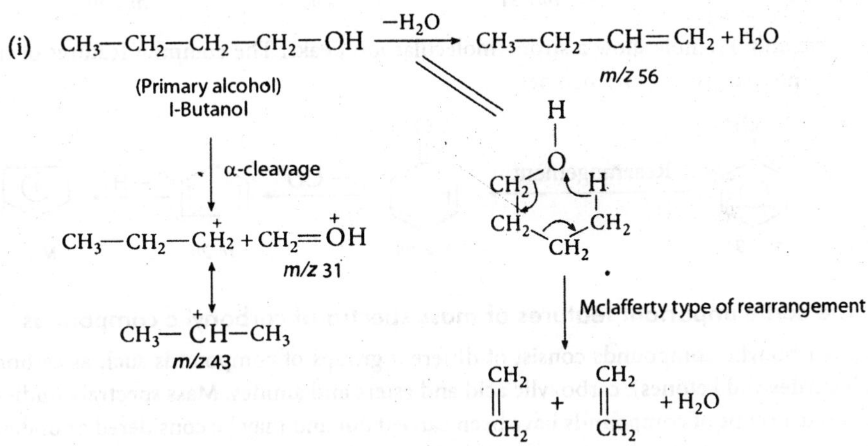

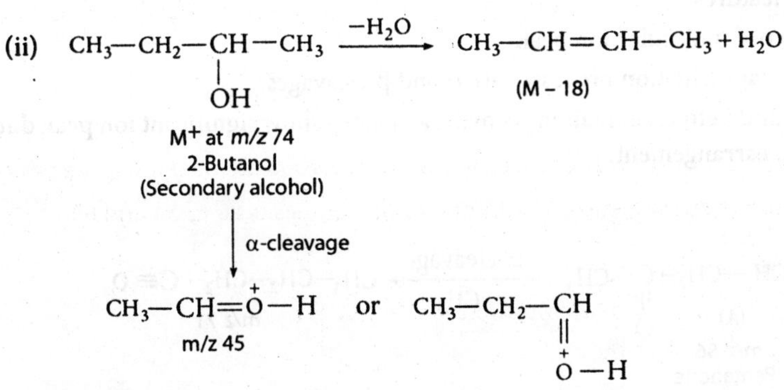

(iii) 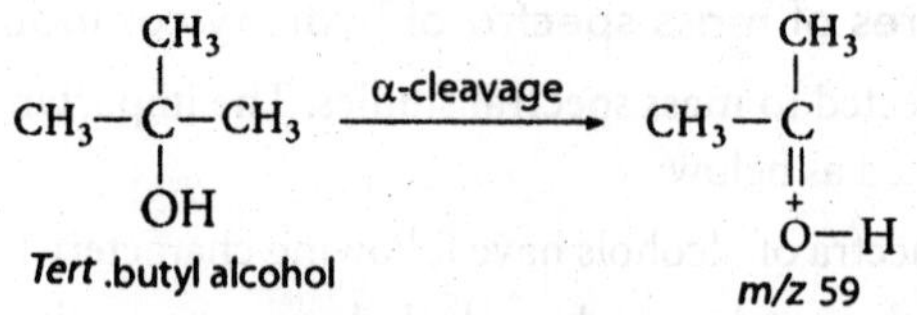

(b) *Aromatic Alcohols*: Benzyl alcohol may be considered as a representative member of this group of alcohols. It exhibits an intense molecular ion peak in its mass spectrum. The characteristic fragmentation that are observed in its mass spectrum are shown below:

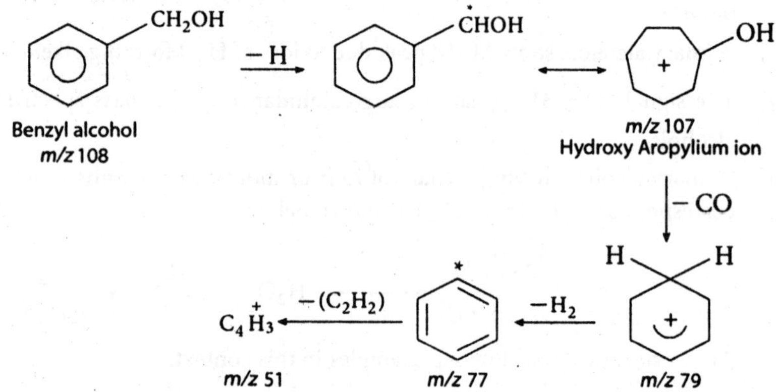

(c) *Phenols*: Phenols show a strong molecular ion peaks. The common features of fragmentation are as described below:

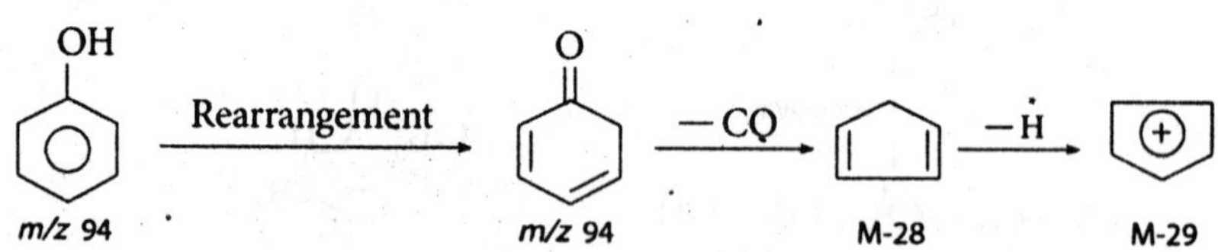

23. Write some important features of mass spectra of carbonylic compounds.

➠ The carbonylic compounds consist of different groups of compounds such as carbonyls (aldehydes and ketones), carboxylic acid and esters and amides. Mass spectrals studies of all these groups of compounds have been carried out and may be considered as under:

(a) *Aldehydes and ketones*: The mass spectra of aldehydes and ketones have the following important features.

(i) Exhibit intense molecular ion peak.

(ii) The major fragmentation processes are α and β cleavages.

(iii) Aldehydes and ketones containing γ-hydrogen atom show significant ion peak due to Mclafferty rearrangement.

Thus,

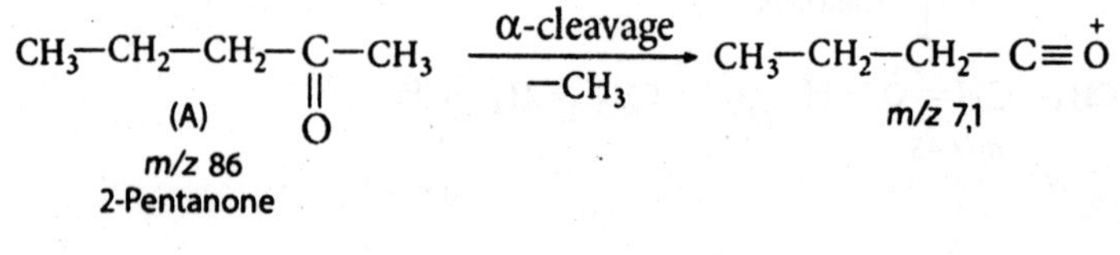

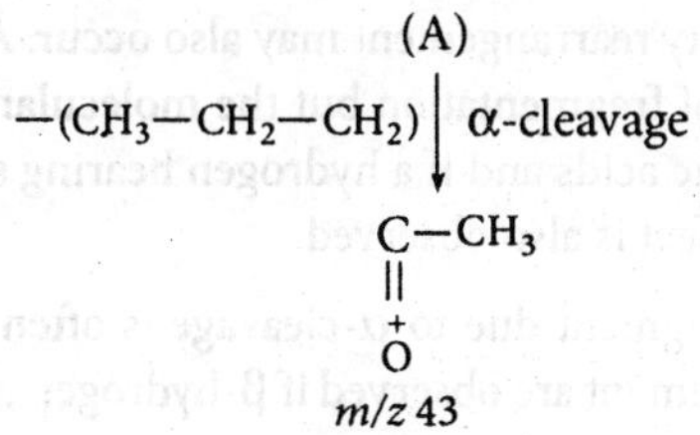

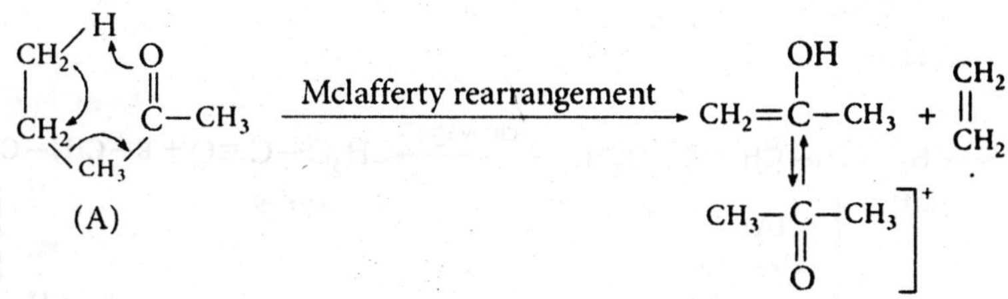

In aldehydes following ion peaks are observed:

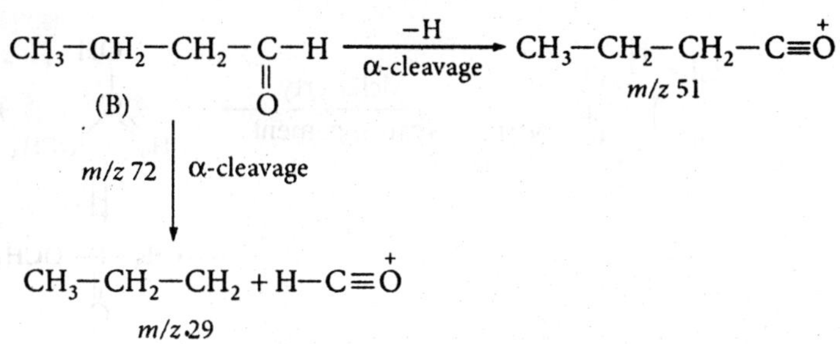

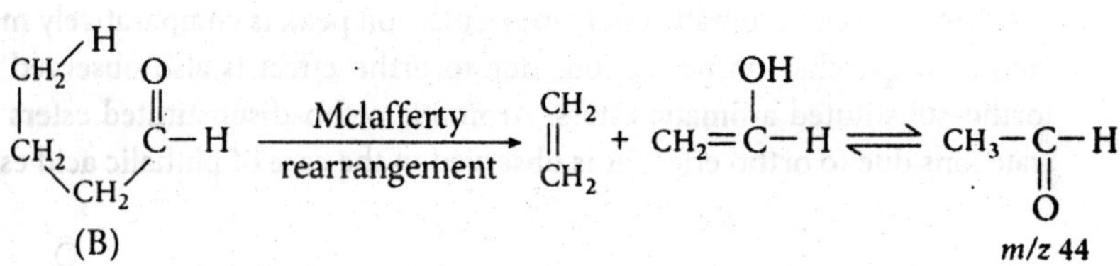

In the case of aromatic aldehydes following ion peaks are often observed.

$$C_6H_5{-}CHO \xrightarrow{-H} C_6H_5{-}\overset{\cdot +}{C{\equiv}O}\ (M\text{-}1) \xrightarrow{-CO} C_6\overset{+}{H}_5\ (m/z\ 78) \xrightarrow{-H} C_6\overset{+}{H}_5\ (m/z\ 77) \xrightarrow{-(C_2H_2)} C_4\overset{+}{H}_3\ (m/z\ 51)$$

(b) *Carboxylic acids*: Modes of fragmentation of carboxylic acids may be elaborated as

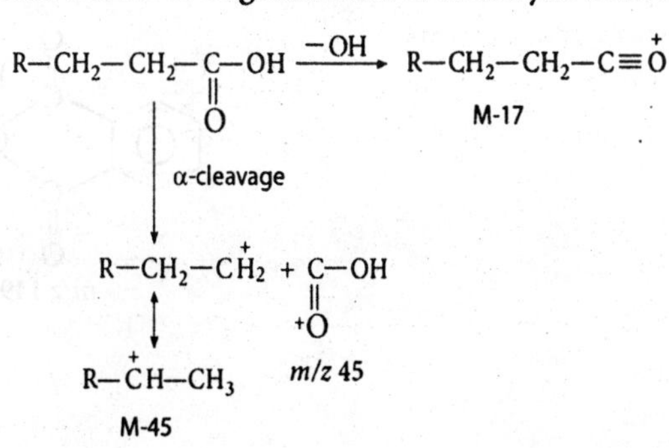

If H-atom is available at γ-position, Mclafferty rearrangement may also occur. Aromatic carboxylic acids show similar mode of fragmentation but the molecular ion peaks are more intense than those of aliphatic acids and if a hydrogen bearing substituent is present at ortho position ortho effect is also observed.

(c) *Esters*: The molecular ion peak is weak. Fragment due to α-cleavage is often observed. Ion peaks due to Mclafferty rearrangement are observed if β-hydrogen atom is available:

Thus,

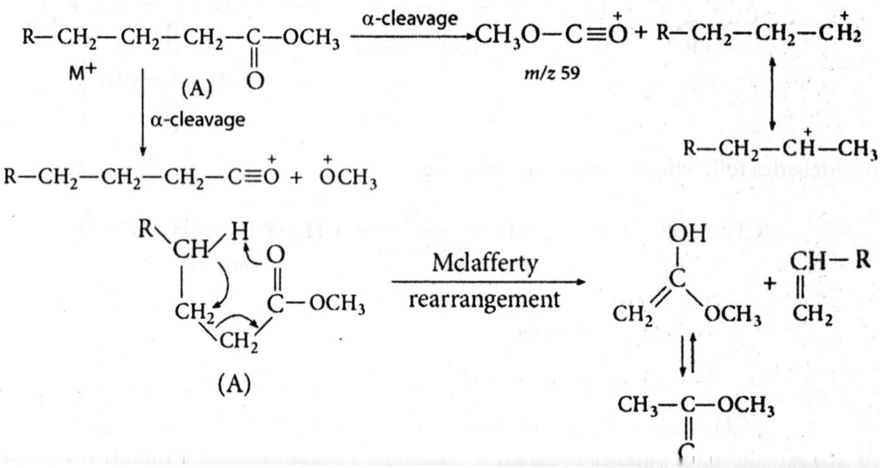

In the case of aromatic esters molecular ion peak is comparatively more intense and show similar ion peaks. Ions due to ortho effect is also observed in suitably ortho-substituted aromatic esters. Aromatic ortho-disubstituted esters may generate ions due to ortho effect as is observed in the case of phthalic acid esters.

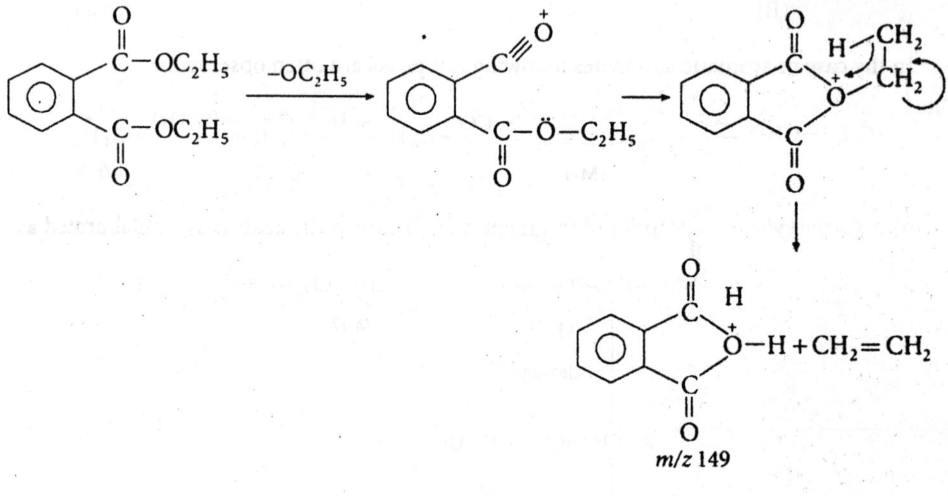

(d) *Amides*: In case of amides, the base peak in straight chain primary amides higher than propionamide, is the Mclafferty rearrangement to generate a peak at $\frac{m}{z}$ 59. The α-cleavage in all primary amides give a peak at $\frac{m}{z}$ 44.

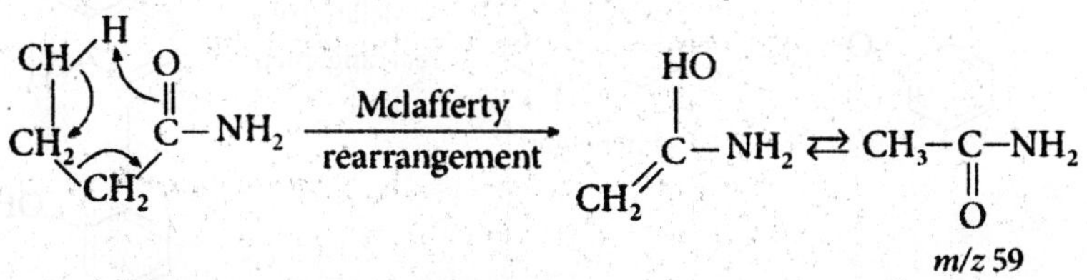

In the case of N-alkyl amide another mode of cleavage predominates. This is the cleavage of N-alkyl group β to the nitrogen atom.

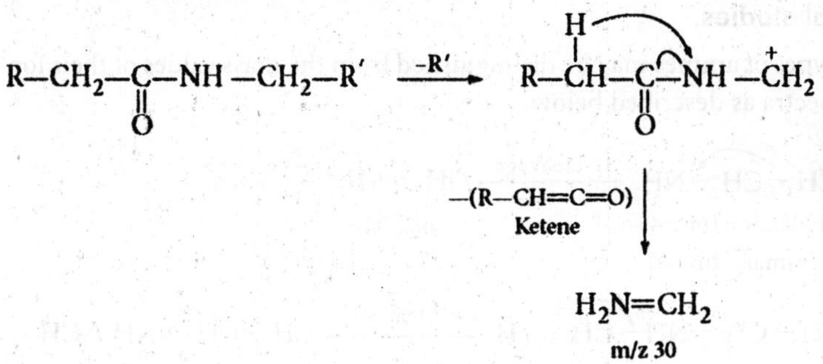

24. Predict the fragmentation pattern of diphenyl ether.

➡ The fragmentation pattern of diphenyl ehter is highly interesting. There is a loss of carbon monoxide due to a rearrangement as shown below.

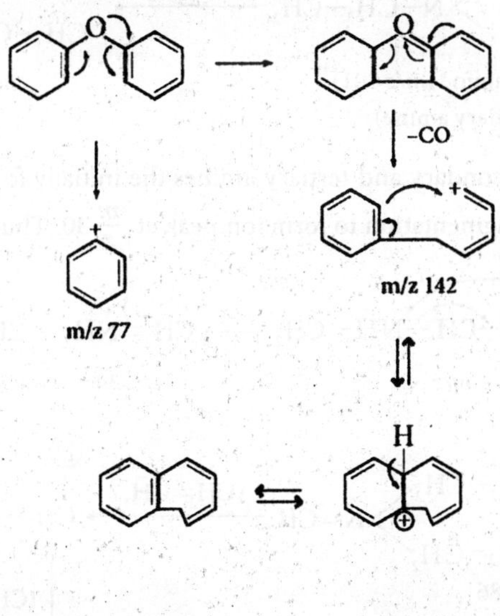

25. Phenetole in its mass spectrum exhibits an ion peak at $\frac{m}{z}$ 94. Explain its genesis.

➜ Formation of an ion peak at $\frac{m}{z}$ 94 from phenetole may be rationalised as under:

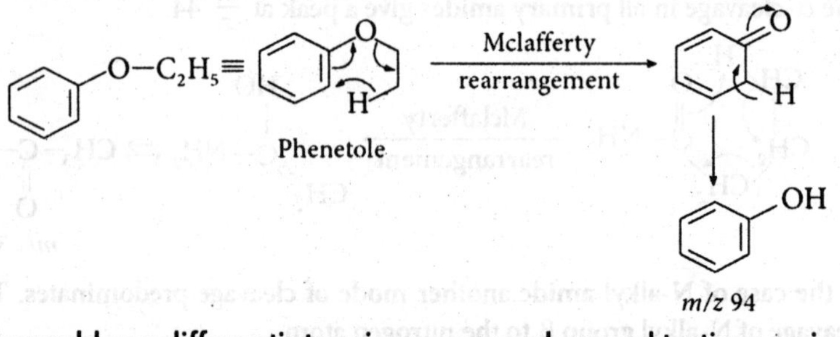

26. How would you differentiate primary, secondary and tertiary amines from mass spectral studies.

➜ Three types of amines may be distinguished from the mass values of their ion peaks in the mass spectra as described below:

$$CH_3-CH_2-NH_2 \xrightarrow{\alpha\text{-cleavage}} CH_2=NH_2$$

Ethylamine (*m/z* 45) (Primary amine) → *m/z* 94

$$CH_3-CH_2-NH-CH_2-CH_3 \xrightarrow{\alpha\text{-cleavage}} CH_3-CH_2-NH=CH_2$$

Diethylamine (*m/z* 73) (Seconday amine) → *m/z* 58 (A)

$$(CH_3-CH_2)_2N-CH_2-CH_3 \xrightarrow{\alpha\text{-cleavage}} (CH_3-CH_2)_2\overset{+\bullet}{N}=CH_2$$

Triethylamine (*m/z* 101) (Tertiary amine) → *m/z* 86 (B)

In the case of secondary and tertiary amines the initially formed fragment ions may undergo further fragmentation to form ion peak at $\frac{m}{z}$ 30. Thus

$$\text{(H)}-CH_2-CH_2-\overset{+}{N}H-CH_2 \longrightarrow CH_2=\overset{+}{N}H_2 + CH_2=CH_2$$

(A) *m/z* 58 → *m/z* 30

$$\text{(H)}-CH_2-CH_2,\ \text{(H)}-\overset{\beta}{C}H_2-\overset{\alpha}{C}H_2 > \overset{+}{N}=CH_2 \xrightarrow{(CH_2=CH_2)} \text{(H)}\overset{\beta}{C}H_2-\overset{\alpha}{C}H_2-\overset{+}{N}H=CH_2$$

(B) *m/z* 86 → *m/z* 58

$$\xrightarrow{(CH_2=CH_2)} \overset{+}{N}H=CH_2$$

m/z 30

It may be mentioned that the formation of ion at $\frac{m}{z}$ 30 is possible only when hydrogen atom at β-carbon is available.

27. Write a short account of the general mode of fragmentation of heteroaromatic compounds.

➜ **The general mode of fragmentation of heteroaromatic compounds is very similar to benzene. They give abundant molecular ions and undergo fragmentation in the same way as in benzene. Thus, when benzene eliminates C_2H_2 from the molecular ion, pyrole and pyridine eliminate HCN, thiophene ejects CHS and furan expels CHO from their molecular ions. Thus,**

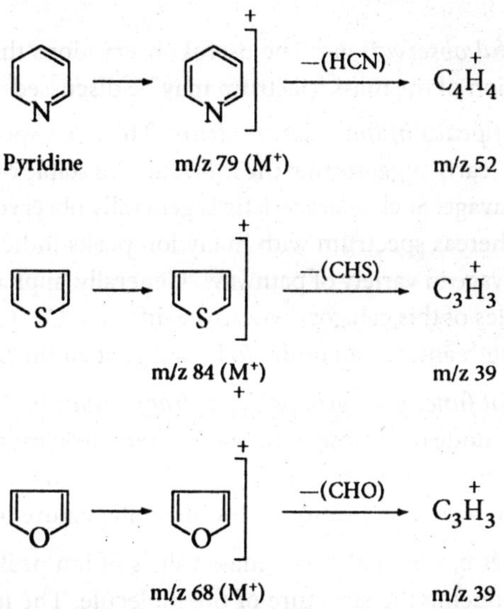

Alkyl substituted heterocycles fragment at the benzylic position.

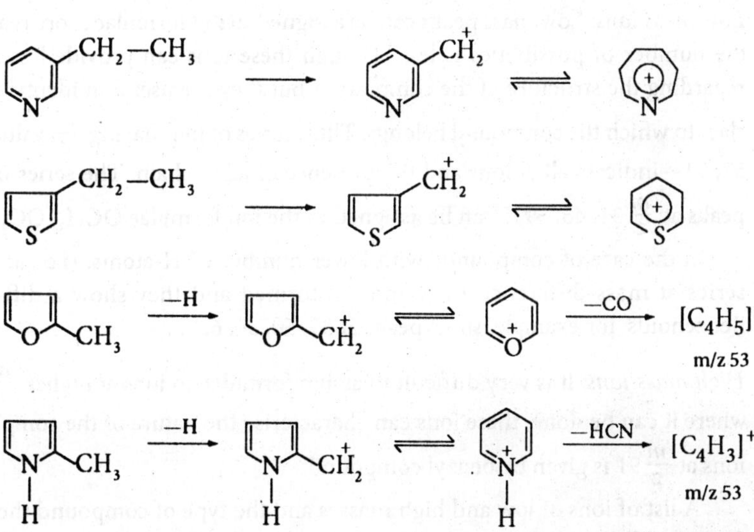

28. Write a short account on how would you proceed to interpret the mass spectral data for the elucidation of the structure of the organic compounds.

➡ With the basic knowledge of different fragmentation processes that may occur during electron impact or in any other ionisation techniques of organic compounds and in conjunction with the consideration of some useful observations mentioned, it is possible to rationalise the mass spectrum for the deduction of the structure of the unknown compound. However, it must be ensured that the compound under consideration is of high purity, otherwise no useful information can be obtained. It may be mentioned that for better interpretation of the mass spectrum of a compound for the assignment of the correct structure, other spectral data such as UV, IR, ^{1}HNMR, ^{13}CNMR, etc. should always be available.

Some useful observations: The useful observations that are to be considered during the interpretation of the mass spectrum may be discussed as under:

1. *Number of peaks in the mass spectrum*: The mass spectrum with a fewer number of the peaks clearly suggests that the molecule is a stable one and has only a few preferred site for cleavage. Such characeristic is generally observed in the case of aromatic compounds whereas spectrum with many ion peaks indicates that the molecule can undergo cleavage in variety of pathways. Generally, aliphatic saturated hydrocarbons are the examples of this category where the different C—C bonds have approximately the same strength and so can undergo breakage at all the bonds.
2. *Influence of functional groups in the fragmentation*: Each functional group has characteristic mode of fragmentation and direct the cleavage process according to its nature. Thus the appearance of peak at M–17 and in some cases at M–18 in the mass spectrum is a clear indication of the alcoholic nature of the compound.
3. *Mass values of ion peaks*: The mass values of ion peaks in the spectrum are of great help in deducing the structure of the molecule. The ion peaks can be classified into two groups—ion peaks with low and ion peaks with high $\frac{m}{z}$ values.

(a) *Low mass ions*: Low mass peaks can be assigned to ion formulae more readily because the number of possiblities is less. Though these ions can provide less information regarding the structure of the compounds but they are useful in identification of the class to which the compound belongs. Thus, series of ions having $\frac{m}{z}$ values 15, 29, 43, 57, 71—indicate alkyl ions and the presence of alkyl chain. The series of significant peaks at $\frac{m}{z}$ 31, 45, 59 ... can be assigned to the ion formulae OCH_3, OC_2H_5, OC_3H_7.

In the case of compounds with lower number of H-atoms, (i.e., aromatics) ion series at mass difference 14 cannot be formed and they show a different series, Benzenoids, for example, show peaks at 39, 50, 51, 63.... .

(b) *High mass ions*: It is very difficult to assign formulae to ions at higher $\frac{m}{z}$ values but where it can be done, these ions can characterise the nature of the compound. Thus, ions at $\frac{m}{z}$ 91 is given by benzyl compound.

A list of ions of low and high masses and the type of compound they indicate is furnished below:

$\frac{m}{z}$	*Nature of the compound*
15, 29, 43 ...	Alkanes, alkyl chains, aldehydes
27, 41, 55 ...	Cyclo-alkanes
63, 77, 91, 105, 119 ...	Aromatic (phenyl)
31, 45, 59, 73, 87 ...	Alcohols, esters
61, 75, 89 ...	Esters
30, 44, 58 ...	Amines
59, 73, 87 ...	Acids

4. *Loss of common masses during ionisation*: The common masses lost from molecular ion or from other ions can be easily identified by the difference in mass of the parent and daughter ions. These common masses have usually low $\frac{m}{z}$ values but their identification can play a significant role in determining the structure of the molecule under consideration. For example, CH_3 (15 mass units), H_2O (18 mass units), $CH \equiv CH$ (26 mass units), etc., are lost from many molecular ion. From the lost fragment nature of the compound may be ascertained. Alcohols generally lose H_2O, phenyl compounds lose $CH \equiv CH$, etc.

   A list of a few such common masses, their compositions and the type of compound that such a loss indicates is given below.

Mass $\frac{m}{z}$	*Composition*	*Type of Compound*
15	CH_3	Large number of compounds
16	O	$ArNO_2$
16	$-NH_2$	Amide $\left(R-\overset{\overset{O}{\parallel}}{C}-NH_2\right)$
17	–OH	–COOH, alcohol
18	H_2O	Alcohols
26	$CH \equiv CH$	Aromatics
27	HCN	Heterocyclics
28	CO, C_2H_4	Ketones, esters, acids, cyclo-alkenes
29	CHO	Aldehydes
	C_2H_5	Ester, Ether
30	CH_2NH_2, CH_3NH,	Primary, Secondary amines
	CH_2O	Ether
31	OCH_3	O-methyl
	CH_2OH	Primary alcohol
32	CH_3OH	$R-OCH_3$
42	$CH_2{=}C{=}O$	$R-CH_2-\underset{\underset{O}{\parallel}}{C}-NH-CH_2-R$, $CH_3-\underset{\underset{O}{\parallel}}{C}-O-R$

43	$CH_3{-}C(=O)$	Keto methyl
	C_3H_7	$(CH_3)_2{-}CH$ containing compound
44	CO_2	Acids
45	$-COOH$	Acids
	$-OC_2H_5$	Esters, ethers
59	$-COOCH_3$	$RCOOCH_3$
60	CH_3COOH	CH_3COOR

Keeping in mind the above useful observations the general approach mentioned below may be followed to interpret the mass spectrum for the elucidation of the structures of the molecules. However, it may be remembered that every spectrum presents its own challenge and so no rigid procedure for structure elucidation can be specified.

(a) Detection of accurate molecular ion.

(b) Determination of the elemental composition of the compound.

(c) Deduction of suitable conclusion from the general appearance of the spectrum.

(d) Proposal of probable structure on the basis of these observations and deduction and finally comparison of spectral behaviour of the compound with that of related reference compounds to confirm the proposed structure. Some appropriate problems as depicted below may be cited to show the applicability of the above approach.

(i) An organic compound having molecular formula, C_7H_{12} (M^+ at $\frac{m}{z}$ 96) shows intense peaks at $\frac{m}{z}$ M−15 and 54. What structure can be assigned to the compound?

Solution: The compound contains seven carbon atoms. So for saturated hydrocarbon, (C_2H_{2n+2}), 16 hydrogen atoms are necessary. But it contains 12 hydrogen atoms. So the deficiency of hydrogen atoms is 16 – 12 = 4. Hence, the compound contains two double bonds or one ring and one double bond. The presence of M−15 mass peak suggests the presence of a branched methyl group. The peak at $\frac{m}{z}$ 54 may be due to C_4H_6, i.e., butadiene. It is known that retro-Diels-Alder cleavage eliminates charged butadiene fragment at $\frac{m}{z}$ 54 from cycloalkene. The hydrogen deficiency is also in favour of the presence of cycloalkene moiety in the compound. The structure of the compound, thus, may be proposed as 4-methyl cyclohexene.

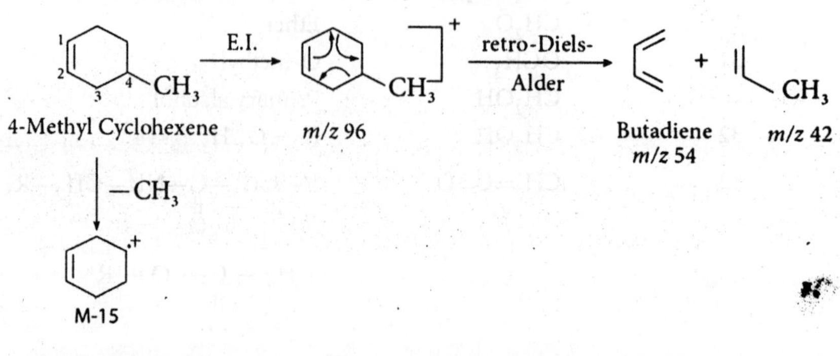

(ii) An organic compound, $C_5H_{10}O_2$ displays the significant ion peaks at $\frac{m}{z}$ 102, 87, 74, 71, 59, 43 (base peak) and 53.8. Deduce its structure.

Solution: The molecular ion has mass value 102 as required by its elemental composition. The peak at $\frac{m}{z}$ 87 is due to M–CH_3 whereas peak at $\frac{m}{z}$ 74 is due to M–28. Now the metastable peak indicates that $\frac{m}{z}$ 74 is directly formed from the molecular ion by the loss of 28 mass units.

$$\frac{74^2}{102} \approx 53.7$$

The fragment lost can be C_2H_4 or CO. The compound contains two oxygen atoms. So it may be assumed that the compound may be either acid or ester. It is known that acid and ester can undergo Mclafferty rearrangement readily and in such reaction CO cannot be lost; so the mass lost 28 is C_2H_4, i.e., ethylene. So the probable structure of the compound stands as noted below from which the formation of other fragments can be explained.

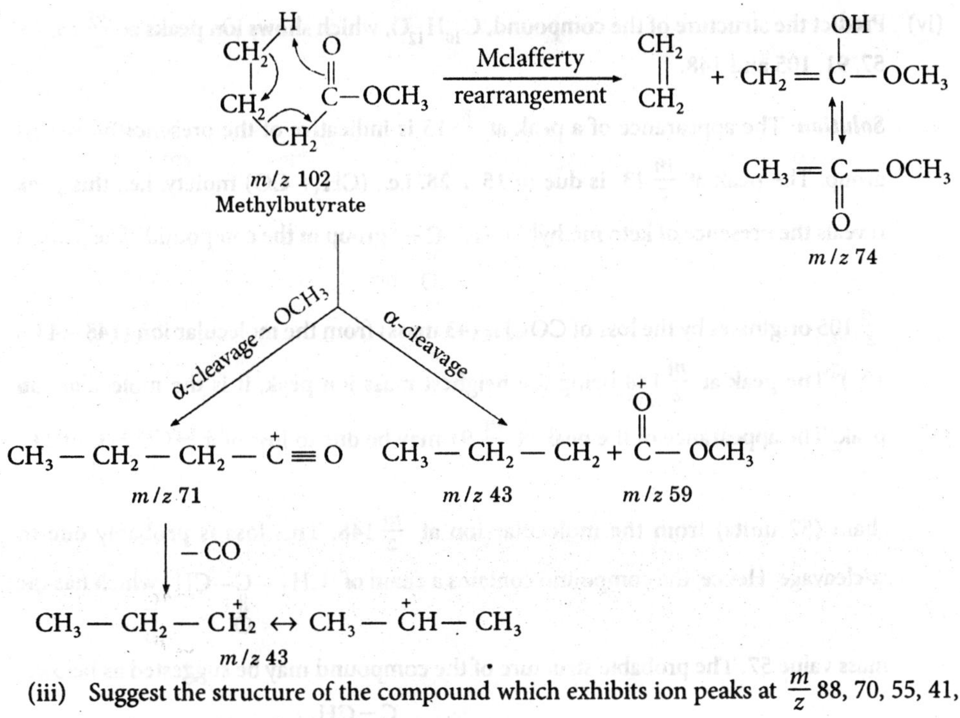

(iii) Suggest the structure of the compound which exhibits ion peaks at $\frac{m}{z}$ 88, 70, 55, 41, 31 (base peak) and 29 in the mass spectrum.

Solution: The peak at $\frac{m}{z}$ 88 is the heighest mass ion peak. So it is most probably the molecular ion peak. The peak at $\frac{m}{z}$ 70 is due to loss of 18 mass units (H_2O). This suggests that the compound is alcohol, probably primary alcohol. The appearance the peak at $\frac{m}{z}$ 31 shows the α-cleavage of primary alcohol. The signal at $\frac{m}{z}$ 55 is due of (M–18—CH_3). This also supports the primary alcoholic nature of the compound. The peak at $\frac{m}{z}$ 42 is due to ($M-18-C_2H_4$). On the basis of the above, the structure of the compound may be suggested as under:

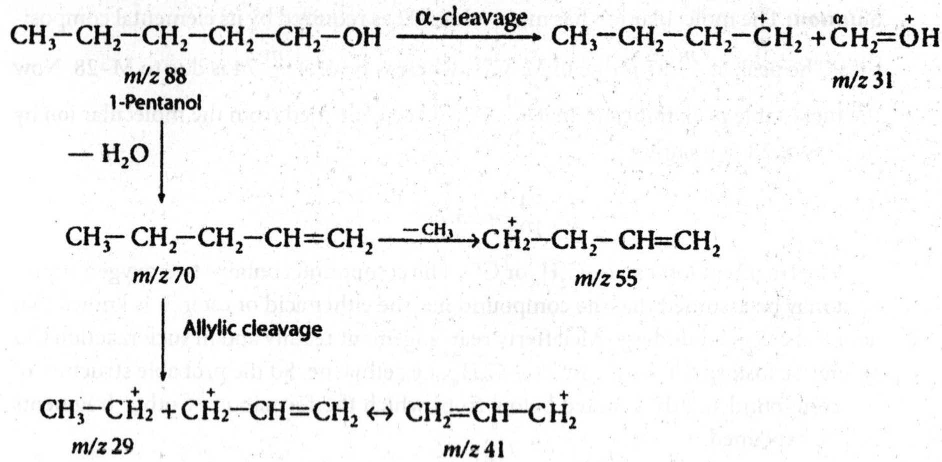

(iv) Predict the structure of the compound, $C_{10}H_{12}O$, which shows ion peaks at $\frac{m}{z}$ 15, 43, 57, 91, 105 and 148.

Solution: The appearance of a peak at $\frac{m}{z}$ 15 is indicative of the presence of methyl group. The peak at $\frac{m}{z}$ 43 is due to 15 + 28, i.e., (CH_3+ CO) moiety, i.e., this peak reveals the presence of keto methyl ($CH_3-\underset{\underset{O}{\|}}{C}-$) group in the compound. The peak at $\frac{m}{z}$ 105 originates by the loss of $COCH_3$ (43 mass) from the molecular ion (148 – 43 = 105). The peak at $\frac{m}{z}$ 148 being the heighest mass ion peak, it is the molecular ion peak. The appearance of the peak at $\frac{m}{z}$ 91 may be due to loss of a $-CH_2-\underset{\underset{O}{\|}}{C}-CH_3$ chain (57 units) from the molecular ion at $\frac{m}{z}$ 148. This loss is probably due to α-cleavage. Hence, the compound contains a chain of $CH_3-\underset{\underset{O}{\|}}{C}-CH_2$ which has the mass value 57. The probable structure of the compound may be suggested as below:

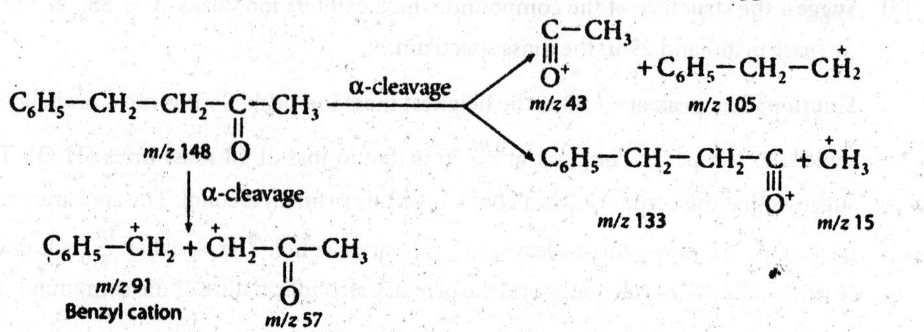

(v) An organic compound displays peaks at $\frac{m}{z}$ 182, 183, 152, 105 (base peak), 77, 76.5, 77.5 and 51 in the mass spectrum. Deduce its structure.

***Solution*:** The spectrum contains few peaks. This suggests that the molecule is stable and probably it is aromatic. The presence of peaks at $\frac{m}{z}$ 77 and $\frac{m}{z}$ 51 suggests a phenyl ring.

$$\underset{m/z\ 77}{\overset{+}{C_6H_5}} \xrightarrow{-(C_2H_2)} \underset{m/z\ 51}{\overset{+}{C_4H_3}}$$

The base peak is at $\frac{m}{z}$ 105. It is probably $C_6H_5-C\overset{+}{O}$ or $C_6H_5C_2\overset{+}{H}_4$. The molecular ion is at $\frac{m}{z}$ 182 (peak at $\frac{m}{z}$ 183 being M + 1 peak).

The molecular ion ($\frac{m}{z}$ 182) undergoes simple cleavage to form $C_6H_5C\overset{+}{O}$ or $C_6H_5C\overset{+}{H}_2$ at $\frac{m}{z}$ 105 which further undergoes cleavage forming $C_6\overset{+}{H}_5$ ($\frac{m}{z}$ 77). So the compound may be (105 + 77). The probable structure may be

$C_6H_5COC_6H_5$ (A) or $C_6H_5-CH_2-CH_2-C_6H_5$ (B)

Out of (A) and (B), (A) is more probable because the presence of 14C atoms is unlikely on the basis of the presence of (M+1) peak. Again, (B) would be expected to give a base peak at $\frac{m}{z}$ ($C_6H_5C\overset{+}{H}_2$) rather than less stable $C_6H_5-CH_2-C\overset{+}{H}_2$ ($\frac{m}{z}$ 105). Hence the structure of the compound is (A).

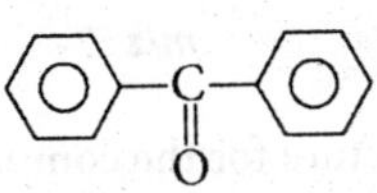

Benzophenone

(vi) Deduce the structure of the compound which shows peaks at $\frac{m}{z}$ 102, 85, 60 (base peak), 57 and 45. The compound on heating with an alcohol in presence of concentrated sulphuric acid gives fruity smell.

***Solution*:** Since the compound gives a fruity smell on heating with an alcohol, most probably it is an acid (The fruity smell is due to ester formation). Hence it must contain a COOH group. The presence of peaks at $\frac{m}{z}$ 85 by the loss of 17 mass units (OH) from molecular ion (182), and $\frac{m}{z}$ 45 (COOH) supports this view.

Further, it is known that carboxylic compound can undergo Mclafferty rearrangement if γ-hydrogen with respect to carboxylic group is available. The formation of the base peak at $\frac{m}{z}$ 60 may be explained on the basis of Mclafferty rearrangement as shown below.

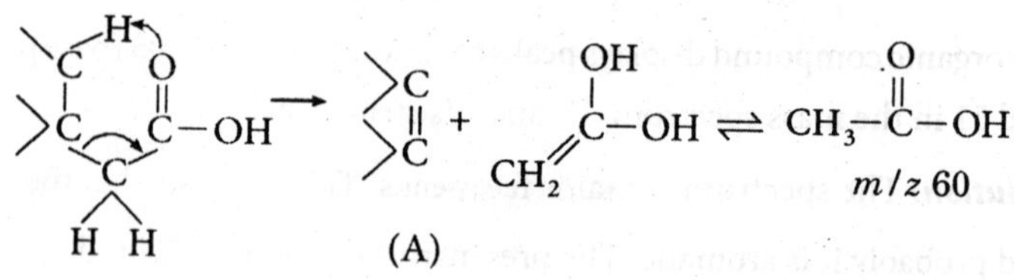

Therefore, the alkene portion (A) should have mass value of 42. This suggests that (A) is $CH_3-CH=CH_2$. Thus the probable structure of the compound is

$$CH_3-CH_2-CH_2-CH_2-COOH$$

n-Pentanoic Acid

The formation of different ion peaks may be rationalised as below:

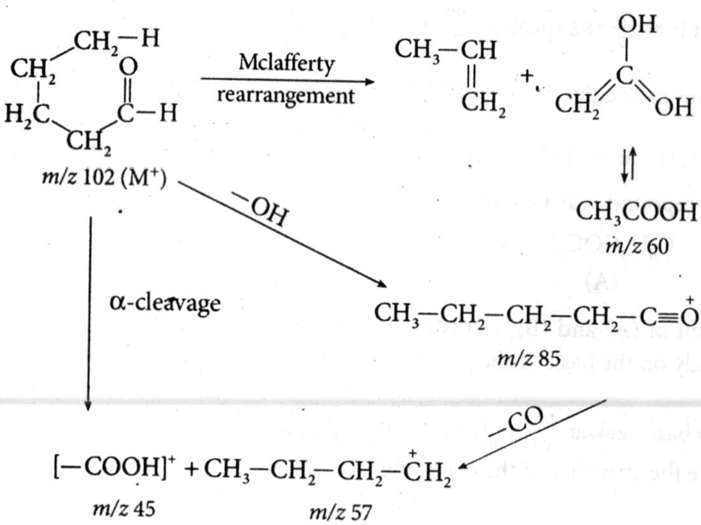

(vii) Deduce the probable structure for the compound with molecular formula, $C_{11}H_{14}O_2$ which exhibits in its mass spectrum the peaks at $\frac{m}{z}$ 178, 122, 105, 77 and 56.

***Solution*:** Since the heighest mass peak is at $\frac{m}{z}$ 178, so the molecular ion peak is at $\frac{m}{z}$ 178 which is in accordance with the molecular formula of the compound. The compound contains low content of hydrogen atoms as required by 11 carbon and two oxygen atoms. Yet the molecule is stable (replacing two oxygen atoms by four hydrogen atoms, the molecular formula for corresponding hydrocarbon becomes $C_{11}H_{18}$). So the compound may be aromatic in nature. Further, the presence of peak at $\frac{m}{z}$ 77 suggests the phenyl ring in the molecule. The peak at $\frac{m}{z}$ 105 may be due to $C_6H_5C\equiv\overset{+}{O}$ and it is formed by the elimination of a fragment with mass value 73. This mass value may be due to OC_4H_9 fragment. So if the fragments at $\frac{m}{z}$ 105 and $\frac{m}{z}$ 73 are combined, the tentative structure stands as $C_6H_5-\underset{\underset{O}{\|}}{C}-OC_4H_9$, i.e., butyl (or its any isomer) ester of benzoic acid. The genesis of different mass ion peaks may be rationalised on the basis of butyl benzoate as shown below:

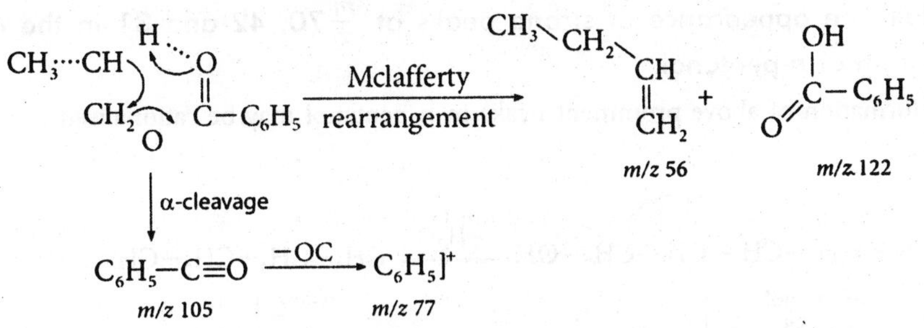

29. Alkylanilines exhibit the base peak at $\frac{m}{z}$ 106. Which ion is responsible for it? Explain its formation.

The three isomeric alkyl anilines lose a hydrogen atom to form an ion at $\frac{m}{z}$ 106. The genesis of this ion may be rationalised as below:

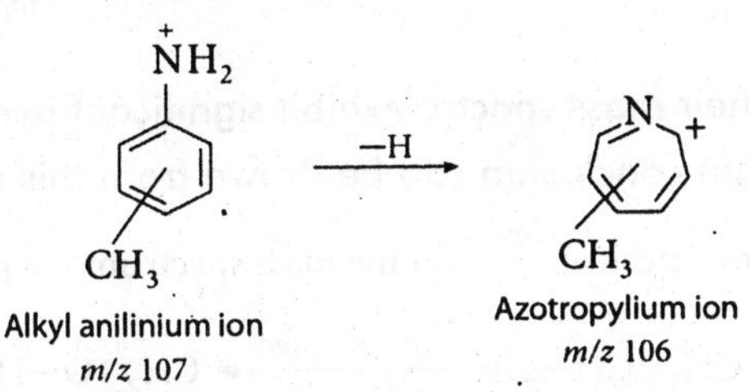

30. What is the structural formula of the ion which is responsible for the formation of the appearance of the strong peak at $\frac{m}{z}$ 149 in the mass spectrum of all ester of phthalic acid? Explain its genesis.

The strong peak at $\frac{m}{z}$ 149 in all ester of phthalic acid is due to the formation of the following ion.

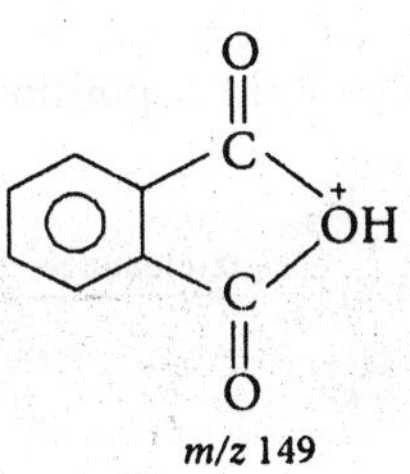

m/z 149

It is originated by the ortho effect as discussed below:

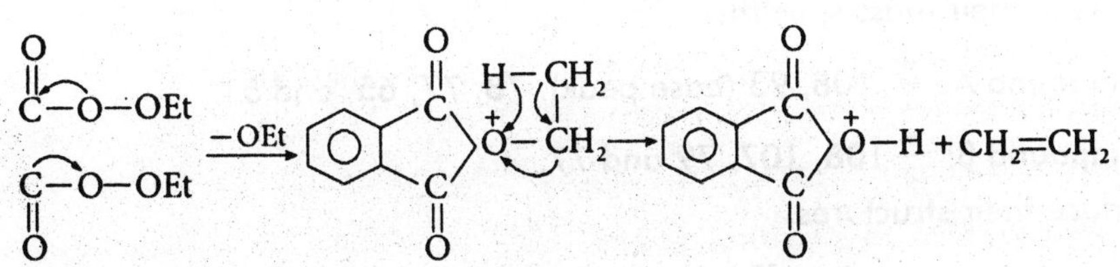

m/z 149

31. Explain the appearance of strong peaks at $\frac{m}{z}$ 70, 42 and 31 in the mass spectrum of *n*-pentanol.

➜ The formation of above prominent peaks in *n*-pentanol may be rationalised as shown under:

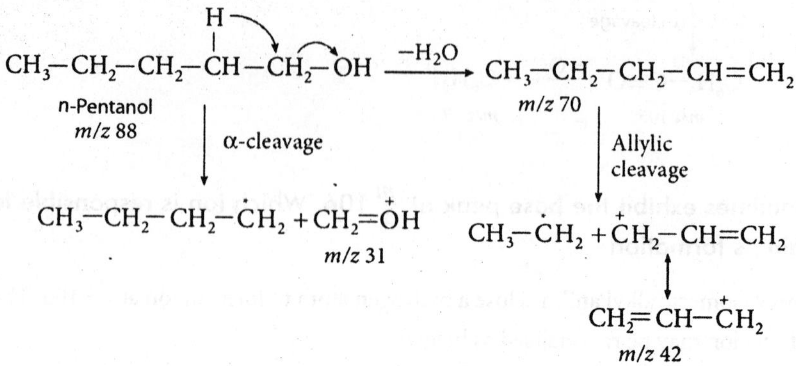

32. Three alcohols in their mass spectra exhibit significant peaks at $\frac{m}{z}$ 31, 45 and 59 respectively. What conclusion can be drawn from this observation.

➜ The alcohol which shows peak at $\frac{m}{z}$ 31 in the mass spectrum is a primary alcohol.

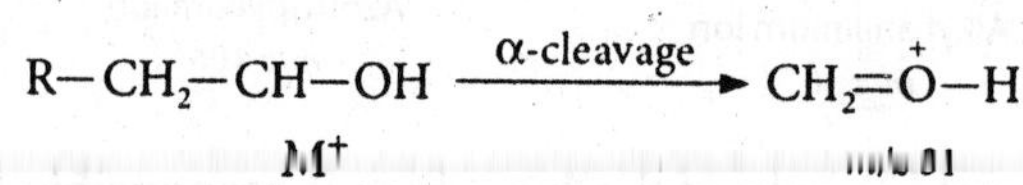

The peak at $\frac{m}{z}$ 45 is given by secondary alcohol with CH_3 group at α-position.

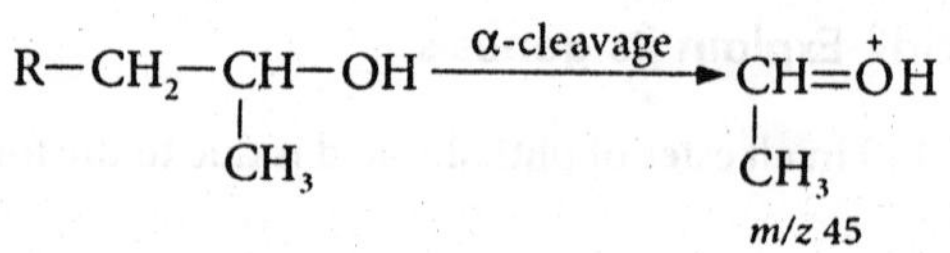

Tertiary alcohol with two CH_3 groups at the α-position generates the peak at $\frac{m}{z}$ 59.

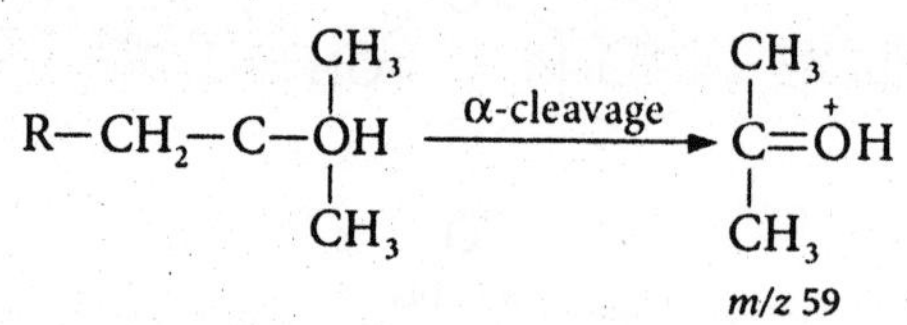

33. Two isomeric compounds with molecular formula, C_9H_8O, display the following peaks in their mass spectra.

Compound A: $\frac{m}{z}$ 108, 93 (base peak), 78, 77, 65, and 51.

Compound B: $\frac{m}{z}$ 108, 107, 79 and 77.

Deduce their structures.

➜ The presence of ion peak at $\frac{m}{z}$ 77 in both the isomers is a clear indication of the presence of phenyl ring in both the compounds. Again, both the isomers contain one oxygen atom, and their moelcular ion has mass value 108. On these basis the structures of these two may be suggested as below:

(A) Anisole (B) Benzyl alcohol

The formation of other ions in their mass spectra may be explained as below:

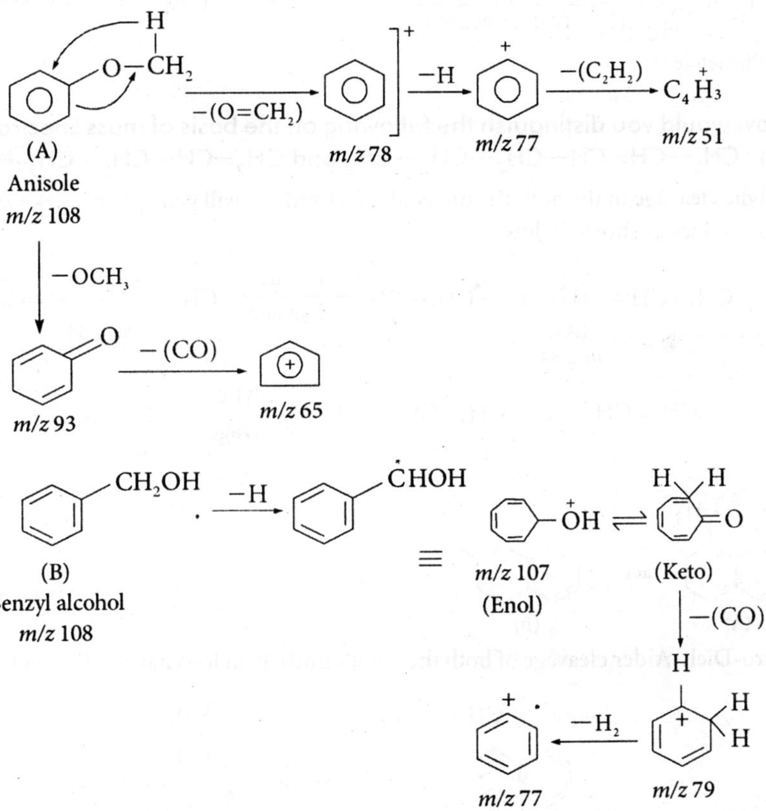

34. Do all the bonds in a molecular ion undergo fission?

- All the bonds in the molecular ion do not undergo cleavage. Some bonds break rapidly. This is due to the availability of insufficient energy at 70 eV which is the normal electrovolt supplied.

35. What is expected to happen if the energy of the ionisation electrons is increased beyond 70 eV?

- It the energy of the ionising electron beam is increased beyond 70 eV no significant changes occur in the mass spectrum. Eventually second ionisation energy is reached and M^{2+} ions are generated.

36. Discuss some unusual aspects of amines which are helpful in identifying them.

- Amines exhibit following characteristic features which help in characterising them.

  (i) They show an odd number of molecular mass.

(ii) They have a high tendency of capturing proton and so normally low intense (M+1) ion peak is observed.

37. Explain the formation of a mass ion at 94 in the mass spectrum of phenetole.

➜ The ion peak at $\frac{m}{z}$ 94 in the mass spectrum of phenetole may be explained as below.

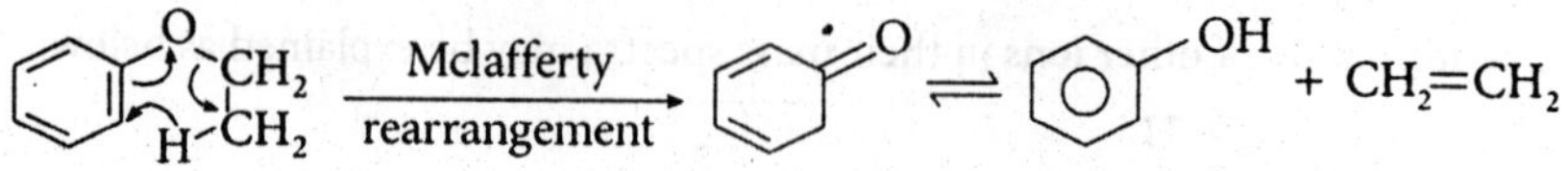

Phenetole

38. How would you distinguish the following on the basis of mass spectral analysis?

(a) $CH_3—CH{=}CH—CH_2—CH_2—CH_3$ and $CH_2{=}CH—CH_2—CH_2—CH_2—CH_3$

➜ Allylic cleavage in the both the molecules (A) and (B) will generate in peaks with different mass values as shown below:

$$\underset{\substack{(A)\\ m/z\ 84}}{CH_3-CH{=}CH-CH_2-CH_2-CH_3} \xrightarrow[\text{cleavage}]{\text{Allylic}} \underset{m/z\ 55}{CH_3-\overset{+}{C}H-CH{=}CH_2}$$

$$CH_2{=}CH-CH_2-CH_2-CH_2-CH_3 \xrightarrow[\text{cleavage}]{\text{Allylic}} \underset{m/z\ 41}{\overset{+}{C}H_2-CH{=}CH_2}$$

(b) 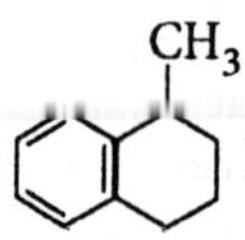 and 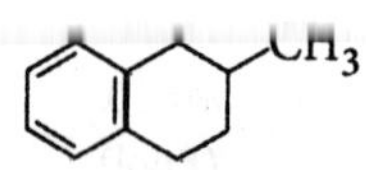

(A) (B)

➜ retro-Diels-Alder cleavage of both the compounds form ions having different mass values.

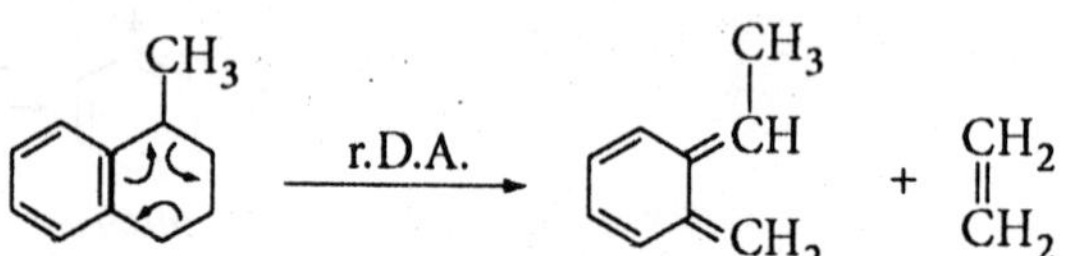

m/z 146 m/z 118 m/z 28

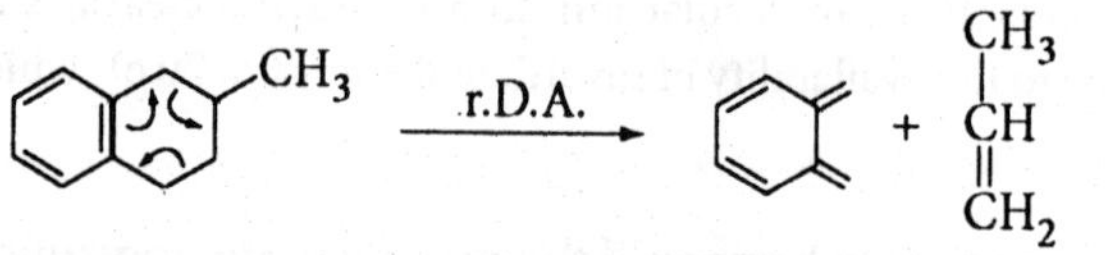

m/z 104 m/z 42

(c) 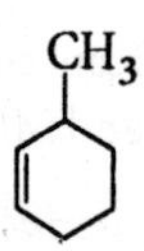 and 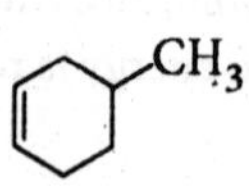

(A) (B)

➜ Both (A) and (B) will undergo retro-Diels-Alder cleavage to generate mass ions with different mass values.

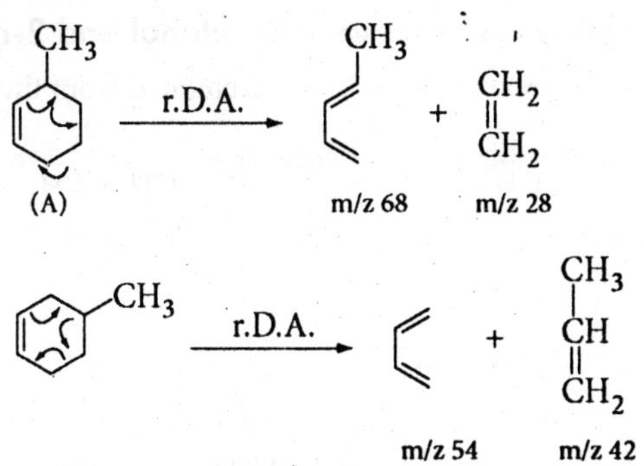

(d) *o, m,* and *p*-dimethoxy benzenes.

➜ All these three dimethoxy benzenes exhibit the following ion peaks in their mass spectra.

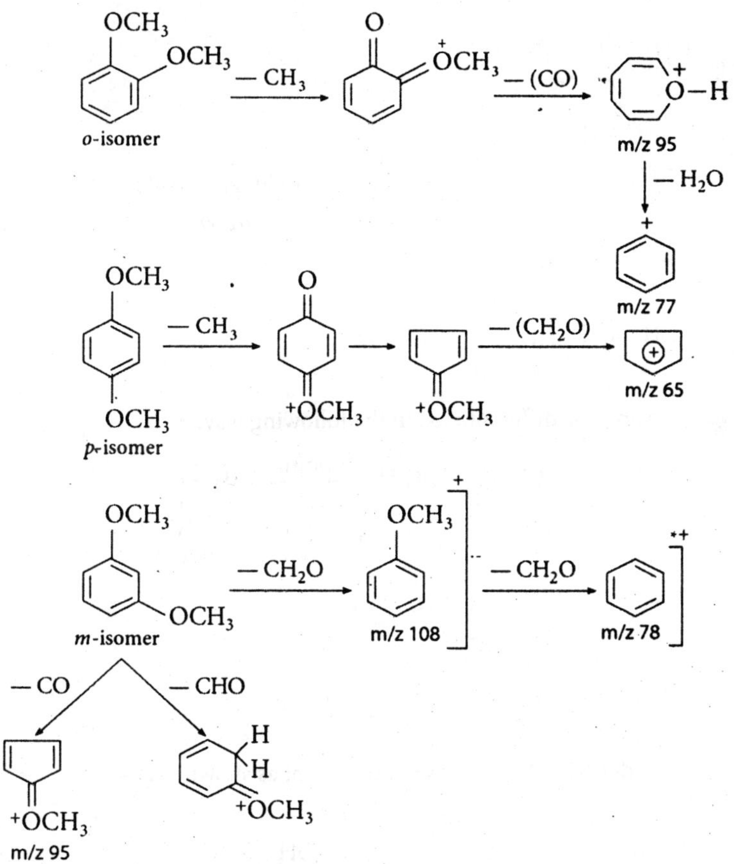

The *o* and *p*-dimethoxy benzene show very little loss of CHO. M–CH_3 ion is very abundant in them, but in *m*-dimethoxy benzene loss of CHO is prominent and successive loss of two molecules of formaldehyde along with the elimination of a molecule of CO are also observed.

(e) Three isomeric butanols, l-butanol, 2-butanol and 2-methyl-2-propanols.

➜ These three isoemeric butanols may be differentiated from their following mass peaks.

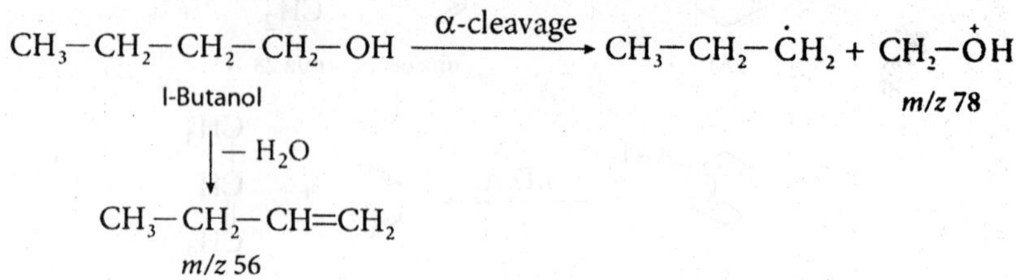

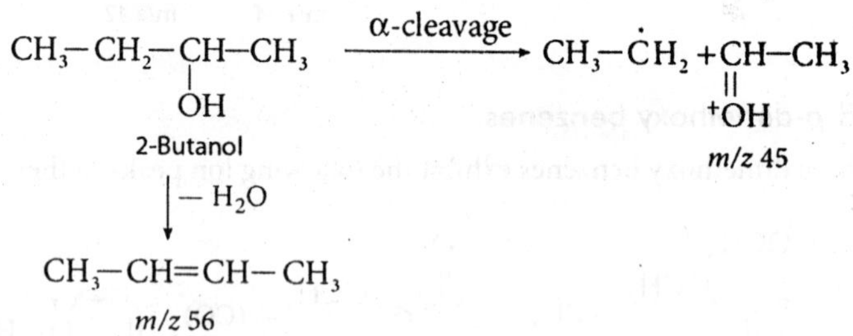

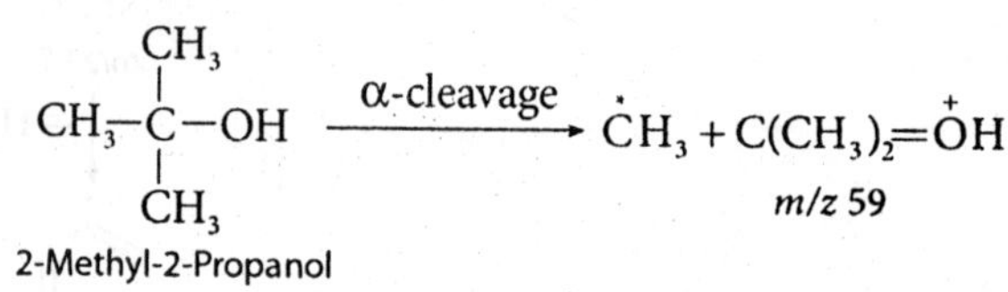

(f) $CH_3{-}C({=}O){-}OC_2H_5$ **and** $C_2H_5{-}C({=}O){-}OCH_3$

➜ These two esters may be differentiated in the following way:

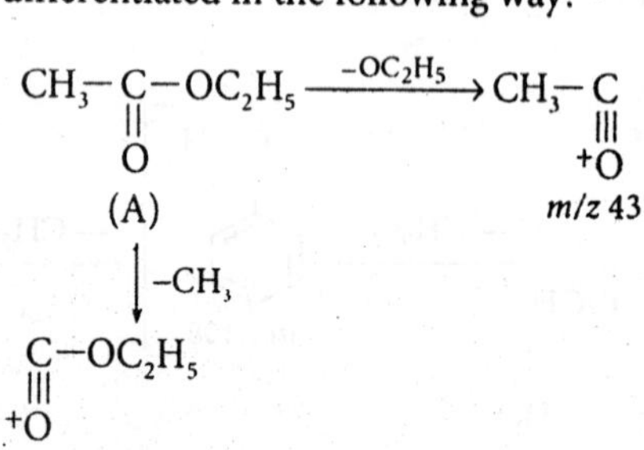

This ester also undergoes Mclafferty rearrangement as shown below:

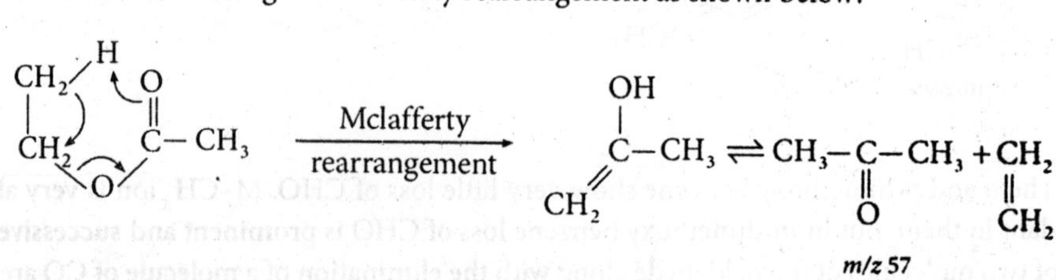

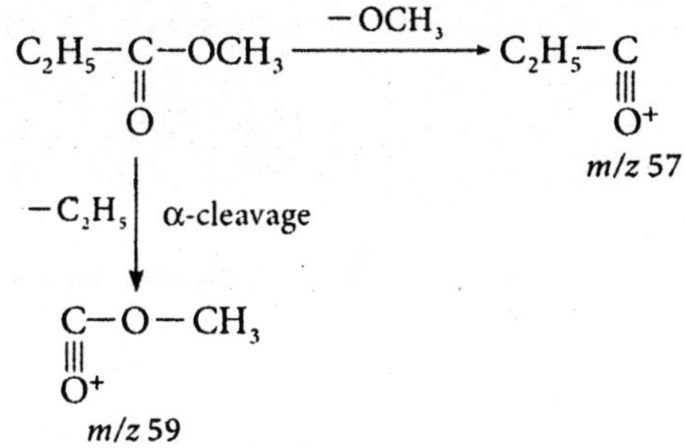

The ester (B) does not undergo Mclafferty rearrangement as there is no γ-hydrogenation.

(g) 3- and 4-methyl-2-pentanone.

➜ The structures of these two ketones may be written as below:

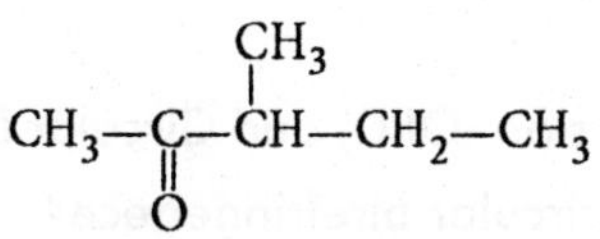

3-Methyl-2-pentanone

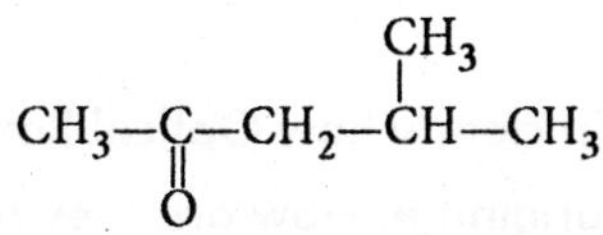

4-Methyl-2-pentanone

Both the ketones contain keto methyl $(CH_3-\underset{\underset{O}{\|}}{C}-)$ grouping and hence both will form peak at $\frac{m}{z}$ 43 through α-cleavage. Again, both the ketones contain γ-hydrogen atom and hence Mclafferty rearrangement will occur producing ion peaks of different mass values as shown under.

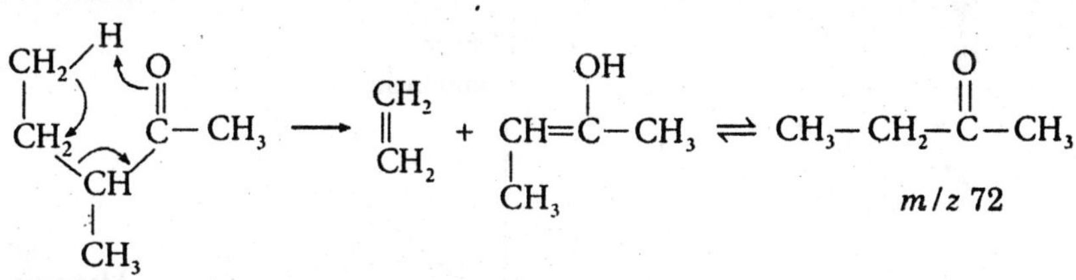

3-Methyl 2-pentanone

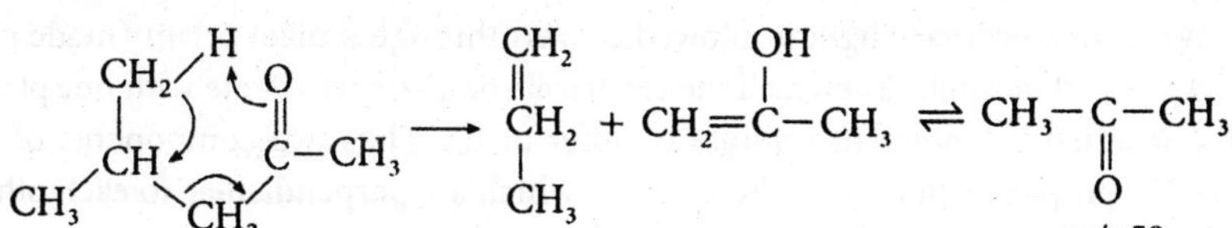

4-Methyl-2-pentanone

CHAPTER 7

OPTICAL ROTATORY DISPERSION (ORD) AND CIRCULAR DICHROISM (CD)

1. Explain clearly how Optical Rotatory Dispersion (ORD) and Circular Dichroism (CD) originate. How are they related with circular birefringenece?

➜ Both Optical Rotatory Dispersion (ORD) and Circular Dichroism (CD) are the optical phenomena and both originate from the differential absorption of circularly polarised light by optically active medium. Origin of these two may be rationally discussed as below:

It is known that light is an electromagnetic radiation and is an wave phenomenon. It consists of two component forces—(i) electrical and (ii) magnetic which act perpendicular to each other and propagates through vibration. At any point of propagation these two component forces of ordinary light act in all possible directions at random as shown in the figure.

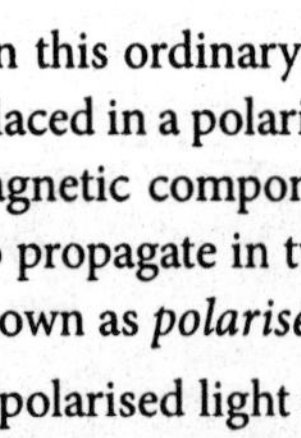

Ordinary Light

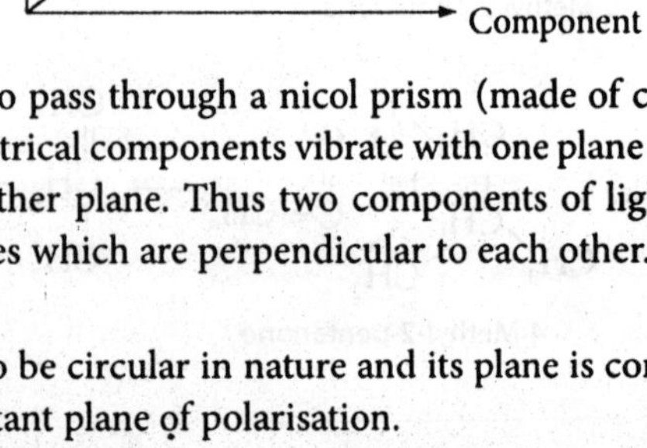

When this ordinary light is allowed to pass through a nicol prism (made of calcite, $CaCO_3$) placed in a polarimeter, all the electrical components vibrate with one plane while all the magnetic components vibrate in other plane. Thus two components of light are allowed to propagate in two different planes which are perpendicular to each other. Such light is known as *polarised light.*

This polarised light has been found to be circular in nature and its plane is continuously rotating across the plane of the resultant plane of polarisation.

Another very interesting and important characteristic of circularly polarised light is that each of the components (electrical and magnetic) is made of two parts—one is right circularly polarised part and the other is left circularly polarised part. In absence of any medium these two parts of each component move in the resultant direction. Thus, there are two resultants—one for electrical and other for magnetic components. These two resultants are mutually perpendicular to each other and the final resultant of these two resultant forces propagate linearly through only one plane known as *plane of polarisation*. That is why this light is known as *linearly plane polarised light*. This statement may be expressed by the following diagram.

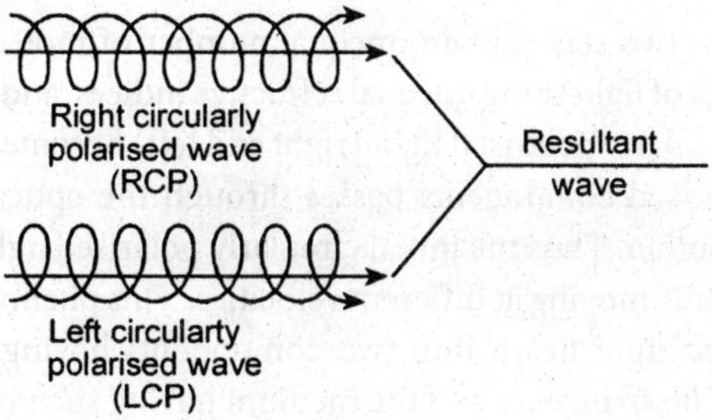

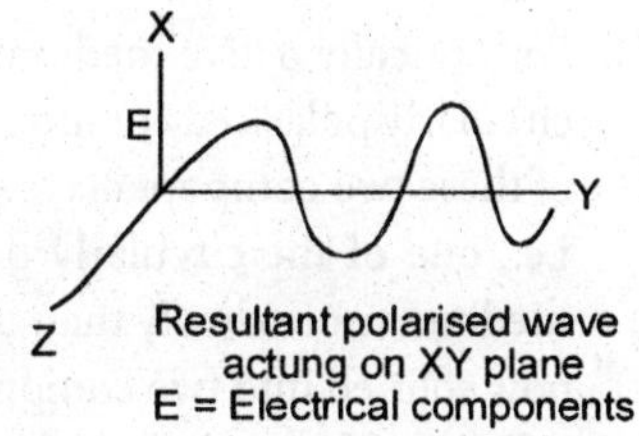

It may be mentioned that the magnetic field of electromagnetic wave lies perpendicular to the electric field and acts along YZ-plane. Now when the plane polarised light is allowed to pass through a chemical compound there will be polarisation (induced polarisation) of the electrons of the molecule by the electrical component of light. As a result of such random polarisation in a gram-mole of the substance containing *N* number of molecules where *N* is the Avogadro number—a few of the molecules will be polarised by the right circularly polarised part of the electrical component while remaining number of molecules may be polarised by the left circularly polarised part or in other words it may be concluded that during passing through the molecule (medium) aggregates (*N* bumber of molecules) are distributed into two sets or the original medium is divided into two media—one due to left and other due to right circularly polarised parts of the plane polarised light.

Again, it is logical to conclude that the thickness (density) of any medium depends upon the number of molecules. So refractive indices of the medium may have relation with the polarisation [Polarisation $(\alpha) = \frac{\mu^2 - 1}{\mu^2 + 2} \cdot \frac{M}{D}$ where μ is the refractive index of the medium, *M* is the molecular weight and D is the density of the medium]

Further it has been established that refractive index is related to the velocity as

$$v = \frac{c}{\mu},$$

where *c* is the velocity of light in vacuum and *v* is the velocity of the light in the medium. Thus, from the above discussion following deductions may be made.

I. Since two polarised sets contain the same molecules, one set may be regarded as an object and the other as its image. In this situation two cases may arise—

(a) When both polarised sets contain same number of molecules, one set (say, object) will be exactly identical (superimposable) with the other set (say, image).

(b) But if both the sets contain unequal number of molecules, then one set (say, object) is not exactly superimposable with other set, i.e., image.

It is an established fact that the superimposability or non-superimposability of the object with its image is related to optical activity or inactivity. When the object (molecule) and its image are non-superimposable to each other the object (molecule) is said to be dissymmetric and it exhibits optical activity, i.e., the object is optically active. On the other hand when the object (molecule) and its image are superimposable, the molecule is said to be *symmetric* and it does not show any optical activity, i.e., the molecule is optically inactive.

II. For optically active medium, as two sets contain unequal number of molecules, two circularly polarised components of light show unequal refractive indices, and velocities of these two components of circularly polarised light (right and left) become different, i.e., one of the circularly polarised components passes through the optically active medium more slowly than the other. Thus the initial circularly polarised light beam is now splitted into two components moving at different velocities. This phenomenon of splitting of circularly polarised light beam into two components having different velocities is termed as *circular birefringence* and the medium having such property is known as *circularly birefringent*.

Again, as a result of splitting of circularly polarised light into two components (circular birefrignence), the plane of the incident polarised light is rotated by an angle α. In practice, this angle is measured in terms of either specific rotation, [α], as

$$[\alpha] = \frac{\alpha \times 100}{l \times c}$$

where α is in degrees, l is the length of the tube in which the solution of the medium taken in decimetres, c is the number of grams per 100 ml of the solution, or we may measure as molecular rotation defined as

$$\text{Molecular rotation} = \frac{[\alpha] \times M}{100}.$$

M is the molecular weight of the optically active medium. If the direction of rotation of the plane of polarisation is to the right, the substance (medium) is called dextrorotatory (+); if to the left, laevorotatory (–). Thus the cause of optical rotation (manifestation of optical activity) is circular birefringence.

It has been observed that the refractive index of a medium is not a constant but depends upon the wavelength as is evident from the following expression

$$\mu \text{ (refractive index)} = I + \frac{k\lambda^2}{\lambda^2 - \lambda_0^2},$$

where k and λ_0 are constants for a given medium and μ is the refractive index at a particular wavelength λ of the light ray. It follows, therefore, that the angle of rotation α and hence specific rotation or molecular rotation will change with the wavelength of the incident polarised light. This phenomenon of change of optical rotation with wavelength is designated as ***optical rotatory dispersion*** (ORD). We can, thus, conclude that ORD is nothing but a form of manifestation of circular birefringence.

Now due to circular birefringence of the optically active medium, as we have seen, there will be unequal rates of propagation of left and right circularly polarised light. As a result, these two circularly polarised components not only are out of phase but are of unequal amplitudes, i.e., the components are absorbed to differnt extents in the region of absorption and hence the optically active medium has a unequal molar absorptivity coefficients (ε) for right and left circularly polarised components of the light ray. This difference in molar absorption coefficients (Δε) is termed as *differential dichronic absorption* and the phenomenon of creation of difference in molar absorptivity by optically active medium is known as *circular dichroism* (CD).

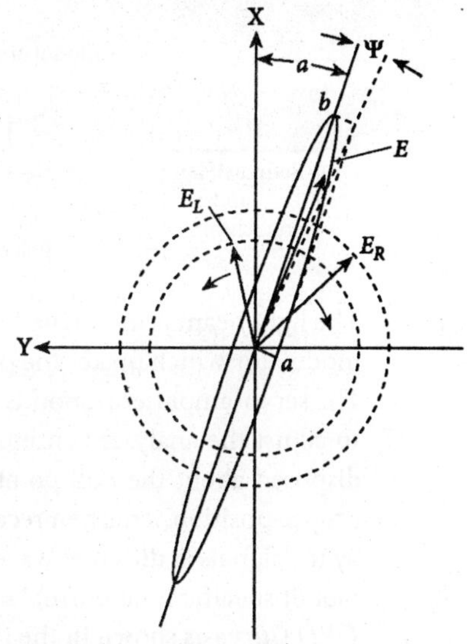

Further, due to unequal absorption by left and right circularly polarised components of the light beam passing through optically active medium, an imbalance is created between the strength of two circularly polarised rays of emergent beam. As a result the emergent beam is not truly linearly polarised but traces out an ellipse as shown in the figure, and the transmitted light becomes elliptically polarised. Similar to circularly polarised ray, this ray may be left and right handed elliptically polarised. The ellipticity of the ellipse is defined by an angle Ψ given by

$$\Psi = \frac{\pi}{\lambda}(k_L - k_R)$$

where k_L and k_R are absorption coefficients of left and right circularly polarised components. The specific ellipticity [Ψ] may be expressed as

$$[\Psi] = \frac{\Psi}{l.c.} \quad \text{(i)}$$

where l is the path length in centimetres and c is the concentration of the solution. The molar ellipticity [θ] in terms of extinction coefficients may be written as

$$[\theta] \approx 3300\ \Delta\ t \quad \text{(ii)}$$

The value of specific ellipticity (i) or molar ellipticity (ii) cannot be directly determined but can only be determined through CD.

It is thus revealed from the above discussion that the only cause of both these phenomena is the circular birefringence of the optically active compound and is therefore they are very closely similar and has been found to occur simultaneously. For this reason these chiroptical phenomena are always written as ORD-CD or CD-ORD and appear as if a single phenomenon.

2. How ORD and CD are measured?

- Both ORD and CD are measured with instruments—ORD is measured by *spectropolarimeter* while CD is measured by conventional spectrophotometers.

(a) *Spectropolarimeter*: There are several types of spectropolarimeters for recording ORD. They, in general, are modified single-beam spectrophotometers. The sechematic representation of single-beam spectropolarimeters may be shown as in the figure (p. 230):

In this type of spectropolarimeter the light beam from a monochromator passes seqeuentially through polariser, the sample and an analyser to a photomultiplier tube.

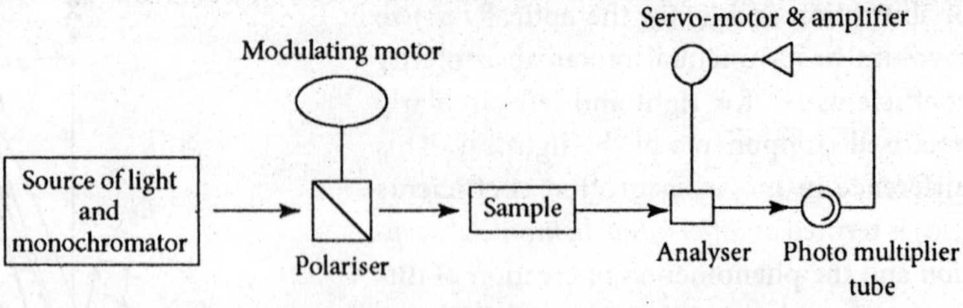

The light beam is modulated with the state of polarisation at 12 Hz by a motor-driven modulator which makes the polariser to rock back and forth through an angle of ±1°. The servo-amplifier responds only to the 12 Hz frequency and causes the servomotor to adjust the analyser continuously to the point where 12 Hz signal is symmetrically disposed about the null point. The servomotor also keeps the recording pen in the proper position. This pen records the rotation of the plane of polarised light obtained by the signals at different wavelengths and the result is expressed in a graph which is a plot of specific rotation/molar rotatin against the wavelength. This graph is known as *ORD Curve* as shown in the figure below:

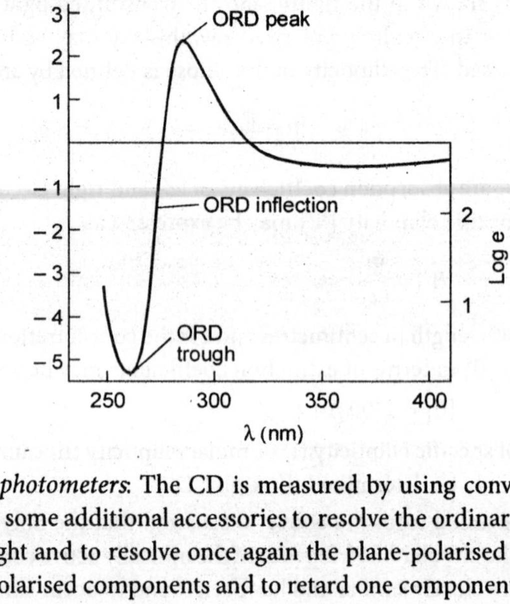

(b) CD *Spectrophotometers*: The CD is measured by using conventional spectrophotometers with some additional accessories to resolve the ordinary light beam into plane-polarised light and to resolve once again the plane-polarised light into right and left circularly polarised components and to retard one component relative to other by exactly one-quarter wavelength. The important circular resolvers may be of two types, viz., *Fresnel rhomb* and *Pockels electro-optical modulator*. Fresnel rhomb has limited wavelength range while in Pockels type resolver can be programmed to give continuously correct output as the wavelength is changed. A general schematic representation of rhomb is as shown in the figure.

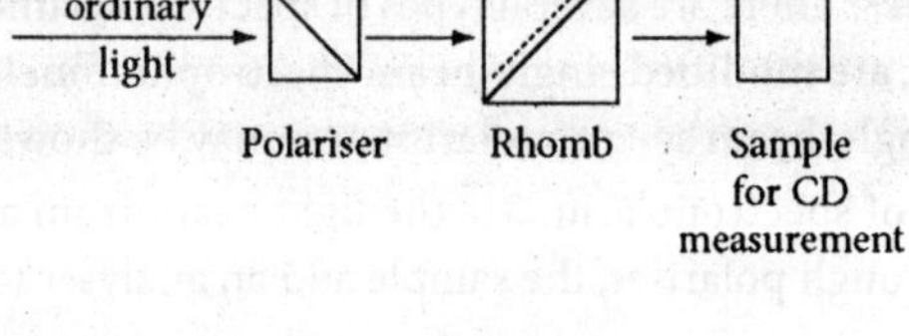

This entire unit of the rhomb is fitted into the sample chamber of a standard spectro photometer. The result is expressed by the curves known as CD-curves. These curves are the plot of $\frac{\Delta\varepsilon}{[\theta]}$ (difference in molar absorptivity ($\Delta\varepsilon$)) or molar ellipticity $[\theta]$ against wavelength as shown below:

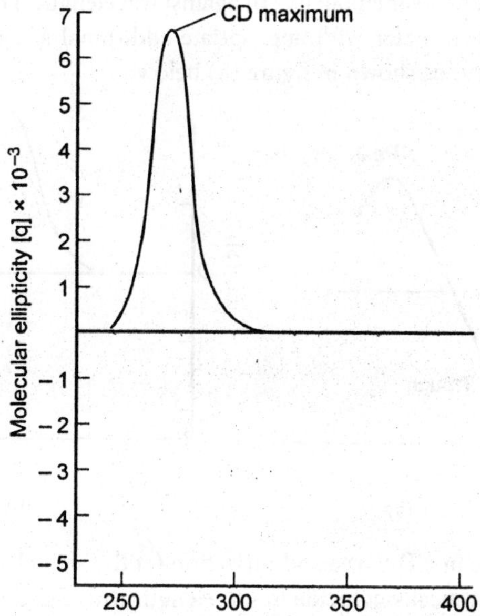

It may mentioned in this connection that it is not easy to measure CD alone in the ultraviolet region- the only region where it can be important for colourless compounds.

3. Explain why is the study of ORD and CD also termed as chiroptical spectroscopy?

→ It is known that circular birefringence, circular dichroism and their combination (Cotton effect) as well as optical rotatory dispersion are chiroptical properties of chiral compounds. That is why the study of ORD and CD is also termed as chiroptical spectroscopy.

4. (a) Define the term Cotton effect. (b) How is it studied? (c) Why most of the Cotton effect studies has been carried with carbonyl chromophore? (d) What do you mean by positive and negative Cotton effects?

→ (a) It has been observed that any medium which exhibits circular birefringence, i.e., optical rotatory dispersion (ORD) also exhibits circular dichroism (CD). The combination of these two effects in the region, in which, optically active absorption bands are observed gives rise to a phenomenon known as Cotton effect after the name of French physicist Aime Cotton who first observed this phenomenon along with ORD and CD. It may be mentioned that the Cotton effect is observed precisely for the compounds which absorb in the ultraviolet region but for compounds that absorb only in the far ultraviolet region, i.e., well below 220 nm no Cotton effect is observed, because the effect occurs only near an absorption maximum.

(b) Cotton effect is generally studied by plotting the change of optical rotation with wavelength. This plot as we know is known as ORD Cotton effect curve. Such plot is readily obtained by means of spectropolarimeter. In general, ORD curve showing Cotton effect has a peak (high point) and a trough (low point), and is known as *anomalous* ORD cotton effect curve [figure (a) below.]

However, Cotton effect may be investigated by plotting difference in molar absorptivity, $\Delta\varepsilon$ or molar ellipticity $[\theta]$ against wavelength. This plot obtainable by the use of spectrophotometer with appropriate additional accessories is known as CD *Cotton effect* curve as shown in figure (b) below:

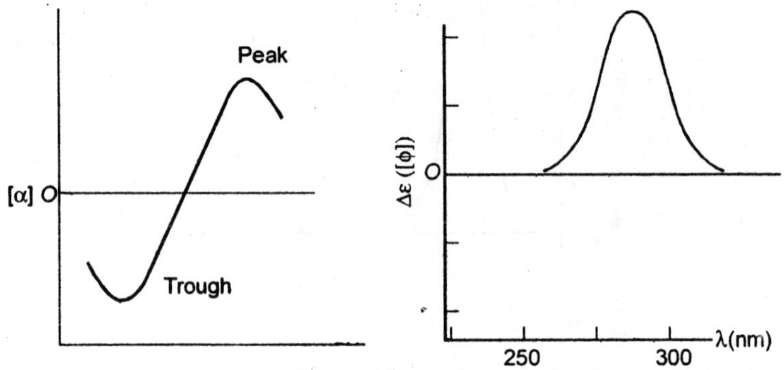

(a) (b)

The maxima in CD curve and inflection (at the sign change cross-over point) of ORD curve theoretically coincide in wavelength with the absorption maxima i.e. the region of Cotton effect.

(c) A significant point which needs to be mentioned in this context that the measuerment of Cotton effect in ORD curve is satisfactory only when the functional group (chromophore) of the compound under study has an absorption band with very low extinction coefficient so that it cannot interfere with measurement of optical rotation through the region of absorption band. The carbonyl group which acts as a chromophore due to excitation of an electron in a non-bonding orbital *n* (lone pair on oxygen) into low energy antiboding orbital (π-bond) absorbs uv-light in the region of $n \rightarrow \pi^*$ transition (280–290 nm) and its absorption is not very strong. This makes carbonyl group ideal for ORD Cotton effect study and that is why most of the studies on ORD cotton effect have been carried with ketones.

(d) The Cotton effect has been found to be of two types—positive (+) and negative (–). Consequently, curves for Cotton effect may be of two types designated as positive (+) *Cotton effect curve* and *negative* (–) *Cotton effect curve.* The ORD Cotton effect curve (anamalous curve) is called positive ORD Cotton effect curve when to approach the region of Cotton effect (absorption maxima) from long wavelength, rotation first increases, i.e., peak is first reached with the decrease in wavelength [figure (a)] but when in such approach rotation decreases, i.e., trough is reached first with the decrease in wavelength the ORD curve is called negative ORD Cotton effect curve as shown in the figure (a):

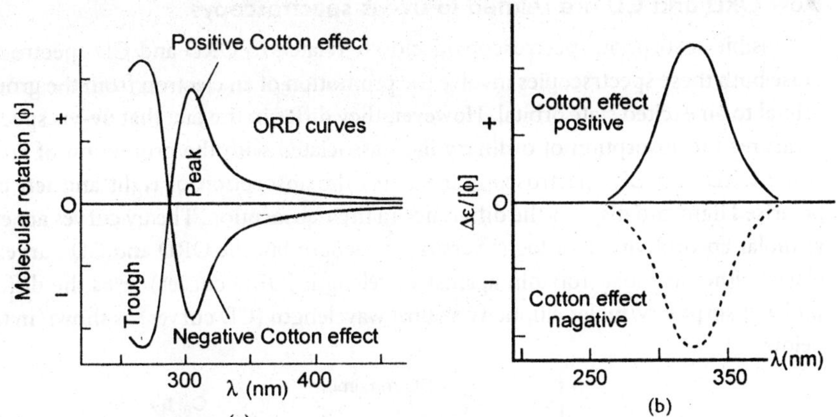

(a) (b)

Similar is the case with CD-Cotton effect as is evident from the figure (b).

5. Discuss how CD and ORD are related to each other.

➜ The relationship between CD and ORD may be discussed as below.

ORD involves the measurement of optical rotation and, therefore, can be detected over all the wave lengths whereas CD pertains to the measurement of absorption, i.e., the differential absorption of left and right circularly polarised light. Thus CD occurs in the vicinity of absorption band. The relationship between ORD and CD may be expressed in terms of integral transform known as *Kramers relation* and corresponding CD curve may be calculated from the experimentally determined ORD curve and vice-versa. The equation is known as *Kroning-Kramers relation* which relates the amplitude of an ORD Cotton effect to $\Delta\,\varepsilon$ of the corresponding CD peak as given by the following expression.

$$\alpha = 40.28\,\Delta\,\varepsilon = (1.22 \times 10^{-2}) \times [\theta]$$

where α is the angle of rotation of the plane polarised light, $\Delta\,\varepsilon$ is the difference in molar absorption and $[\theta]$ is the molar ellipticity.

Both ORD and CD curves have equal significances in organic chemistry with respect to stereo chemistry. The curves of both the phenomena may be positive or negative. A positive ORD curve usually gives a positive CD and a negative ORD gives a negative CD curve.

In general, CD curve is more simple than ORD because ORD is always associated with a background curve due to transitions at distant wavelength which interfere with the determiantion of the sign of the Cotton effect while CD curves are, in general, free from such background effects and if the compound under study has more than one absorption maximum, CD maxima are well separated.

It may be mentioned that for a CD Cotton effect of a compound there also exists a corresponding ORD Cotton effect as shown in the figure because both the phenomena reflect the interaction of polarised light with the same chromophore.

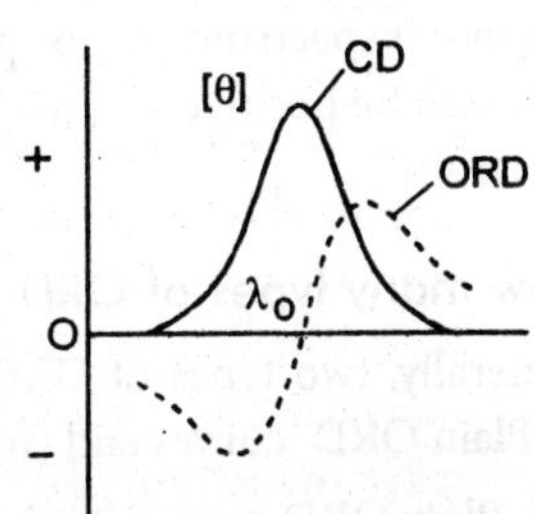

Positive Cotton effects

6. How ORD and CD are related to uv-vis spectroscopy?

➜ UV-visible absorption spectroscopy is closely related to ORD and CD spectroscopy because both these spectrscopies involve the promotion of an electron from the ground state orbital to an excited state orbital. However, they differ in the fact that uv-vis spectrsocopy measures the absorption of ordinary light associated with the promotion of an electron while ORD and CD spectroscopy measures the absorption of right and left circularly polarised light and displays the difference in their absorption. The uv curves are expressed as molar absorptivity (ε or log ε) versus wavelength but the ORD and CD curves are displayed either as molar rotatoin against wavelength (ORD curves) or as the difference in molar absorptivity/molar ellipticity against wavelength (CD curves) as shown in the figure below:

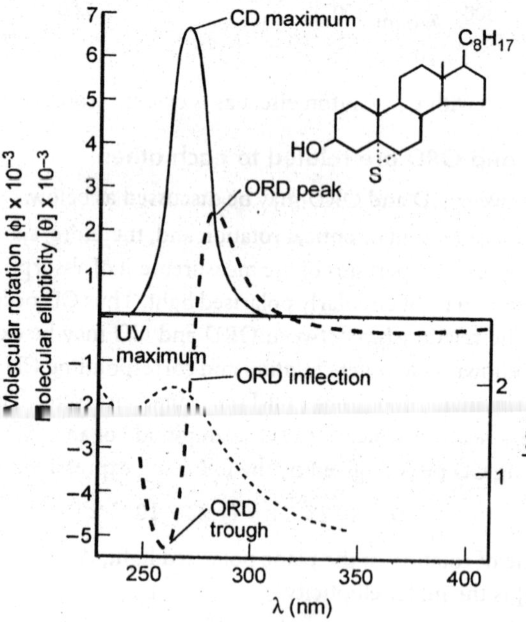

Theoretically, maxima of CD and uv-visible curves and the inflection of ORD curve coincide in wavelength but the superimposed effects of chromophores may cause shifts in the observed maxima in uv-vis curve. CD curves will often pinpoint "hidden absorption maxima" (uv-vis-maxima) with less ambiguity than the corresponding ORD curves.

A CD curve has very close similarities in appearance with uv-vis curve, however, unlike uv-vis spectrum may be positive and negative as is evident from the figure. ORD curve may also be positive and negative in addition to dissimilarities in appearance with uv-vis spectrum.

7. How many types of ORD curves are observed. Discuss them in short.

➜ Generally, two types of ORD curves are observed for the organic compounds, such as (i) Plain ORD curves and (ii) anomalous ORD curves.

(i) *Plain* ORD *curves*: It is known that for the compounds that absorb only in the far ultra violet region, i.e., well below 220 nm normally no Cotton effect is observed because Cotton effect occurs near the absorption maximum.

However, if the measurement is carried out below 220 nm a steady increase or decrease in optical rotation with decrease of wavelength is observed. The nature of the resulting plot becomes as shown in figure.

Such type of plot has been termed as *plain ORD curve* by Djerassi. This plot may or may not cross the zero rotation line but it lacks inflections and extrema. It differs from *normal dispersion* curve in the fact that the normal dispersion curve is always devoid of all the three features, viz., inflection, extrema and crossing of zero rotation axis. Plain curve may be positive or negative in nature. When the molecular rotation increases in the positive direction, the curve is referred to as *positive plain dispersion curve* as seen in the curve 'A' of the figure above while the curve 'B' of the figure is called negative dispersion curve in which molecular rotation increases in the negative direction.

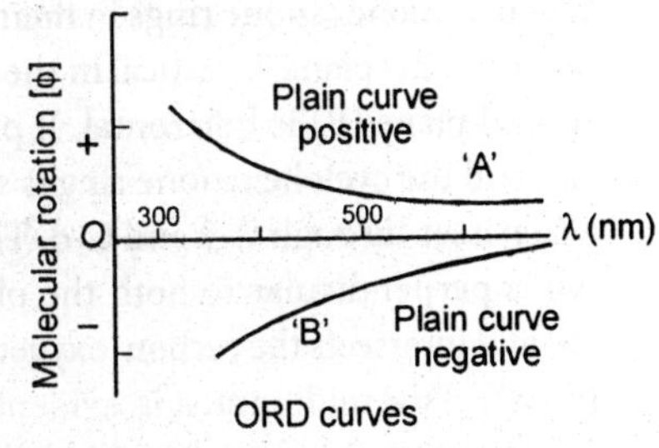

ORD curves

The plain dispersion curve obeys Drude equation, $[\alpha]_D = \frac{k}{(\lambda^2 - \lambda_0^2)}$, where k is an empirical constant, λ is the wavelength at which the molecular rotation $[\alpha]_D$ is measured and λ_0 corresponds to the wavelength due to absorption maximum.

(ii) *Anomalous* ORD *curves*: The ORD curve in which there is a high point known as 'peak' and a low point termed as 'trough' (the terms maxima and minima are avoided to prevent confusion with ultraviolet absorption maxima and minima) are present. This type of ORD curves are known as *anomalous* ORD *curves* and are obtained for compounds which have absorption maxima (λ_{max}) in or near the region under study (220–650 nm) and exhibit Cotton effect. This curve is S-shaped and this curve crosses the zero axis of rotation which closely corresponds to uv-maxima. They do not obey Drude equation. This type of curve may be positive or negative. When passing from the long wavelength if peak is first reached it is called positive anomalous ORD curve as in the curve 'A' of the figure but if the trough is reached first as in the curve 'B' of the figure, the curve is called negative anamalous ORD curve.

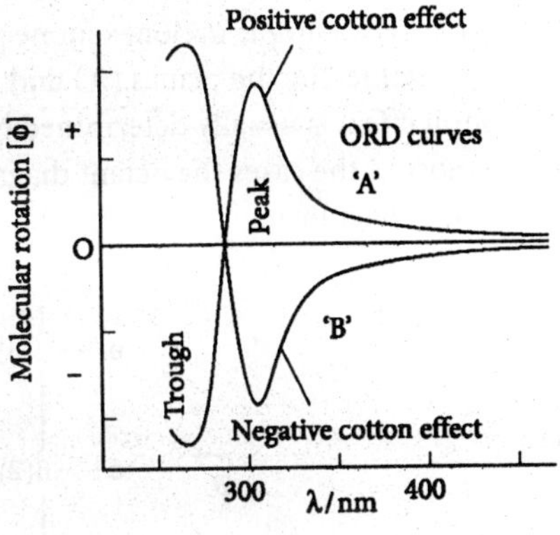

8. What is octant rule? What information can be obtained by the use of this rule to cyclohexanone?

➜ Moffiti and his associates first proposed an empirical rule for deducing the sign of Cotton effect of chiral cyclohexanone derivatives from their structure, conformation and configuration. This rule may also be useful in deducing the configuration or conformation when the other variables are known. Later this rule has been found to be applicable for a large number of keto carbonyl compounds.

In order to deduce the sign of Cotton effect, the space around the keto-carbonyl group is divided into eight sectors (octants) around the coordinate axes X, Y, and Z with the help of three planes (A), (B) and (C). The first plane (A) passes through the carbonyl group C–1 and C–4 of cyclohexanone rings in their chair conformation. This plane is vertical in the diagram. The second plane (B) is horizontal. It passes through C-1 and the cyclohexanone ring is so titled that it also passes through C-2 and C-6. The third plane (C) is perpendicular to both the planes (A) and (B) and intersects the carbon-oxygen double bond (C=O) at the midpoint. It is, evident from the figure thus that the two planes (A) and (B) divide the space around the carbonyl group into four quadrants designated as upper right (UR), upper left (UL), lower right (LR) and lower left (LL) for the purpose of *octant rule*. These four quadrants constitute the set of *rear octants* (away from the observer) and contribute to the sign of Cotton effect.

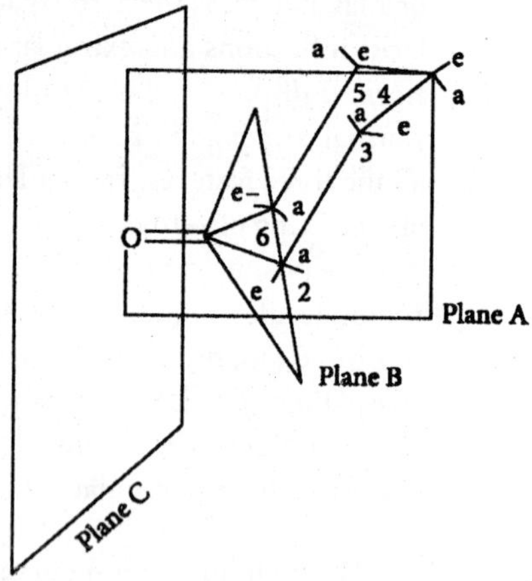

It may be mentioned that there is another set of four *front* quadrants which are the mirror image of the rear quadrants. The contribution of the substituents in these front quadrants to the sign of the Cotton effect is exactly opposite to that by the substituents in the rear quadrants. However, scrutiny of the literature reveals that in the case of vast majority of cyclohexanone derivatives, substituents rarely appear in the front quadrants and therefore for simple cyclohexanone plane (C) may be ignored and the only four rear quadrants defined by the planes (A) and (B) are to be considered. Thus, the overall sign of the cotton effect is usually determined by the substituents in these four rear quadrants. Thus, in most of the cases the octant diagram may be simplified as shown in the figure (Planar octant Diagram)

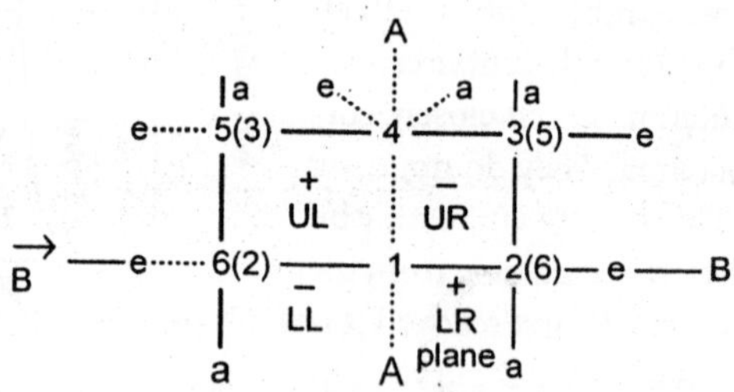

Rear Octants (Planar Octant Diagram)

From the above discussion, the statements of octane rule towards the determination of sign of Cotton effect stand as under:

(i) The substituents lying in the coordinate planes (A) and (B) have practically no controbution to n $\rightarrow \pi^*$ Cotton effect. These are 2e, 4a, 4e and 6e substituents.

(ii) The substituents in the sectors UL and LR make positive contribution to Cotton effect, i.e., axial substituents at C-2 and axial and equatorial substituents at C-5 have positive contribution to the Cotton effect.

(iii) Substituents present in sectors UR and UL, i.e., 6a and 3a, 3e make a negative contribution.

Let us consider an example to illustrate the operation of the octant rule. The cyclohexanone derivative; 3-Methyl cyclohexanone can exist in two conformation—one having methyl group at axial position (upper right) as seen in figure (a) and another with methyl group at equatorial position (upper right) as in figure (b). As there is no other substituents, the equatorial conformer (b) should predominate. This conformer is identical to the conformer (c). In this conformer (c) methyl group is at upper left (UL) sector. Hence, it should have positive Cotton effect. In fact, its Cotton effect has been found to be positive and it has R-configuration.

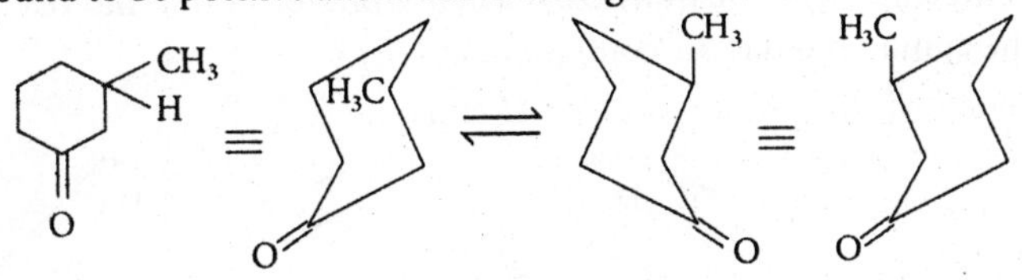

(a) (b) ©

9. (a) State and explain haloketone rule. How is it related to Cotton effect?
(b) Mention some of its applications in organic stereochemical problems.

(a) The haloketone rule is applicable to halogenated cyclohexanone derivatives. This rule is concerned with the effect of introducing halogen atom next to keto group, i.e., at α-position. This rule states that introduction of an equatorial halogen at α-position does not alter the sign of Cotton effect of the halogen-free ketone whereas axial halogen atom in α-position alters the sign of Cotton effect of the parent ketone. The cotton effect of α-halocyclohexanone may be predicted by looking along O = C axis, if halogen is on the left of line of view [figure (a)], the compound exhibits a negative Cotton effect while if the halogen is on the right as in the [figure (b)] Cotton effect will be positive.

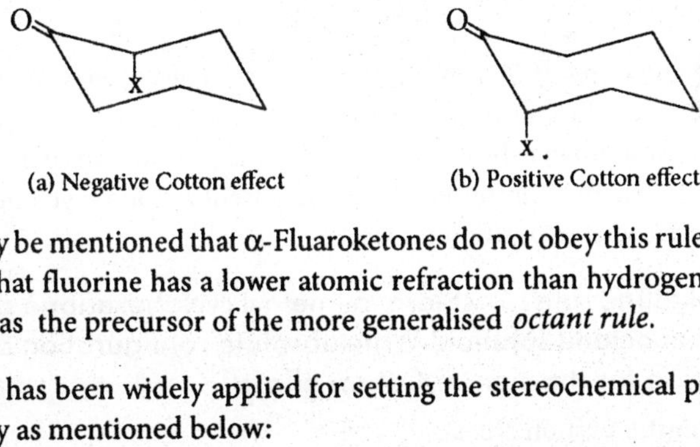

(a) Negative Cotton effect (b) Positive Cotton effect

It may be mentioned that α-Fluaroketones do not obey this rule. This is because of the fact that fluorine has a lower atomic refraction than hydrogen. This rule may be regarded as the precursor of the more generalised *octant rule.*

(b) This rule has been widely applied for setting the stereochemical problems in organic chemistry as mentioned below:

(i) *Determination of absolute configuration*: When conformation of the ring system and the site of axial α-halogen atom is known, the absolute configuration may be settled by the application of this rule. Thus, (–)-trans–1–decalone (A) on bromination forms (+)–2–β–bromo–*trans*–1–decalone (B) which shows strong positive Cotton effect. According to axial-haloketone rule the dextrorotatory bromoketone has configuration (B) and so its precursor (–)-*trans*–1–decalone must have configuration as shown in (A).

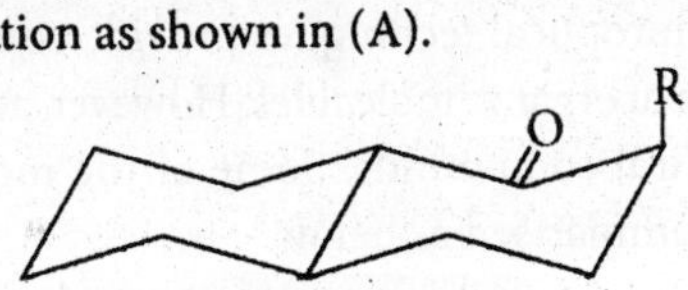

A; R=H, B : R=Br

(ii) *Assignment of conformation*: The conformation of the ring system may be settled if the other parameters, viz., configuration, position of the halogen are known and the nature of halogen be axial then the conformation of the ring system may be ascertained. The example cited in this connection is the chlorination of (+) 3–methyl cyclohexanone (A). The product formed was found to be 2–chloro–5–methyl cyclohexanone (B). In octane solution IR spectrum of the compound reveals the axial nature of halogen and its Cotton effect is negative which is in conformity with the trans-diaxial configuration of the chloroketone (B). Hence, the compound has the configuration (A).

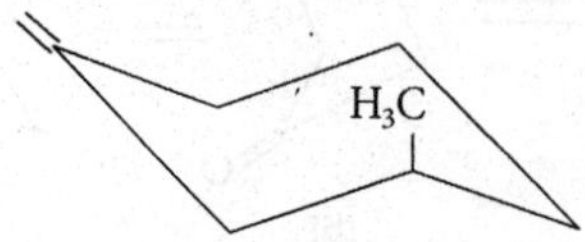

(A)

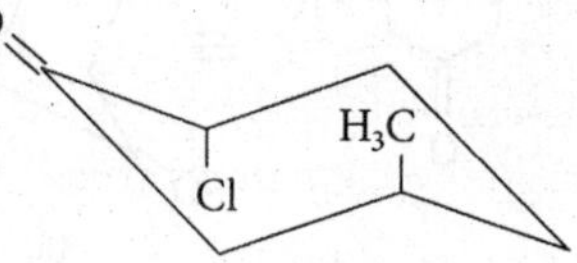

(B)

(iii) *Determination of the location of the axial halogen*: From the known configuration and conform of the parent ketone and if the α-halogen is axial, its location may be settled by the application of haloketone rule. Thus, 2,3–seco–cholestan–6–one–2,3–dioc acid on bromination gives an axial bromo derivative which exhibits negative Cotton effect. The negative Cotton effect clearly suggests that the bromine is present at 5-position.

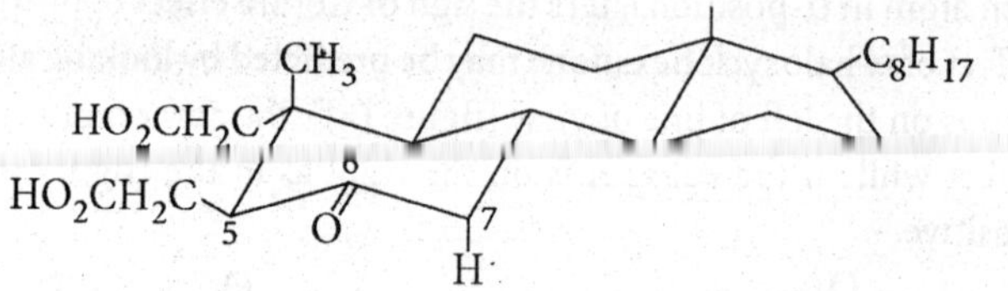

(iv) *Determination of the axial nature of the halogen atom in cyclohexanone ring*: When cyclohexanone is brominated it forms α-bromo cyclohexanone. The bromine atom may be axial or at equatorial position. In such situation, if there is reversal of the Cotton effect of the parent ketone, cyclohexanone, bromine introduced at α-position is axial in nature.

10. Define 'Haloketone rule'. 2–chloro–5–methyl cyclohexanone shows a negative Cotton effect in octane solution. Write absolute configuration of the compound and give reason for the sign of Cotton effect.

➜ See answer against Question No. 9.

11. State and explain Haloketone rule. How this rule can be used to determine the configuration and conformation of the compound?

➜ See answer against question no. 9.

12. Discuss some significant applications of ORD and CD.

➜ ORD and CD, two chiroptical techniques provide complimentary information regarding chiroptical properties of organic molecules. However, at present CD has practically replaced ORD in studying chiral compounds. Some of the most significant applications of these techniques may be summarised as below:

(i) *Study of absolute configuration and stereochemistry*: From the sign of Cotton effect stereochemistry in the vicinity of the carbonyl group can be easily ascertained by the comparative optical rotatory dispersion studies of two similar compounds. Thus compounds (A) and (B) are different. But they show almost mirror image relationships, i.e., they show Cotton effects of opposite sign as seen in the figure below. This observation reveals that the two compounds have enantiomeric (mirror image relationship) stereochemistry of the groups in the immediate vicinity of the carbonyl chromophore which is *A/B* ring juncture. So, if the stereo configuration of one is known, other can be ascertained easily.

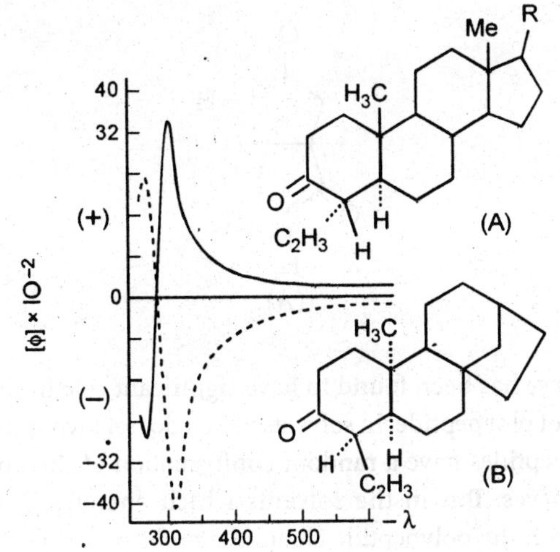

(ii) *Location of functional group*: In some cases traditional spectral measurements fail to differentiate isomeric oxo compounds like I- and 3-oxo-5α steroids. However, their CD curves clearly distinguish such type of compounds. Thus 3-oxo-compound shows a positive Cotton effect while 1-oxo-derivative exhibits a negative Cotton effect as is evident from the figure.

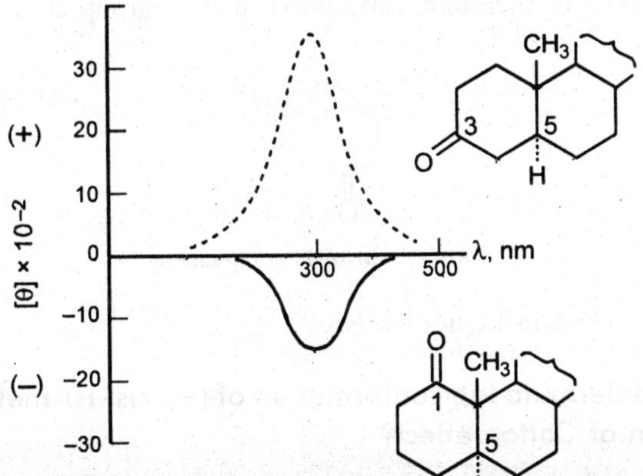

(iii) *Usefulness of plain* ORD *curves*: Plain ORD curves (when no Cotton effect is observed) can be very helpful in ascertaining important chiroptical properties. Thus,

(a) specific rotation of the chiral compounds may be determined with lesser amount of the substance even when normal method fails to record the rotation. Thus rotations of 4–cholestene and 5–cholestene which differ by less than 10% at ordinary measurement differ doubly at lower wavelength (at 300 nm).

(b) Another significant application of ORD where plain curve is obtained is in deciding whether a compound is optically active or not. Thus the compound (A) below shows no optical rotation in sodium D line (ordinary method) but records an optical rotation of –165° in ORD.

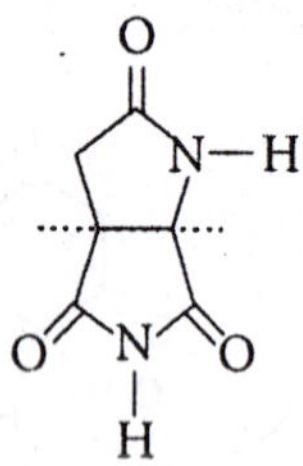

(A)

(c) Plain ORD curve has been found to have significant role in the determination of the conformation of ploypeptide. In solvents which do not favour intramolecular hydrogen bonding polypeptides have a random conformation of the polymer chain and show plain ORD curves. But in the solvents which encourage intramlecular H-bond formation within the polypeptide chain, the molecules coil up in helical form. Thus, ORD throws light on the helical formulation of polypeptide molecule.

(iv) *Determination of conformation*: If the absolute configuration of a compound is known, its preferred conformation may be settled by the ORD study. Let us consider (+)–3–methyl cyclohexanone which is known to have R–configuration. It is expected that it will exist in chair form with energetically favourable equatorial methyl group. By applying octant rule after projecting it properly, it has been found that it exhibits positive Cotton effect. It is, therefore, certain that methyl group at C–3 position should be equatorial.

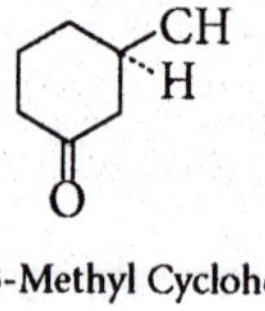

(R)-(+)-3-Methyl Cyclohexanone

See also answer against Question No. 8.

13. How will you determine the conformation of (+)-cis-10-methyl-2-decalone by the application of Cotton effect?

➔ (+) *cis*–10–methyl–2–decalone may exist in either of two conformations (A) and (B). Both of them will exhibit Cotton effect differently and so can be easily identified.

The application of Cotton effect suggests that the conformation (A) will display negative Cotton effect while (B) will show a positive Cotton effect. The compound actually displays a negative Cotton effect and hence its conformation is as shown in (A).

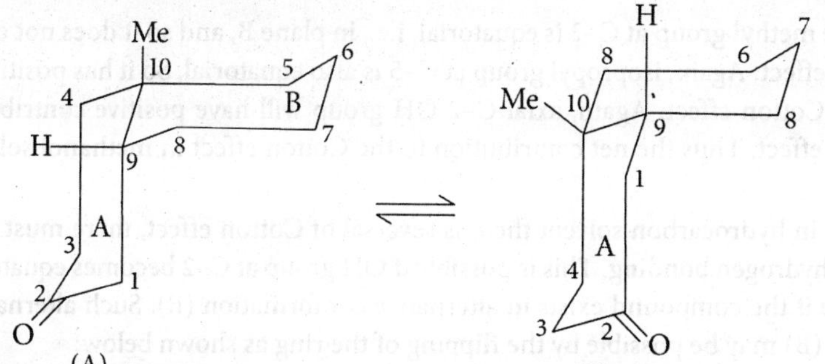

(A) (B)

14. How will you determine the absolute configuration of (–) trans–1–decalone?

➜ The absolute configuration of the above compound may be determined by the application of haloketone rule. When brominated, it gives (+) *trans*-2–bromo–1–decalone (B, R=Br) with bromine atom in axial position and it gives negative Cotton effect. Hence, the configuration of (–) *trans*–1–decalone is as shown in the structure. (A, R=H).

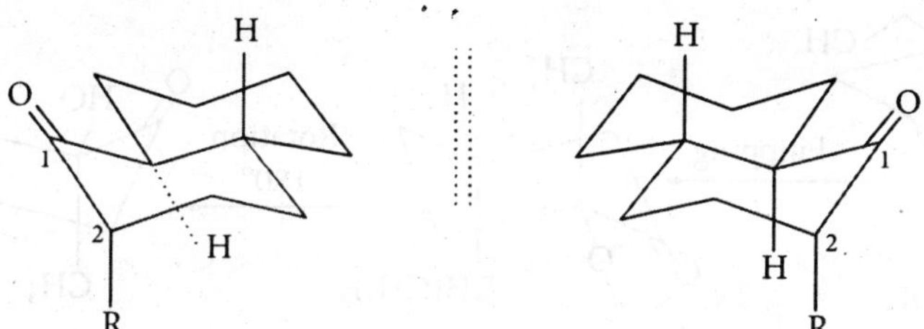

(CE +ve) (A, R=H) (CE -ve) (B, R=Br)

15. Define 'Heloketone rule'. Discuss with example how this rule can be applied for the determination of absolute configuration of the organic compound.

➜ See answer against Question No. 7.

16. Explain: The following compound exhibits positive CD Cotton effect in methanol but displays negative Cotton effect in hydrocarbon solvent.

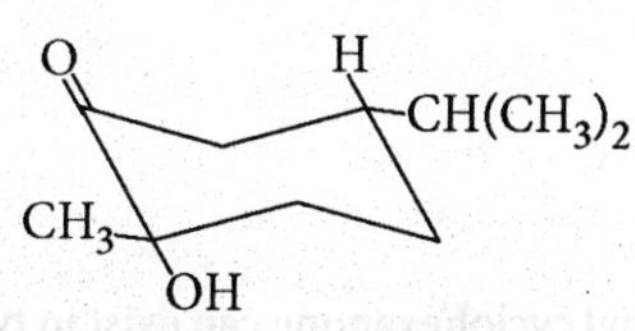

➜ The Cotton effect may be predicted by the proper application of octant rule. For such purpose, the structure of the compound may be written as (A).

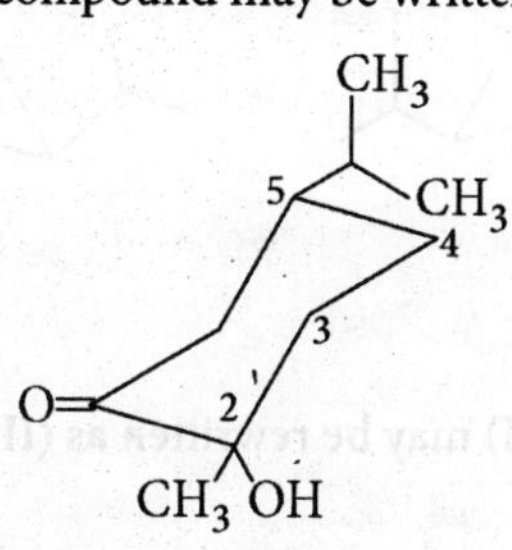

(A)

The methyl group at C–2 is equatorial, i.e., in plane B, and so it does not contribute to Cotton effect. Again, isopropyl group at C–5 is also equatorial. So it has positive contribution to Cotton effect. Again, axial C–2 OH group will have positive contribution to the Cotton effect. Thus the net contribution to the Cotton effect in methanol solvent is positive.

As in hydrocarbon solvent there is reversal of Cotton effect, there must be intramolecular hydrogen bonding. This is possible if OH group at C–2 becomes equatorial. This is possible if the compound exists in alternative conformation (B). Such alternative conformation (B) may be possible by the flipping of the ring as shown below:

Now from the conformation (B) it is evident that C–2 equatorial methyl now becomes C–6 axial which has negative contribution to Cotton effect but axial OH group at C–2 now becomes equatorial at C–6 which will have no contribution to the Cotton effect. Again, C–5 isopropyl group now (in B) becomes axial at C–3. This will have negative contribution. So the net Cotton effect is negative.

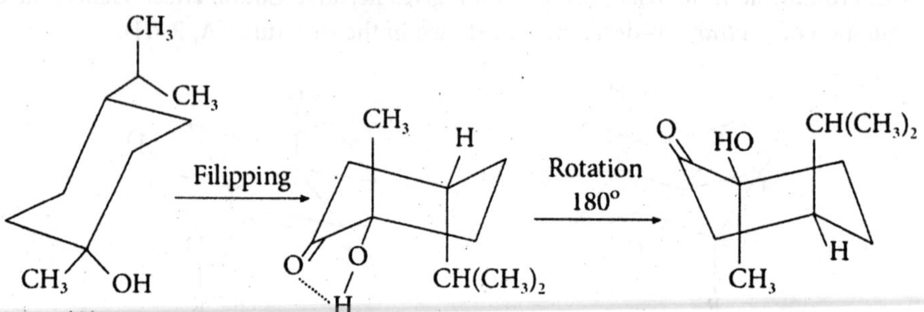

(A) (B)

17. Predict the Cotton effect of the following compounds.

(a) cis-2-Methyl-4-isopropyl cyclohexanone

(b) (+) cis-10-Methyl-2-decalone

(c) (+) trans-10-Methyl-2-decalone

(d) 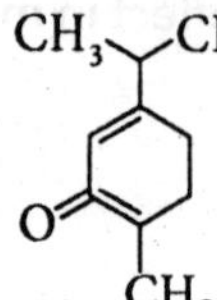

(a) *cis*-2-Methyl-4-isopropyl cyclohexanone can exist in two stereoisomeric forms (I) and (II).

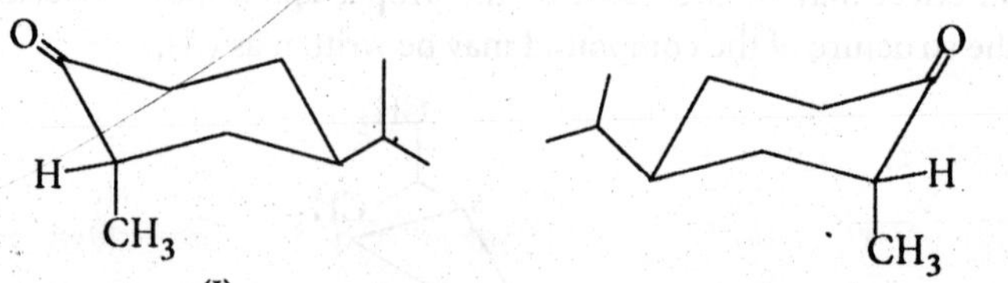

(I) (II)

For applying octant rule, (I) may be rewritten as (III) and (II) as (IV).

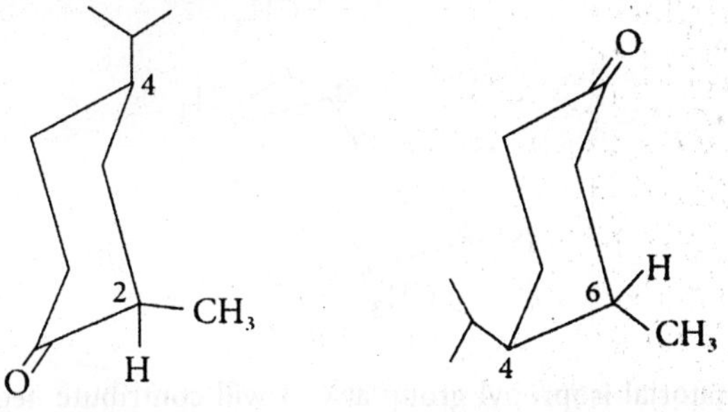

(III) (IV)

In both the structures there is an isopropyl group at C-4 position which will have no contribution to cotton effect. So only methyl substituent is responsible for the Cotton effect. In (III) this group (methyl) is at C-2 and axial and it will show ***positive Cotton effect*** while in (IV) methyl group at C=6 is axial. So it will display ***negative Cotton effect.***

(b) (+)-*cis*-10-Methyl-2-decalone may exist as either (I) or (II).

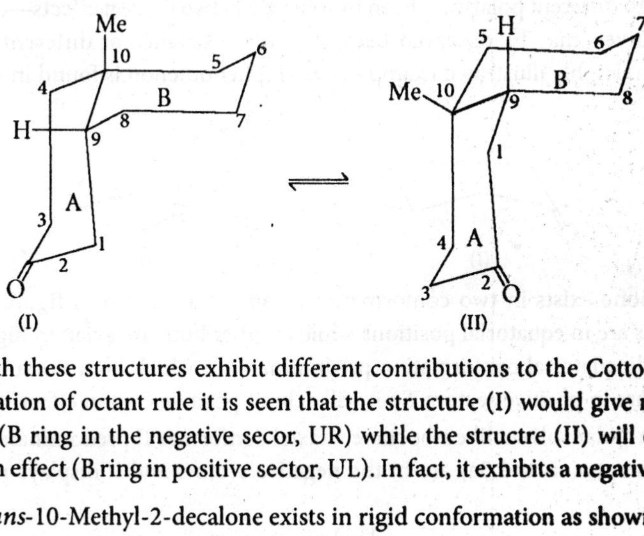

(I) (II)

Both these structures exhibit different contributions to the Cotton effect. By the application of octant rule it is seen that the structure (I) would give negative Cotton effect (B ring in the negative secor, UR) while the structre (II) will display positive Cotton effect (B ring in positive sector, UL). In fact, it exhibits a negative Cotton effect.

(c) (+) ***trans***-10-Methyl-2-decalone exists in rigid conformation as shown below:

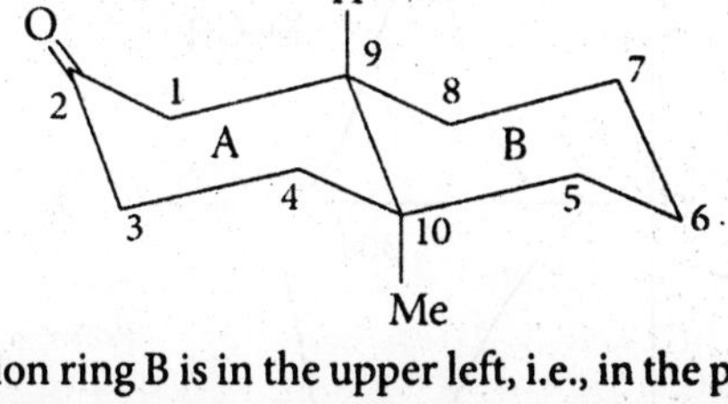

In this conformation ring B is in the upper left, i.e., in the positive sector. (substituents at C-5 and C-6 with respect to cyclohexane ring). So it will display positive Cotton effect. Its mirror image projection will show negative Cotton effect.

(d) The compound, 3-isopropyl-6-methyl cyclohexanone will preferably exist in chair with both substituents at equatorial position as shown in the figure.

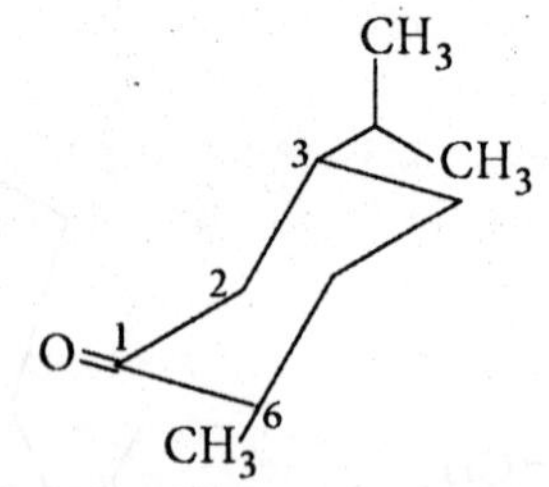

The equatorial isopropyl group at C-3 will contribute negative to Cotton effect while equatorial methyl at C-6 will have no contribution. Hence net Cotton effect shown by the compound is *negative*.

18. Explain citing an example how the sign and magnitude of the Cotton effect change with the change of solvent polarity and temperature.

➜ It has been observed that when a molecule exists in more than one conformations in solution, change in sign and magnitude may occur with the change of polarity of the solvent. This is because each conformer has its own ORD or CD curve and so the sign and magnitude of the Cotton effect will change with the change of population of the conformer in different solvents with different polarities. Even in some cases two Cotton effects—one positive and other negative—may be observed because of the existence of different conformers in equilibrium. Highly illustrated example of such phenomenon is found in (–) menthone.

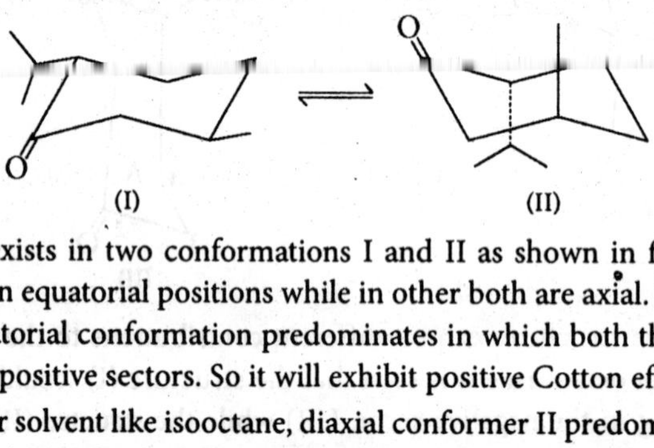

(I) (II)

Menthone exists in two conformations I and II as shown in figure. In I, both the substituents are in equatorial positions while in other both are axial. In high polar solvent like water diequatorial conformation predominates in which both the substituents (C-2e and C-5e) are in positive sectors. So it will exhibit positive Cotton effect.

In non-polar solvent like isooctane, diaxial conformer II predominates in which both the substitutents (6a and 3a) are in the negative sectors and it displays negative Cotton effect.

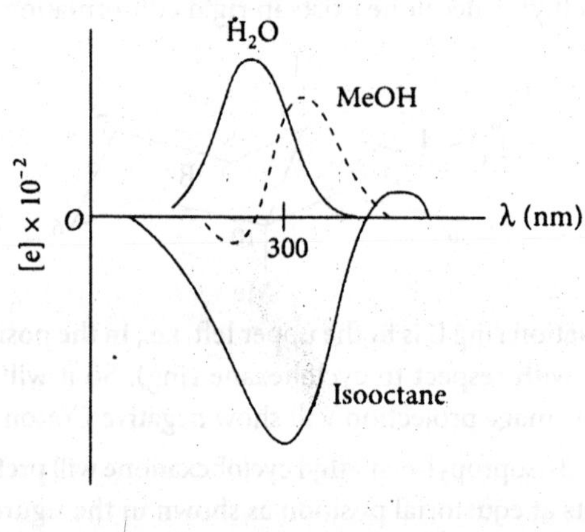

But in methanol, a solvent of moderate polarity, both the conformers exist in equilibrium in different proportions. Hence both the Cotton effects—positive and negative are observed. This type of curves with two CD maxima of opposite sign is known as *bisignate*. It may be noted that in bisignate curves, the individual maxima are widely separated.

It has also been observed that similar effects is displayed by some compounds with the variation of temperature. At low temperature the diequatorial conformation of menthone predominates but its population gradually decreases with the increase in temperature. Thus, at different temperature Cotton effects of different sign and magnitude may be observed as in the case of variation of solvent polarity.

19. Write a brief note on Benzoate rule and Benzoate sector rule.

➜ Freudenberg in 1933 first showed that the benzoyl derivative of secondary alcohols like (A) is more dextrorotatory than the parent alcohol.

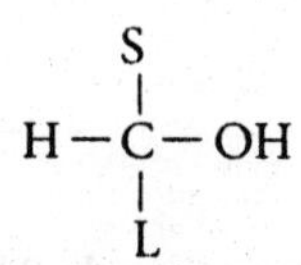

(A)
S = Small group, L = Large group

This is because in Newman projection the planar benzoyl group due to its high bulk is placed between Hand S as shown in (B).

Since S has more polarisability than H, a clockwise asymmetric screw pattern results in the increased dextrorotation. The reverse effects takes place in the enantiomeric configuration. This is an empirical rule and is known as *benzoate rule*.

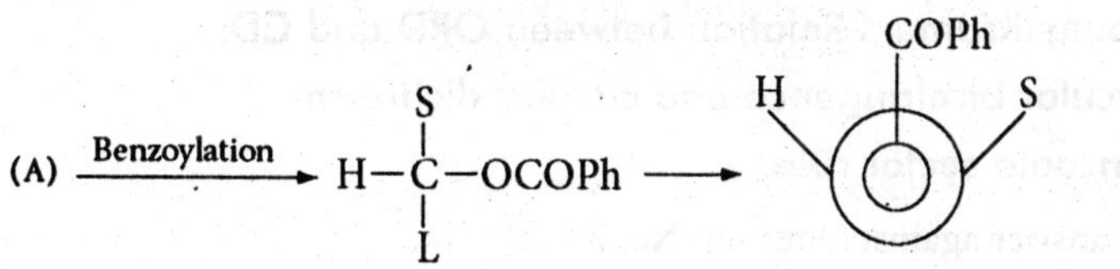

(B)

Based on this empirical rule Nakanishi and Harada in 1969 developed an empirical rule for the determination of the absolute configuration of 1,2-glycols through the CD spectra of their dibenzoyl derivative. It may be mentioned that the absorption band at 230 nm in unsubstituted dibenzoate is due to an intramolecular charge transfer transition in which the electric transition moment is approximately equivalent to the C–O of the ester group. In the case of dibenzoyl derivatives of 1, 2-glycols, the two chromophores form a chiral screw pattern either right-handed or left-handed like (C) or (D).

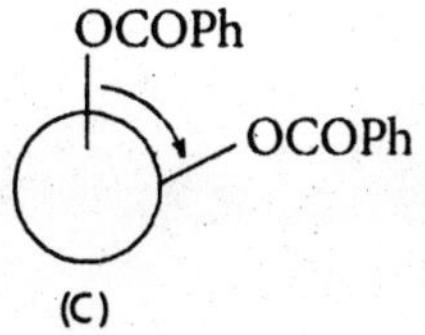

(C)

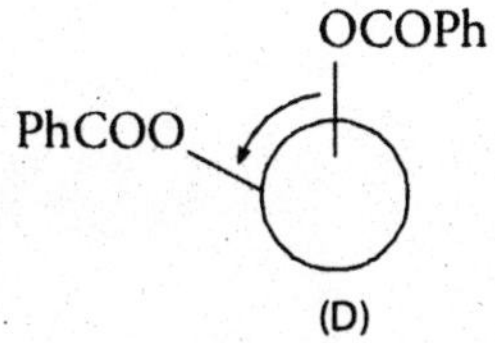

(D)

As a result of interaction of two chromophores, there is splitting of Cotton effect into two components—positive and negative. This has actually been observed in the case of

2α, 3β—dibenzoyloxy cholestane (E). In this case the first Cotton effect appears at 233 nm and the second one at 219 nm. The sign of cotton effect at 233 nm corresponds to right-handed screw pattern and shows positive Cotton effect while the second one at 219 nm corresponds to left-handed screw pattern and gives a negative Cotton effect. As a whole the combined CD Cotton effect curve resembles an anomalous ORD curve as shown in the figure below (F).

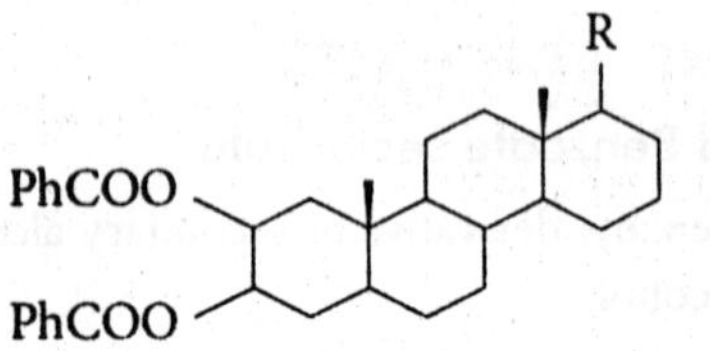

(E)

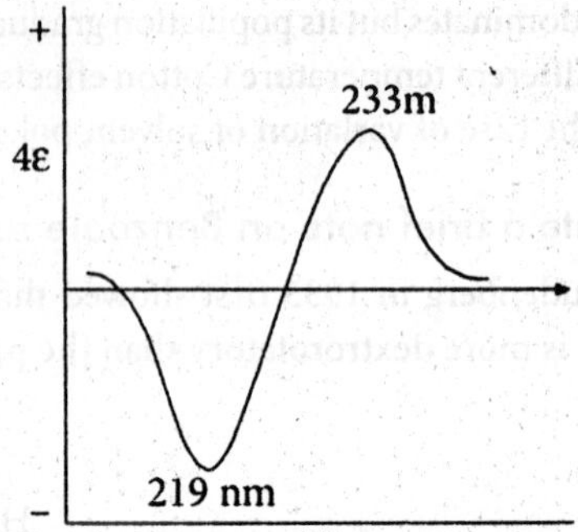

(F)

This rule is known as *benzoate sector rule.* This rule has been found to be applicable to the benzoates of 1,2-glycols and OH groups may be alcoholic or phenolic in nature. It is also applicable to benzoates of 1,2,3-triols. The principle holds good for any pair of chromophores provided the direction of electric moments is known. It may be mentioned that this rule has subsequently been designated as *the exciton chirality method.*

20. Write notes on.

(a) Plain ORD and anomalous ORD curves.

(b) Kronig–Kramers Relation between ORD and CD.

(c) Circular birefringence and circular dichroism.

(d) Benzoate sector rule.

(a) See answer against Question No. 7.

(b) See answer against Question No. 5.

(c) See answer against Question No. 1.

(d) See answer against Question No. 19.

CHAPTER 8

SOLUTIONS OF SOME COMPOSITE SPECTRAL PROBLEMS

1. An organic compound, $C_6H_{12}O$ gave the following spectral information:
(i) IR: 1715 cm^{-1}, (ii) ^{1}HNHR: δ 2.0 (3H, S), 1.0 (9H, S).
Suggest the structure of the compound.

➜ The compound contains one oxygen atom. The IR spectrum exhibits a band at 1715 cm^{-1} which is characteristic for a carbonyl group (>C=O). The ^{1}HNHR spectrum shows only two singlets indicating that it has only two types of non-equivalent protons. The singlet at δ 2.0 for three protons suggests the presence of a ketomethyl $(-\underset{\underset{O}{\|}}{C}-CH_3)$ group. The remaining nine equivalent protons at δ 1.0 (9H, S) is a clear indication of the presence of a tertiary butyl group $[-C(CH_3)_3]$. Hence, the most probable structure of the compound is as under.

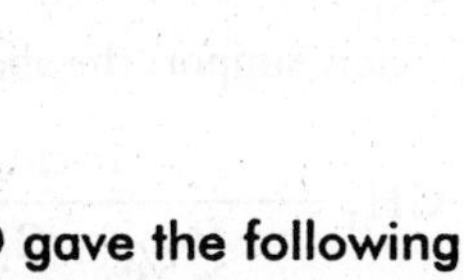

3, 3-Dimethyl-2-butanone

2. An organic compound, $C_5H_{10}O$ exhibits the following spectral characteristics
(i) UV (λ_{max}): 280 nm
(ii) IR (ν_{max}): 1715 cm^{-1}
(iii) ^{1}HNMR (δ): 0.90 (3H, t), 1.60 (2H, m), 2.20 (3H, s), 2.40 (2H, t)
(iv) Mass $\left(\frac{m}{z}\right)$: 86, 71, 58, 43 (100%).
Deduce the structure of the compound and explain the formation of different ions in mass spectrum.

➜ Molecular formula, $C_5H_{10}O$, suggests the presence of only one oxygen atom. The IR spectrum shows a band at 1715 cm^{-1} which is typical of a carbonyl group. The UV band at λ_{max} 280 nm is in agreement with this conclusion because the band at λ_{max} 280 nm may be assigned to n → π* transition of a >C=O group. A singlet signal of three protons at δ 2.20

may be assigned to a ketomethyl group ($-\underset{\underset{O}{\|}}{C}-CH_3$) while a triplet of two protons at δ 2.40 may be due to a methylene group attached to carbonyl carbon ($CH_2-\underset{\underset{O}{\|}}{C}-CH_3$). The signal at δ 1.60 for two protons appearing as a multiplet is most probably due to another methylene group attached to previous methylene group and the triplet **signal of three protons** at δ 0.90 is due to another methyl group attached to the **methylene group (δ 2.40, t).** Thus the most tentative structure may be written as

$$CH_3-CH_2-CH_2-\underset{\underset{O}{\|}}{C}-CH_3$$

Methyl Propyl Ketone

The mass spectral data completely support the above structure. Thus

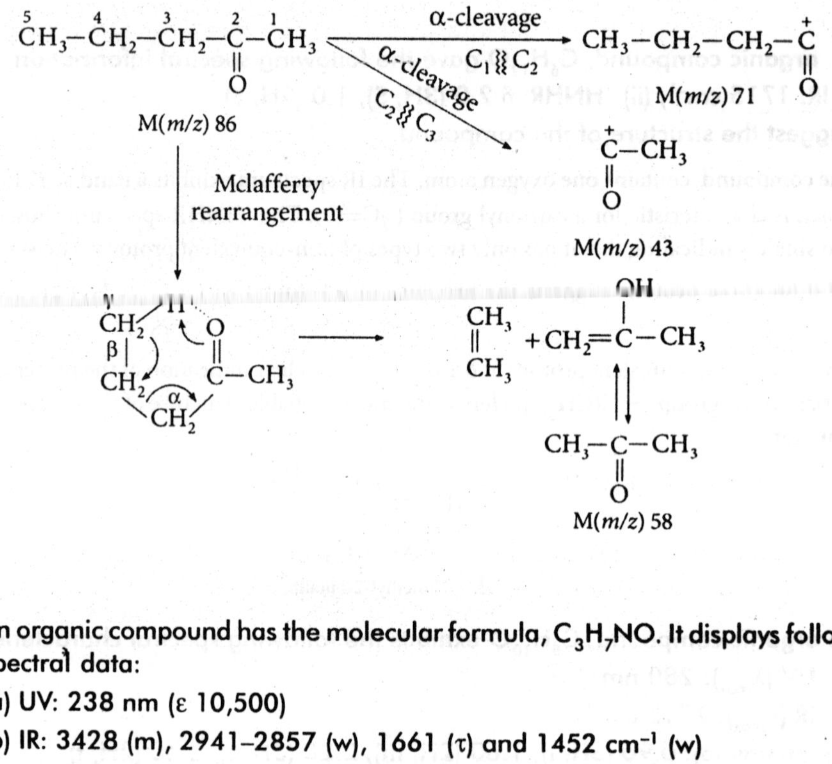

3. An organic compound has the molecular formula, C_3H_7NO. It displays following spectral data:

(a) UV: 238 nm (ε 10,500)

(b) IR: 3428 (m), 2941–2857 (w), 1661 (τ) and 1452 cm^{-1} (w)

(c) ^{1}HNMR: τ1.87 τ(1H, s), 7.30 τ(3H, s) and 8.1 τ(3H, s)

Suggest the structure of the compound.

➜ The appearance of a band at 238 nm with ε 10,500 is characteristic of conjugation. The IR band at 2941–2887 cm^{-1} (w) is due to C–H stretching. The band at 3428 cm^{-1} (m) suggests N–H stretching and the band at 1681 cm^{-1} (s) is assignable to amide carbonyl which is in conjugation with the lone pair of electrons on nitrogen atom.

The ^{1}HNMR spectrum exhibits three signals. The three-proton singlet at 8.1 τ suggests a CH_3 group adjacent to nitrogen atom. The other three proton-singlet at 7.30 τ shows the presence of another methyl group. A one-proton singlet appearing at 1.87 τ is due to N–H proton. On the basis of the above, following structure may be written for the compound.

$$CH_3-\underset{\underset{O}{\|}}{C}-NH-CH_3$$

N-Methylacetamide

4. An organic compound exhibits the following spectral properties:

MS: $\frac{m}{z}$ **164, 149, 91, 73.**

IR: ν_{max} **1730 cm^{-1}**

^{1}HNMR: δ **2.00 (3H, s) 2.93 (2H, t) 4.30 (2H, t) 7.30 (5H, s)**

Suggest the probable structure.

➜ The presence of an absorption band at 1730 cm^{-1} in the IR spectrum suggests the presence of an ester group in the molecule. The ^{1}HNMR signal at δ 2.00 for 3H which appears as a singlet is due to ketomethyl ($-\underset{\underset{O}{\|}}{C}-CH_3$). The appearance of a signal as a triplet at δ 4.30 for two protons is due to methylene ($-CH_2$) protons. Its appearance at a very low field is indicative of its presence as $-CH_2-O-\underset{\underset{O}{\|}}{C}-CH_3$ which is in agreement with the presence of an ester group in the compound. Again, its appearance as a *triplet* suggests its attachment to another CH_2 group. So the ester group is most probably of the type $-CH_2-CH_2-O-\underset{\underset{O}{\|}}{C}-CH_3$. Lastly the singlet signal at δ 7.30 for five protons is characteristic of aromatic protons. Hence, the compound contains a benzene ring. Thus, the tentative structure of the compound is 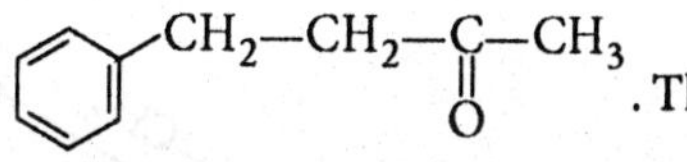. The attachment of CH_2 group to benzene ring is quite in conformity with its chemical shift value (δ 2.93). The mass spectral data are in complete accord with above structure as shown below.

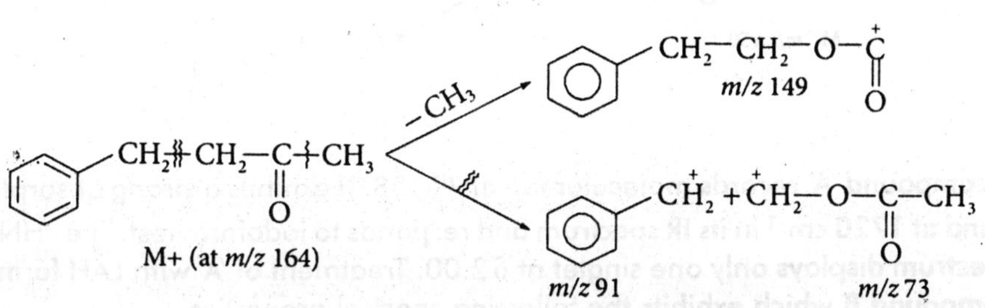

5. An organic compound having molecular formula, $C_9H_{10}O_2$, gave the following spectral data:

UV: λ_{max} 274 nm (ε = 2050)
IR: ν_{max} 3031, 2941, 1724, 1608, 1504, 1060 and 830 cm^{-1}.
^{1}HNMR δ: 2.35 (3H, s), 3.82 (3H, s) and 7.20–7.85 (4H, m)

Ms $\left(\frac{m}{z}\right)$: 150, 145, 119

What is the probable structure of the compound?

➡ The UV absorption maximum at 274 nm (ε = 2050) may be ascribed to benzene ring substitued by groups which causes bathochromic shift. The IR band at 3031 cm^{-1} is due to C—H stretching of aromatic ring and the band at 2941 cm^{-1} is due to C—H stretching from alkyl group. The band at 1724 cm^{-1} is probably due to ester carbonyl streching which is supported by the presence of a band for C—O streching at 1060 cm^{-1}. The bands at 1609 and 1504 cm^{-1} are characteristics of C—C stretching of aromatic ring. The band at 830 cm^{-1} indicates a disubstituted benzene ring.

The presence of aromatic ring is also supported by the appearance of a multiplet of four protons in the regions 7.20–7.85 δ. Two three-proton singlets at 2.35 and 3.82 suggest the presence of two methyl groups one of which is connected to aromatic ring while the other is due to carbomethoxy (—CO—O—CH_3) group. On the basis of above evidences the structure of the compound stands as under:

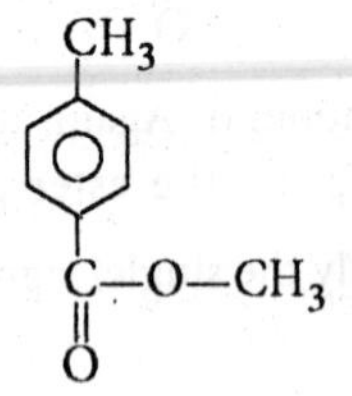

The mass spectral data are also in favour of the above structure.

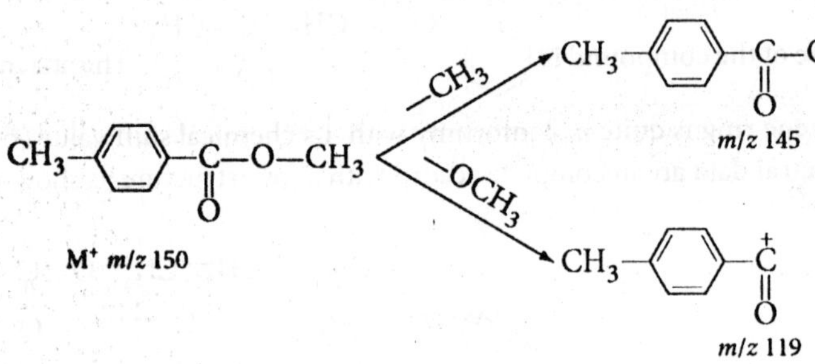

6. A compound 'A' records molecular ion at $^m/_z$ 58. It exhibits a strong absorption band at 1720 cm^{-1} in its IR spectrum and responds to iodoform test. The ^{1}HNMR spectrum displays only one singlet at δ 2.00. Treatment of 'A' with LAH forms a compound B which exhibits the following spectral properties.

MS: M^+ at $^m/_z$ 60, IR: 3400 cm^{-1}, ^{1}HNMR: δ 1.2 (6H, d), 3.9 (m, 1H) and 4.7 (1H, s) Signal at δ 3.9 disappears on shaking with D_2O.
Suggest the structures of A and B.

➜ The IR absorption band at 1720 cm^{-1} is due to a carbonyl group. The positive response of iodoform is an indication of the presence of ketomethyl group (—$COCH_3$). The presence of only singlet at δ 2.00 is due to ketomethyl group.

Thus the part structure of the compoud is CH_3—CO—R

The molecular weight of the compound is 58. The mass value of CH_3CO is 43. Hence, the mass value of R should be 15 which must be due to CH_3 group. Thus the compound is acetone, CH_3—C(=O)—CH_3. This is supported by its reduction with lithium aluminium hydride. Thus,

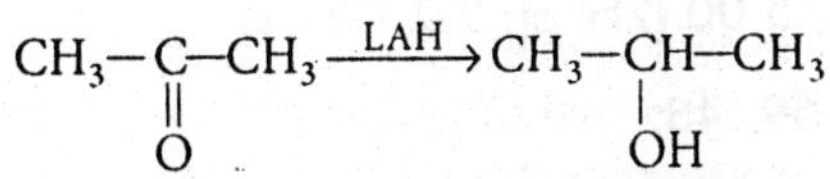

Acetone Isopropanol

m/z 60; two CH_3 groups at δ 1.20 (6H, d), carbinol proton >CH at δ 4.7 (1H, *m*) and
OH
3400 cm^{-1} is due to OH group

7. An organic compound contains 85.7% carbon, and 14.3% hydrogen. It is transparent above 210 nm in the UV spectrum. In IR spectrum it shows bands at 3022, 1676 and 965 cm^{-1}. Two signals appear in ^{1}HNMR spectrum at δ 1.6 (6H, d) and 4.55 δ (2H, q). Determine its structure.

➜ From the percentage composition the molecular formula of the compound stands as C_4H_8. As it is transparent in UV above 210 nm it has no conjugation.

The IR band at 3022 cm^{-1} is due to C—H streching and the band at 1676 cm^{-1} is due to carbon-carbon double (C=C) streching.

This suggests that the compound is unsaturated hydrocarbon. The ^{1}HNMR signal at 1.60 δ of six protons may be due to two methyl groups attached to unsaturated carbon, = C—CH_3 and as it appears as a doublet the chain like =CH—CH_3 may be present in hydrocarbon. This is also supported by the appearance of a quartet at 4.55 δ.

Again, the IR band at 965 cm^{-1} is the characteristic of an alkene in which hydrogen atoms are *trans*. Thus, the structure of the hydrocarbon stands as

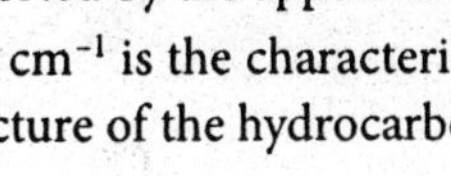

Trans-Butene-2

8. A colourless compound, $C_9H_{10}O_2$, records the following spectral data:
IR: 1720, 1602, 1581, 1270, 1105 cm^{-1}
^{1}HNMR: δ 1.29 (3H, t, J = 8 Hz), 4.35 (2H, q, J = 8 Hz), 7.40 (3H, m), 8.81 (2H, m)
Suggest the structure of the compound.

➜ The IR band at 1720 cm^{-1} indicates a carbonyl function. The bands at 1270 and 1105 cm^{-1} are compatible with an aromatic ester while the bands at 1602 and 1581 cm^{-1} suggest the presence of an aromatic ring.

The triplet at 1.29 (3H, $J = 8$ Hz) and quartet at 4.55 (2H, $J = 8$ Hz) suggest the presence of $-O-CH_2-CH_3$ moiety.

Again, two signals at δ 7.40 (3H, *m*) and 8.81 (2H, *m*) are due to five aromatic protons. On the basis of the above, the structure of the compound may be written as below:

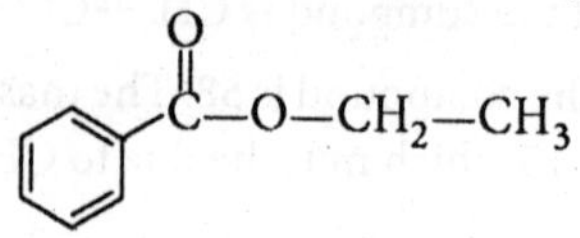

9. An organic compound, $C_9H_{10}O_2$, exhibits the following spectral characteristics.
IR: 1740, 1220, 750, 700 cm^{-1}

^{1}HNMR: δ 1.96 (3H, s); 5.00 (2H, s); 7.2 (5H, s)

MS: $\frac{m}{z}$ 150, 135, 91, 59, 43.

Determine the structure of the compound.

➜ The IR bands at 1740 and 1220 cm^{-1} are suggestive of an acetate grouping ($O-CO-CH_3$). The ^{1}HNMR signal at δ 1.96 (3H, *s*) for three protons is suggestive of a ketomethyl group ($-COCH_3$). The signal at δ 5.00 (2H, *s*) is due to $-CH_2$ group and its chemical shift is indicative of its attachment with benzene ring. ^{1}HNMR spectrum shows a five proton singlet at δ 7.2 which suggests the presence of a monosubstituted benzene ring. On the basis of the above evidences the structure of the compound may be written as

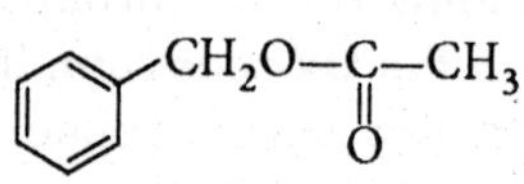

The mass spectral data are in agreement with the above structure as explained below:

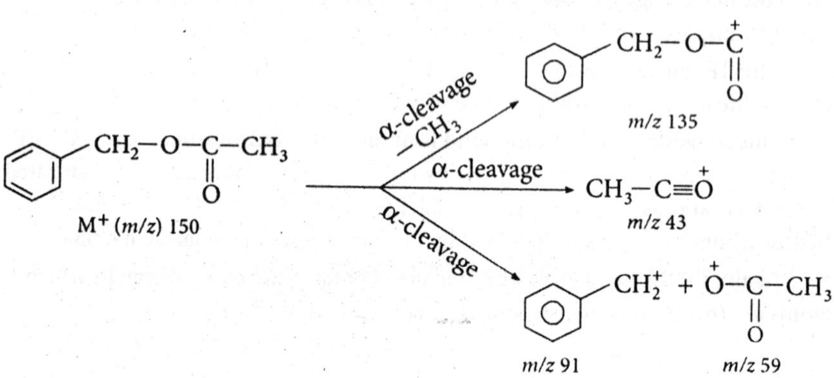

10. A compound A exhibits only one singlet in ^{1}HNMR spectrum at δ 2.17 and in the mass spectrum records mass peaks at $\frac{m}{z}$ 58, 43, 39, 27 and 15; when A is treated with PhCHO in presence of ethanolic NaOH forms B. The IR spectrum of B displays bands at 3025, 1665, 1630, 1600, 763 and 753 cm^{-1} and its ^{1}HNMR spectrum shows signals at δ7.82 (d 1H, J = 18); 7.60 (5H, m) and 7.05 (1H, d, J = 18). What are the structures of A and B?

➜ The molecular ion of A has mass 58. This suggests that A can accommodate maximum 4C atoms. So the possible molecular formulae are C_4H_{10}, C_3H_6O, $C_2H_2O_2$, $C_2H_6N_2$.

The base at $\frac{m}{z}$ 43 (58 – 15) is a clear indication that A contains at least one CH_3 group. ^{1}HNMR shows only one signal. So the formulae, C_4H_{10} and $C_2H_2O_2$, are rejected. So the possibilities for A are C_3H_6O and $C_2H_6N_2$. These two molecular formulae can be rewritten as

$$CH_3-\underset{\underset{O}{\|}}{C}-CH_3 \text{ and } CH_3-N=N-CH_3$$

Of these two, only $CH_3-\underset{\underset{O}{\|}}{C}-CH_3$ can react with PhCHO in the presence of base. Hence,

A is $CH_3-\underset{\underset{O}{\|}}{C}-CH_3$

Acetone

The structure of B should be

$$CH_3-\underset{\underset{O}{\|}}{C}-CH_3 \xrightarrow[\text{Base}]{\text{PhCHO}} Ph-CH=CHCOCH_3$$

$$\downarrow \text{PhCHO}$$

$$Ph-CH=CH-\underset{\underset{O}{\|}}{C}-CH=CHPh$$

^{1}HNMR spectrum of B does not show any signal for CH_3 group. Hence B is

$$Ph-CH=CH-\underset{\underset{O}{\|}}{C}-CH=CH-Ph$$

B

The IR bands and other ^{1}HNMR signals are quite in accord with B.

11. Deduce the structure of the compound from the spectral data given below:

UV λ_{max}: 286 nm

IR: 2970, 2940, 2870, 2820, 2700, 1735 (cm^{-1})

^{1}HNMR: δ 1.08 (d, J = 10 Hz), 2.3 (septet), 9.6 (d)

MS $\left(\frac{m}{z}\right)$: 72, 43 and 29.

➔ λ_{max} at 286 nm suggests the transition of $n \rightarrow \pi^*$. This indicates the presence of a carbonyl function. The IR bands at 2820, 2700 and 1735 cm^{-1} suggest the presence of an aldehydic function, —CHO. Further, ^{1}HNMR signal at 1.08 δ appearing as a doublet shows the presence of CH_3—CH< grouping. The septet at 2.3δ is possible if $(CH_3)_2$—CH— grouping is present while signal at δ 9.60 appearing as a doublet reveals the presence of $CH-\underset{\underset{O}{\|}}{C}-H$ grouping. On the basis of the above, probable structure is shown below:

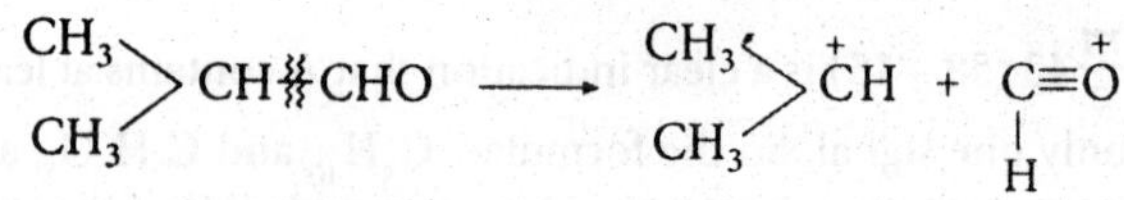

Iso-Butyraldehyde *m/z* 43 *m/z* 29
M^+ (*m/z*) 72

12. An organic compound with molecular weight 130 gave negative iodoform test. It absorbs at 292 nm (ε = 16) in UV spectrum. Its IR spectrum records bands at 3042, 2941, 2862, 2740, 1722, 1605, 1575 and 1462 cm^{-1}. In the ^{1}HNMR spectrum three signals at δ 7.27 (m, 5H); 2.8 (d, 2H) and 0.78 (t, 1H). Determine the structure of the compound.

➜ The UV band at 292 nm (ε = 16) is characteristic of a carbonyl group (n → π*). The presence of a carbonyl group is further supported by strong band at 1722 cm^{-1}. It can only be due to an aldehyde group as there are characteristic C—H streching absorptions at 2862 and 2740 cm^{-1} for —CHO group.

The absorptions at 3042 cm^{-1} is due to aromatic C—H stretching and that at 2941 cm^{-1} is due to C—H stretching from methyl or methylene group. The absorptions at 1605 and 1575 cm^{-1} indicate an aromatic ring which is confirmed by the presence of five protons multiplet at δ 7.27 in ^{1}HNMR spectrum. A triplet of one proton at δ 0.78 is a clear indication of an —CHO group. The other two-protons doublet at δ 2.80 is due to —CH_2 group attached to aromatic ring. Thus, the structure of the compound is

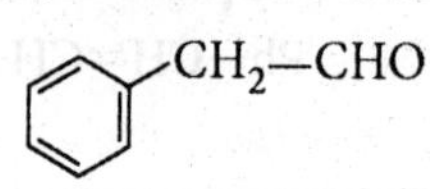

Phenylacetaldehyde

13. The mass of a hydrocarbon records molecular ion peak at $\frac{m}{z}$ 120. The IR spectrum records bands in the aromatic regions whereas ^{1}HNMR spectrum displays signals at δ 1.2 (6H, d), 2.8 (1H, m) and 7.2 (5H, m). Assign structure to this compound.

➜ The compound is an aromatic hydrocarbon as suggested by IR spectrum. ^{1}HNMR signal of six protons appearing as a doublet at δ 1.20 indicates a —$CH(CH_3)_2$ grouping while ^{1}HNMR multiplet signal of one proton at δ 2.80 is due to methine proton of —CH—$(CH_3)_2$. The five-proton multiplet at δ 7.20 is due to aromatic protons. Thus the structure of the compound stands as under which is supported by M^+ at $\frac{m}{z}$ 120.

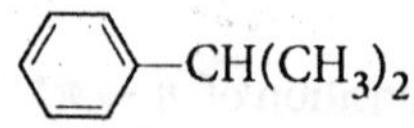

M^+ (*m/z*) 120

14. An organic compound containing C, H, O only showed molecular ion peak at $\frac{m}{z}$ 136 and other fragment ion peaks at $\frac{m}{z}$ 91 and 45. The other spectral data are:

UV: λ_{max} 229, 257 nm

IR: 1710 cm^{-1}

^{1}HNMR: δ 7.2 (5H, s), 3.5 (2H, s)

Predict the structure of the compound.

➻ The appearance of a singlet of five protons at δ 7.2 is indicative of the presence of a mono-substituted benzene ring while the singlet at δ 3.5 for two protons suggests the existence of benzylic CH_2 group. The UV absorption bands at 229 nm ($n \rightarrow \pi^*$) and 257 nm ($\pi \rightarrow \pi^*$) suggest the presence of a carboxylic group and phenyl moiety. The IR absorption band at 1710 cm^{-1} suggests the presence of a —COOH function. On the basis of the above, the probable structure is as under.

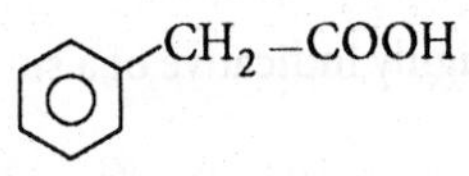

The mass spectral data supports the above structure.

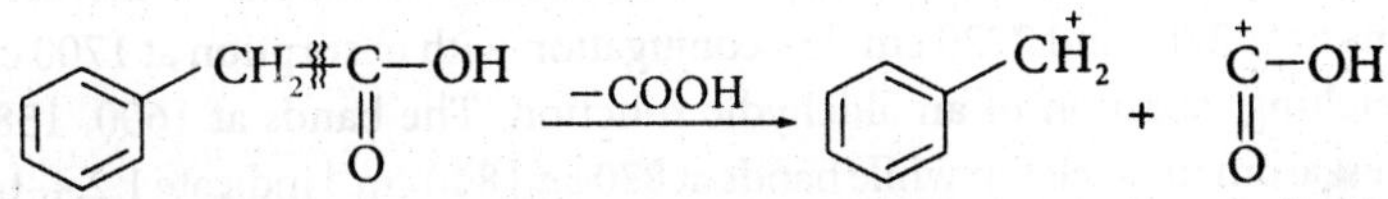

m/z 136 *m/z* 91. *m/z* 45

15. An organic compound contains carbon and hydrogen only. Its spectral properties are as follows:

UV: λ_{max} 270 nm (ε = 300)

IR: 3020, 2920, 2840, 1610, 1585, 1470, 1380 and 835 cm^{-1}

^{1}HNMR: δ 2.1 (9H, s), 6.3 (3H, s)

MS: M^+ at m/z 120, 105, 91

Determine the structure of the compound.

➻ The band at 3020 cm^{-1} is due to aromatic C—H stretching. The bands at 2920 and 2840 cm^{-1} may be due to aliphatic C—H stretching while the bands at 1470 and 1380 cm^{-1} are due to bending vibrations of methyl group. The band at 835 cm^{-1} is indicative of bending vibration due to 1,3,5-trisubstituted benzene ring. The UV bond at 270 nm (ε = 300) also supports this view. The singlet at δ 2.1 for nine protons may be assigned to three CH_3 groups while the singlet at δ 6.3 for three protons indicates symmetrical trisubstituted benzene ring. On the basis of these evidences, the probable structure of the compound appears to be 1,3,5-trimethyl benzene, which is supported by mass spectral fragmentation pattern of the compound.

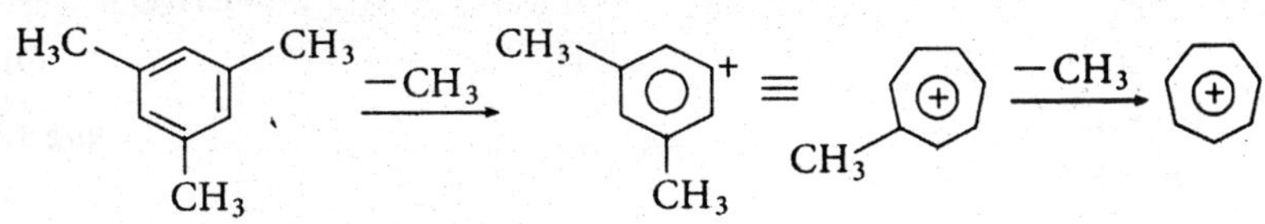

1, 3, 5-Trimethyl Benzene *m/z* 105 *m/z* 91

m/z 120

16. An organic compound, $C_8H_8O_3$, exhibits the following spectral properties.

UV: 282, 295 and 335 nm

IR: 3250, 3060, 2950, 2840, 2760, 2720, 1700, 1600, 1580, 1500, 870 and 840 cm^{-1}

^{1}HNMR: δ 5.75 (3H, s), 7.1 (d, J = 8 Hz), 7.4 (dd, J = 8 and 2H$_2$), 7.35 (1H), 8.05 (1H) and δ 9.80 (1H).

MS: m/z 152, 151, 137, 123, 109, 81

Determine the structure of the compound.

➜ The UV absorption bands are highly indicative of a substituted phenol moiety in the compound.

The IR band at 3250 cm^{-1} indicates an intramolecularly bonded hydroxyl group. The absorption band at 3060 cm^{-1} is due to aromatic C—H stretching vibrations. The bands at 2950 and 2840 cm^{-1} may be due to C—H stretching vibrations of methyl ether while absorptions near 2760 and 2720 cm^{-1} in conjugation with absorption at 1700 cm^{-1} indicate C=O stretching vibration of an aldehydic function. The bands at 1600, 1580 and 1500 cm^{-1} suggest aromatic skeleton while bands at 870 and 840 cm^{-1} indicate 1,2,4–trisubstituted benzene ring. A singlet of three protons at δ 5.75 is due to bonded aromatic $-OCH_3$ group. The signals at δ 7.1, 7.4 and at 7.35 indicate 1,2,4–tribustituted benzene ring. The singlet at δ 8.05 for one proton is due to OH and the singlet at δ 9.80 for one proton indicates aldehydic proton. Thus the structure stands as under:

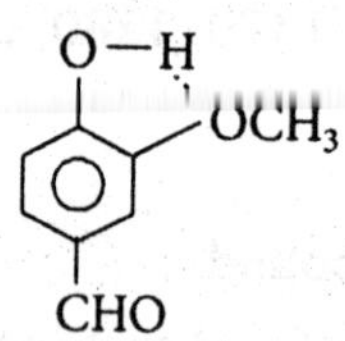

(I)

or

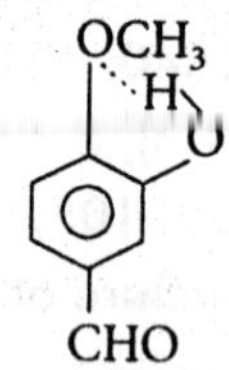

(II)

The UV data favour the structure (II) which has been confirmed by mass spectral fragmentation pattern.

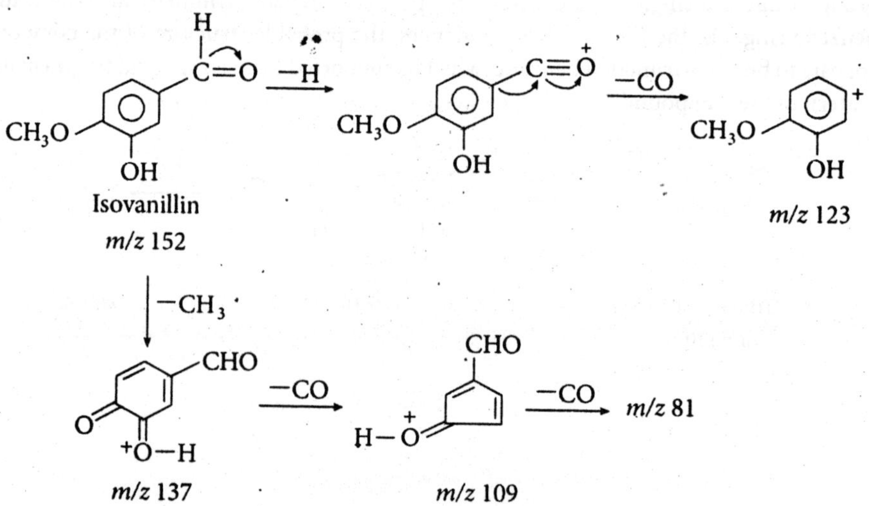

17. An organic compound, C_8H_8O, recorded the following spectral data:
IR: 3040, 2980, 2860, 1690, 1600, 1580, 760 and 690 cm^{-1}.
^{1}HNMR: δ 247 (3H, s), 750 (3H, m) and 8.10 (2H, m)

MS: $\frac{m}{z}$ 120, 105, 77, 51, 43

Predict the structure of the compound.

➜ The different IR bands may be assigned as follows:

(i) 3040 cm^{-1} (aromatic C—H stretching vibration)

(ii) 2950 and 2840 cm^{-1} (C—H stretching vibrations of aliphatic nature)

(iii) 1600, 1580 cm^{-1} (aromatic unsaturation)

(iv) 1690 cm^{-1} indicates >C=O stretching vibration.

(v) 760 and 690 cm^{-1} indicate monosubstituted benzene ring.

The ^{1}HNMR singlet at δ 2.47 for three protons is due to methyl protons attached to —COPh unit. The multiplet at δ 7.50 of three protons is due to two *meta* and one *para* protons while the singlet of two protons at 8.10 δ is due to two *ortho* protons. Thus the compound is acetophenone.

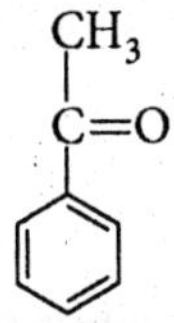

Acetophenone

The mass spectral analysis confirms the above structure.

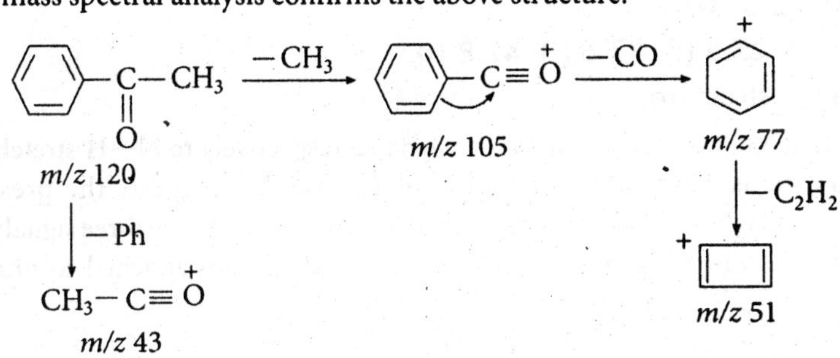

18. Predict the spectral behaviour expected of methyl salicylate.

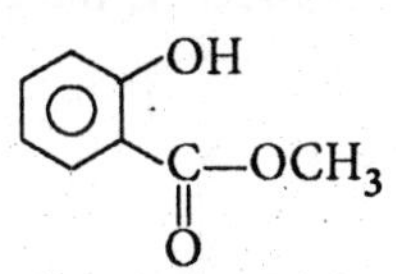

Methyl Salicylate

➜ In this compound, OH group and ester carbonyl group are bonded intramolecularly as shown below:

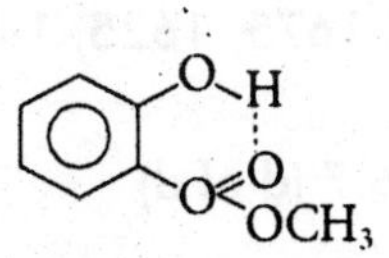

So, in the IR spectrum, compound will show a strong intramolecular hydrogen bonded hydroxyl at about 3200 cm^{-1} and carbonyl absorption band at lower field around 1680 cm^{-1}. Again, due to hydrogen bonding the phenolic proton will exhibit a signal at low field around 12.00 δ in ^{1}HNMR spectrum. Again, the carbomethoxy group ($-OCH_3$) is also deshielded and will appear at low field than expected. In ^{13}CNMR out of six, two carbon atoms are highly deshielded and expected to appear at δ 160 and 170 both as singlets (one due to carbonyl carbon and another due to ring carbon to which carbonyl group is attached). Another ortho proton (C—OH) is expected to appear at δ 120.0. The mass spectral fragmentation is expected to be as discussed below:

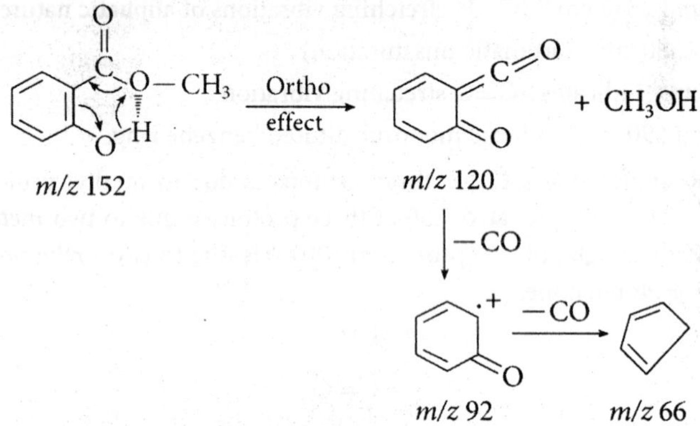

19. An organic compound, $C_5H_{11}N$ exhibits the following properties:

(i) IR: 3300, 1650 cm^{-1}

(ii) ^{1}HNMR: δ 2.8 (4H, s)

(iii) ^{13}CNMR: δ 25.5 (t), 27.6 (t), 47.8 (t)

Determine the structure.

➜ The IR bands at 3300 and 1650 cm^{-1} are assignable respectively to N—H stretching and bending vibrations. The triplet of four protons at δ 2.8 suggests the presence of $-CH_2-NH-CH_2$ linkage which is also supported by ^{13}CNMR. The three signals are due to CH_2 groups of which signal at δ 47.8 is due to CH_2 groups attached to nitrogen as $-CH_2-\underset{|}{N}-CH_2-$.

Now replacing N by CH, the molecular formula stands as $C_5H_{11}CH$, i.e., C_6H_{12}. This formula corresponds to C_nH_{2n}. Hence, the compound is monocyclic. Thus, the expected structure is

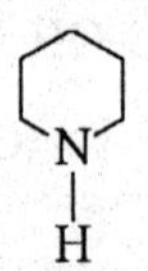

Pipiridine

20. An organic compound, C_9H_8O, has the following spectral properties:

IR: 3060, 3020, 2820, 2750, 1675, 1625, 1450, 750, 680 cm^{-1}

UV: 248 nm

^{1}HNMR: δ 9.7 (d), 7.45 (m), 6.7 (d of d)

^{13}CNMR: δ 128.2 (d), 128.5 (d), 129.0 (d), 131.1 (d), 134.1 (s), 152.5 (d), 193.3 (d) MS: $\frac{m}{z}$ 132, 131, 104, 78, 51

Deduce the structure.

➜ λ_{max} at 248 nm suggests the aromatic conjugations while IR absorptions at 3060, 3010, 1625 and 1450 cm^{-1} are due to C—H stretching vibration of aromatic system. The bands at 750 cm^{-1} and 680 cm^{-1} suggest monosubstituted phenyl ring and band at 1675 cm^{-1} along with 2820 and 2750 cm^{-1} is due to conjugated —CHO group.

^{1}HNMR signal at 7.45 δ indicates aromatic protons while signal at δ 9.7 is due to —CH—CHO and signal at δ 6.7 (d of d) may be assigned to CH—CH.

Assignment of ^{13}CNMR signals may be as follows:

128.2 (d)	128.5(d)	129.0(d)	131.0(d)	134.1(s),	152.5(d)	193.5 (d)
CH	CH	CH	CH	C	CH	CHO

On the basis of the above, the tentative structure is 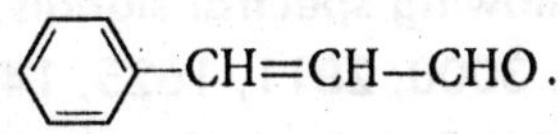

21. Deduce the structure of a compound, C_8H_9NO, having following spectral characteristics.

IR: 3400, 3300, 3100, 1640, 1480 cm^{-1}

^{1}HNMR: δ 7.88, 7.7, 6.7, 6.5, 4.5, 2.5 (S)

^{13}CNMR: δ 27.3 (q), 114.3 (d), 126.8 (s), 132.3 (d), 155.1 (s) amd 196.9 (s)

MS: $\frac{m}{z}$ 135, 120, 92, 65, 40.

➜ Deductions are as follows:

(i) M^+ at $\frac{m}{z}$ 135 i.e. odd number. This suggests the presence of one N atom.

(ii) 3400, 3300 cm^{-1} (—NH_2)

(iii) 3100, 1640, 1480 cm^{-1} (aromatic ring)

^{1}HNMR signals at δ 7.8, 7.7, 6.7 and 6.58 are due to arbmatic protons.

^{13}CNMR assignments are as follows:

27.3 (q)	114.3(d)	126.8(s)	132.3(d)	155.1(s)	196.9(s)
CH_3	CH	C	CH	C	CO

On the basis of the above the probable structure is 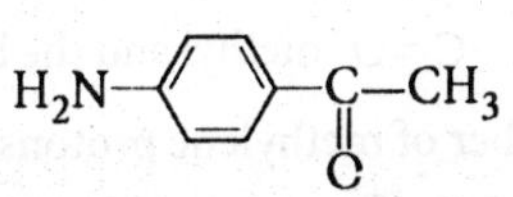

22. Deduce the structure of a compound, $C_6H_5NO_3$ from the following data.

(i) UV: λ_{max} 280 nm (ε 6000)

(ii) IR: 3460, 3035, 1608, 1585, 1510, 1360, 740 cm^{-1}

(iii) ^{1}HNMR: δ 8.00 (1H, s), 7.25–7.39 (unsymmetrical pattern)

➜ The band at 280 nm (ε = 6000) suggests the presence of a benzenoid compound.

The IR band at 3025 cm^{-1} is due to C—H stretching of aromatic system. The band at 3460 cm^{-1} is due to NH or OH stretching vibrations.

One proton singlet δ 8.00 is due to intramolecularly bonded proton of O—H and unsymmetrical pattern at δ 7.25–7.39 clearly indicates aromatic protons.

In the IR spectrum the bands at 1510 cm^{-1} and 1360 cm^{-1} are characteristics of nitro group (—NO_2). The nitro group must be at the ortho position to OH so that intramolecular hydrogen bond can occur. The band at 740 cm^{-1} also supports this view. Hence, the structure of the compound is

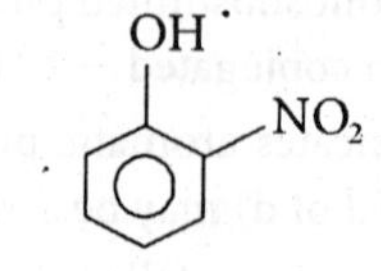

O-Nitrophenol

23. A steroidal alkaloid isolated from *Sarcococca pruniformis*, $C_{26}H_{46}N_2O$, exhibits following spectral signals:

IR: 3000, 2811, 1625, 1404, 1360 cm^{-1}

^{1}HNMR: δ 0.75 (3H, S), 0.76 (3H, S), 1.15 (3H, d, J=7Hz), 1.50 (number of protons, br), 2.10 (3H, s), 2.20 (6H, s), 2.73 (3H, s)

MS: M$^+$ $\left(\frac{m}{z}\right)$ 402, 110, 100, 84, 56

Deduce the structure of the alkaloid and explain the formation of mass spectral data.

➜ The IR bands at 3000 and 2811 are due to C—H stretching vibrations while band at 1625 cm^{-1} indicates the presence of an amide (—CO—N<) carbonyl function in the molecule. The band at 1404 cm^{-1} may be due to N—CH_3 and band at 1360 cm^{-1} may be attributed to C—CH_3 group.

The singlets at δ 0.75 and 0.76 of three protons each in ^{1}HNMR spectrum are suggestive of the presence of two tertiary C—CH_3 groups while the doublet of three protons at δ 1.15 is due to a secondary C—CH_3 (—CH—CH_3) group. The singlet of three protons at δ 2.10 may be assigned to —N—CO—CH_3 group. The six-proton singlet at δ 2.20 may be ascribed to $-N\langle^{CH_3}_{CH_3}$ functions. The singlet of three protons at 2.73 δ may be due to CH_3—N—C=O methyl and the broad signal of a number of protons at δ 1.50 may be due to a number of methylene protons.

The mass fragmentation pattern of the alkaloid is strongly suggestive of a C-3, C-20 amino pregnane like skeleton of the alkaloid. The appearance of the peak at $\frac{m}{z}$ 100 suggests the presence of a substituent of mass 58 on nitrogen atom at C-20. The presence of the mass at $\frac{m}{z}$ 58 indicates that this substituent is probably methyl-N-acetyl —CH_3—N—$COCH_3$ group. The presence of amide linkage has also been suggested from IR and ^{1}HNMR spectral analysis. The appearance of the peak at *m/z* 110 in the mass spectrum is a clear indication of the absence of any double bond between C_5 and C_6 atoms of the steroid nucleus. On the basis of the above spectral evidences the structure of the alkaloid is suggested as shown below and the genesis of the above fragments may rationalised as discussed.

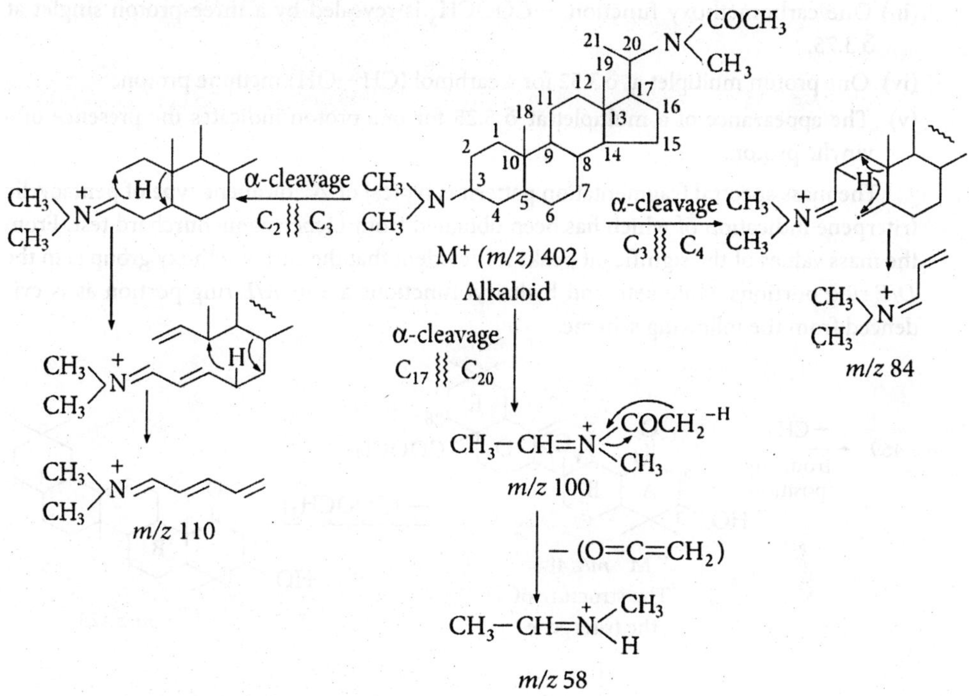

24. A naturally occurring triterpene from *Salvia Coccinea* responds positively to Liebermann–Burchard test for pentacyclic triterpene and exhibits following spectral properties:

IR: 1735, 1720, 1660 cm^{-1}

^{1}HNMR: δ 0.75 (3H, s), 0.80 (3H, s), 0.85 (3H, s), 0.93 (3H, s), 0.98 (6H, s), 1.25 (3H, s), 3.75 (3H, s), 3.75 (lH, br, half band width 15 Hz), 5.25 (1H, m)

MS: $\frac{m}{z}$ 484, 469, 425, 262, 221, 203, 133

^{13}CNHR: δ 218.2 (s), 178.2 (s), 144.1 (s), 123.6 (d), 75.06 (s), 55.4 (d), 51.7 (q), 50.1 (s), 48.4 (s), 46.6 (q), 42.7 (s), 42.6 (q), 37 (S), 36.7 (e), 34.8 (s), 34.1 (ε), 33.7 (ε), 32.6 (q), 31.5 (q), 28.7 (ε), 26.8 (q), 26.3 (q), 24.9 (s), 23.9 (t), 23.8 (q), 21.1 (q), 18.1 (q), 17.5 (q).

Deduce the structure and explain the mass fragmentation pattern of the triterpene.

↔ The IR absorption bands observed in the spectrum may be assigned as follows:

(i) 3440 cm^{-1} for OH group

(ii) 1735 cm^{-1} for ester carbonyl.

(iii) 1720 cm^{-1} for six-membered cyclic carbonyl

(iv) 1660 cm^{-1} for unsaturation.

The different signals displayed in ^{1}HNMR spectrum may be interpreted as:

(i) Seven tertiary methyls suggested by δ 0.75 (3H, s), 0.80 (3H, s), 0.85 (3H, s), 0.93 (3H, s), 0.98 (6H, s) and δ 1.25 (3H, s).

(ii) $-CH_2-\underset{|}{C}=O$ grouping indicated by the two protons multiplet at δ 2.30.

(iii) One carbomethoxy function $-COOCH_3$ is revealed by a three-proton singlet at δ 3.75.

(iv) One proton multiplet at δ 3.52 for a carbinol (CH—OH) methine proton.

(v) The appearance of a multiplet at δ 5.25 for one proton indicates the presence of a vinylic proton.

The mass spectral fragmentation pattern is typical of Δ^{12} oleanene type of pentacyclic triterpene indication of which has been obtained from Liebermann-Burchard test. From the mass values of the significant peaks it is evident that the carbomethoxy group is in the *D/E* ring portions while keto and hydroxyl functions are in *A/B* ring portion as is evidenced from the following scheme.

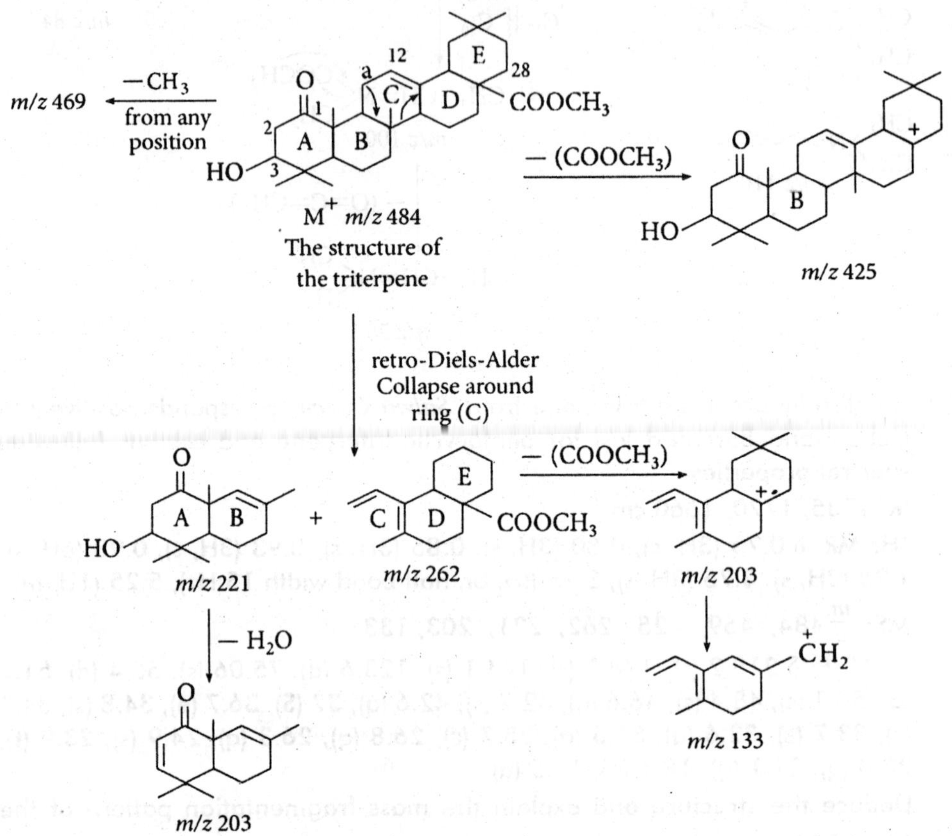

^{13}CNMR spectral data are in complete agreement with the above structure. The signals at δ 218.2 and 178.2 are due to ketocarbonyl carbon at 1 and ester carbonyl carbon of 28 position while the signal at δ 51.7 indicates carbomethoxyl ($-\underset{\underset{O}{\|}}{C}-OCH_3$) group. The signals at δ 144.1 and 123.6 may be assigned to C_{13} and C_{12} carbons. Rest of the signals may be due to carbon atoms expected for different positions.

25. A naturally occurring flavone from *limnophila heterophylla* having molecular formula, $C_{18}H_{16}O_7$ gives positive favonoid test with magnesium hydrochloride and exhibits the following:

(i) λ_{max} (EtOH) (log ε) : 225 (4.25), 280 (4.20) and 335 nm (4.10)

(ii) ν_{max} (KBr): 3400, 1655, 1600, 1575 cm^{-1}

(iii) ^{1}HNMR δ: 3.85 (3H, s); 3.95 (3H, s); 3.98 (3H, s); 7.0 (1H, d, J = 2, Hz); 7.19 (1H dd, J = 2 and 9 Hz); 7.88 (1H, d, J = 2 Hz); 6.95 (1H, s); 13.30 (1H, s).

(iv) $\frac{m}{z}$ 344, 329, 316, 314, 313, 301, 197, 196, 168, 153, 148, 133, 132

(v) ^{13}CNMR δ : 182 (s), 162.4 (s), 158.4 (s), 156.5 (s), 150.5 (s), 148.9 (s), 148.3 (s), 128.3 (s), 122.2 (s), 119.5 (d), 114.6 (d), 107.8 (d), 103.8 (S), 96.0 (d), 61.0 (s), 56.5 (s), 55.8 (s).

Deduce the structure of the flavone.

The UV absorption bands are in the regions, selective for flavone moiety, i.e., band I at 335 nm, band II in the region 280 nm and the band at 225 nm is the band III of flavone moiety.

The IR absorption band at 3400 cm^{-1} is characteristic for O—H stretching vibration of a bonded hydroxyl function. The bands at 1655, 1600 and 1575 cm^{-1} indicate the presence of α, β-unsaturated carbonyl function.

The ^{1}HNMR signals at δ 3.85 (3H, s), 3.95 (3H, s) and 3.98 (3H, s) are due to three methoxyl groups. The proton singlets at δ 7.0, 7.19 and 7.88 are suggestive of three aromatic protons while signal at δ 6.95 for one proton is the C_3—H of flavone nucleus and signal at δ 6.50 (1H, s) may be assigned to C_6–H of ring (B). The signal δ 13.30 (1H, s) is due to chelated phenolic hydroxylic proton.

The mass spectral data are strongly indicative of the presence of a methoxylated flavone skeleton.

The mass ions at $\frac{m}{z}$ 329 and 316 are respectively the conventional M–CH_3 and M–CO peaks characteristic of flavone skeleton while the intense peak at $\frac{m}{z}$ 329 is originated by the loss of 15 mass units from molecular ion peak at $\frac{m}{z}$ 344. The peak at $\frac{m}{z}$ 301 is due to loss of CO and CH_3 group from molecular ion. The fragments at $\frac{m}{z}$ 196 and 148 are generated due to retro-Diels-Alder collapse around hetrocyclic ring. The mass values 196 and 148 clearly suggest the presence of two methoxy and one hydroxyl in ring (A) and one hydroxy and one methoxy in ring B. The genesis of these fragments may be best interpreted in terms of the following structure of the compound.

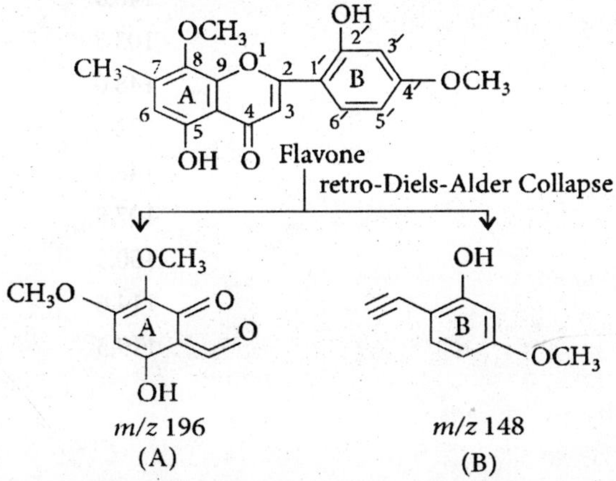

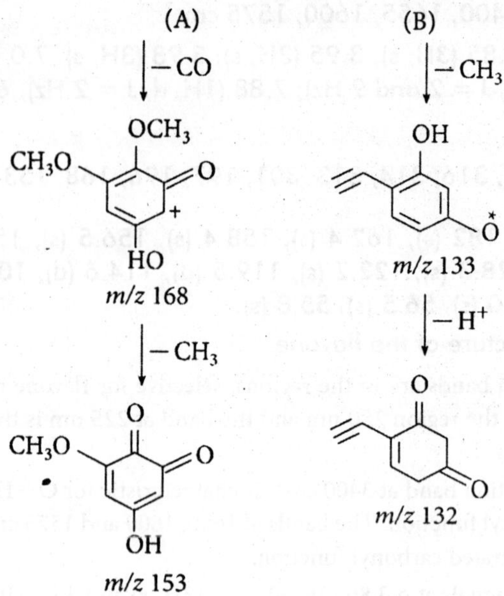

^{13}CNMR spectral data completely favours the above structure and the assignments follow as

C-atom	δ value
7-OCH_3	56.5
8-OCH_3	61.0
4'-OCH_3	55.8
2	162.4
3	108.9
4	182.2
5	156.5
6	96.0
7	158.4
8	128.3
9	148.9
10	103.8
	148.0
1′	122.2
2′	148.3
3′	107.8
4′	150.5
5′	114.6
6′	119.5

26. An organic compound, C_4H_8O, displays the following spectral characteristics:

λ_{max}: 295 nm (ε = 19)

ν_{max}: 2900, 2700 and 1715 cm^{-1}

^{1}HNMR: δ 1.04 (3H, t); 2.4 (2H, q); 2.07 (3H, s); 9.8 δ

^{13}CNMR: 29.37, 36.80, 208.7 δ

MS: $\frac{m}{z}$ at 72 (M+), 57, 43, 29

Suggest the probable structure.

➜ The λ_{max} at 295 nm with low epsilon value is typical of n → π* absorption of a saturated ketone or aldehyde. The presence of a carbonyl group is supported by the appearance of a strong IR absorption band at 1715 cm^{-1}. This carbonyl group is most probably due to keto group as is evidenced from the presence of absorption bands at 2700 and 2900 cm^{-1} in the IR spectrum. The appearance of a signal at δ 9.8 also supports this view. The three proton singlet at 2.09 δ may be assigned to a keto methyl (CH_3—CO) grouping. The three -proton triplet at δ 1.04 and two-proton quartet at 2.4 δ are suggestive of the presence of an ethyl grouping (—CH_2—CH_3). The appearance of a signal at 208.7 δ in the ^{13}CNMR spectrum suggests the presence ot carbonyl group while signals at δ 29.37 and 36.80 may be ascribed due to the presence of methyl (—CH_3) and methylene (—CH_2) carbon respectively. Thus, the structure of the compound stands as $CH_3-\underset{\underset{O}{\|}}{C}-CH_2-CH_3$.

Ethyl methyl Ketone

The above structure is in full agreement with the mass spectral data as shown under.

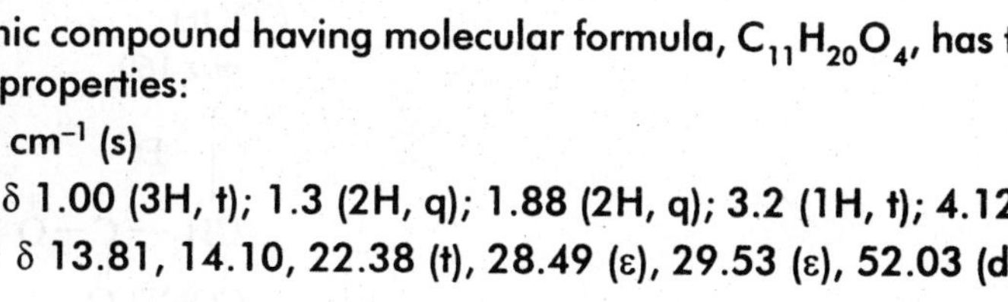

27. An organic compound having molecular formula, $C_{11}H_{20}O_4$, has the following spectral properties:

IR: 1740 cm^{-1} (s)

^{1}HNMR: δ 1.00 (3H, t); 1.3 (2H, q); 1.88 (2H, q); 3.2 (1H, t); 4.12 (2H, q)

C_{13}NMR: δ 13.81, 14.10, 22.38 (t), 28.49 (ε), 29.53 (ε), 52.03 (d), 61.12 and 169.32

MS: $\frac{m}{z}$ at 216, 160, 131, 115.

➜ The strong band at 1740 cm^{-1} in the IR spectrum indicates the presence of a carbonyl group which may be a five-membered ring ketone or a saturated ester.

The signal at δ 4.12 (2H, q) in the ^{1}HNMR spectrum suggests the presence of a —OCH_2—CH_3 grouping in the molecule. The signal at δ 1.3 may also be due to O—CH_2—CH_3. A careful scrutiny shows that the intensity of —OCH_2 signal corresponds to four hydrogen atoms which suggests that there are two identical —O—CH_2—CH_3 groupings in the molecule. Further, the low field appearance of —OCH_2 group (at δ 4.12) indicates that $-OCH_2-CH_3$ groups are actually esters of two-$\overset{\displaystyle O}{\overset{\|}{C}}-O-CH_2-CH_3$ groups.

The signal at δ 3.2 for a single hydrogen atom may be due to $-CH_2-CH(COOEt)_2$ grouping and the two protons signal at δ 1.88 is likely due to methylene hydrogen atoms of $-CH_2-CH\ (COOEt)_2$ and as it appears as a quartet, it is bonded to a CH_2 group to make the total number of vicinal hydrogen atoms three. Thus the molecule contains a moiety $-CH_2-CH_2-CH(COOEt)_2$. The three-proton triplet at δ 1.00 can be due to methyl group of CH_2-CH_3 group. On the basis of above evidence, the structure of the compound has been suggested as under:

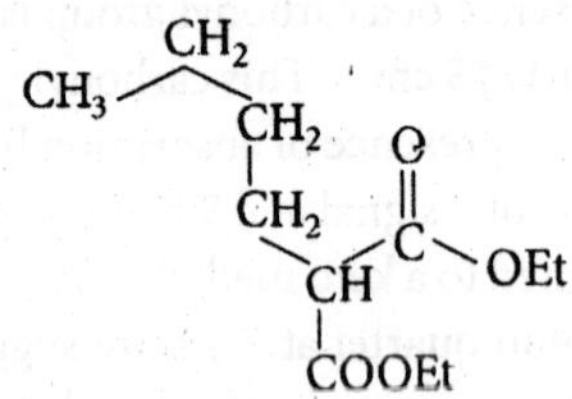

^{13}CNMR spectral data also supports the above structure.

The two methyl carbon atoms appear at δ 13.81 and 14.10. The three triplet signals at 22.38, 28.49 and 29.53 δ are due to $-CH_2-CH_2-CH_2$ chain. The signal at δ 61.12 is due to $-OCH_2$ carbon, 52.03 is due to methine carbon and the carbonyl carbon appears at δ 169.32. The mass spectrum is also in favour of the above structure. Thus,

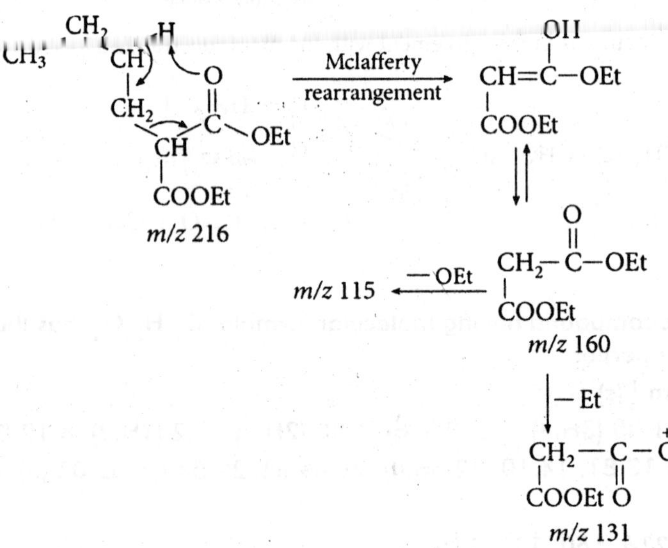

BIBLIOGRAPHY

1. Collin N. Banwell and Elaine M McCash, "Fundamentals of Molecular Spectroscopy, 4th Edition, Tata McGraw-Hill Publishing Company Ltd., New Delhi (1999).
2. Raymond Chang, "Basic Principles of Spectroscopy", International Student Edition, McGraw-Hill Kagakusha Ltd., Tokyo (1971).
3. William Kemp, 'Organic Spectroscopy", 3rd Edition, ELBS with MacMillan, Hampshire (1991).
4. R. K. Prasad, "Teach Yourself Physical Chemistry", Part-2. Bharati Bhawan, Patna (1994).
5. Y. R. Sharma, "Elementary Organic Spectroscopy", 2nd Edition, S. Chand and Company Ltd., New Delhi (1990).
6. V. R. Dani, "Organic Spectroscopy", Tata McGraw-Hill Publishing Company Ltd., New Delhi (1995).
7. Jag Mohan, "Organic Spectroscopy", 3rd Edition, Narosa Publishing House, New Delhi (2007).
8. John R. Dyer, "Applications of Absorption Spectroscopy of Organic Compounds", Prentice-Hall of India Private Limited, New Delhi (1969)..
9. P. S. Kalsi, "Spectroscopy of Organic Compounds", 6th Edition, New Age International (P) Ltd., Publishers, New Delhi (2006).
10. Dudley H. Williams and Ian Fleming, "Spectroscopic Methods in Organic Chemistry", 4th Edition, Tata McGraw-Hill Publishing Company Limited (1988).
11. R. M. Silverstein, G. Clayton Bassler and Terence C. Morill, "Spectrometric Identification of Organic Compounds", 5th Edition, John Wiley and Sons Inc., New York (1991).
12. K. Biemann, "Mass Spectrometry", McGraw-Hill (1960).
13. Budzikiwiez, H., Djeransi, C. and William, D.H.,"Interpretation of Mass Spectra of Organic Compounds, Holder-Day Inc., San Francisco (1964).
14. L. M. Jackman and M. Sternhill, "Nuclear Magnetic Resonance Spectroscopy in Organic Chemistry, Pergamon Press, London (1969).
15. K. S. Mukherjee, B,. Mukhopadhyay and G. Bramhachari, *J. Indian Chem. SOC*, **81** 82(2004).
16. A. Chatterjee, B. Das, C. P. Dutta and K. S. Mukherjee, *Tetrahedron Letters* 67 (1965).

17. K. S. Mukherjee, G Brahamachari, T. K., Manna and P. Mukherjee, *J. Indian Chem. SOC* **75** 260 (1998)

18. T. Morrison and Robert. N. Boyd, "Organic Chemistry" 6th Edition Prentice-Hall of India (Private) Limited, New Delhi (2000).

19. E. L. Elliel, "Stereochemistry of Carbon Compounds", McGraw-Hill Company Inc., Newyork, Kogkausha Company Limited, Tokyo, 1962.

20. D. Nasipuri, "Stereochemistry of Organic Compounds, Principles and Application, Willey Eastern Limited, New Delhi, 2nd Edition, 1992.

21. C. Dejerassi, "Optical Rotatory Dispersion", McGraw-Hill, New York, 1963.

22. Donald L. Pavia, Gary M. Lampman and George S. Kriz, "Introduction to Spectroscopy" Third Edition, Thompson Asia Private Limited, Singapore, 2004.